Science for Sustainable Societies

This series aims to provide timely coverage of results of research conducted in accordance with the principles of sustainability science to address impediments to achieving sustainable societies – that is, societies that are low carbon emitters, that live in harmony with nature, and that promote the recycling and re-use of natural resources. Books in the series also address innovative means of advancing sustainability science itself in the development of both research and education models. The overall goal of the series is to contribute to the development of sustainability science and to its promotion at research institutions worldwide, with a view to furthering knowledge and overcoming the limitations of traditional discipline-based research to address complex problems that afflict humanity and now seem intractable. Books published in this series will be solicited from scholars working across academic disciplines to address challenges to sustainable development in all areas of human endeavors. This is an official book series of the Integrated Research System for Sustainability Science (IR3S) of the University of Tokyo and the Institute for Global Environmental Strategies (IGES).

Julia Bentz • Jelena Ristić Trajković
Editors

Imagining, Designing and Teaching Regenerative Futures: Art-Science Approaches and Inspirations From Around the World

Springer

EUROPEAN COOPERATION IN SCIENCE & TECHNOLOGY

Funded by the European Union

Editors
Julia Bentz
Center for Ecology, Evolution and Environmental Changes (CE3C)
University of Lisbon
Lisbon, Portugal

Jelena Ristić Trajković
University of Belgrade – Faculty of Architecture
Belgrade, Serbia

ISSN 2197-7348 ISSN 2197-7356 (electronic)
Science for Sustainable Societies
ISBN 978-981-96-9028-2 ISBN 978-981-96-9029-9 (eBook)
https://doi.org/10.1007/978-981-96-9029-9

This publication is based upon work from COST Action CA21166, supported by COST (European Cooperation in Science and Technology).

COST (European Cooperation in Science and Technology) is a funding agency for research and innovation networks. Our Actions help connect research initiatives across Europe and enable scientists to grow their ideas by sharing them with their peers. This boosts their research, career and innovation.

www.cost.eu.

This Springer imprint is published by the registered company Springer Nature Singapore Pte Ltd.
The registered company address is: 152 Beach Road, #21-01/04 Gateway East, Singapore 189721, Singapore

Foreword

It is easy to feel overwhelmed and even hopeless amid ongoing ecological, social, political, and economic global crises. We are living the results and priorities of an industrial, extractive, and isolationist way of imagining humankind's relationship with/in the natural world. In this context, this book is a gift; it prioritizes hope to combat the despair and passivity creeping throughout our communities. It acknowledges how current times call for bold and daring people and actions, communities of accomplices working toward a better future.

For many years, my scholarship has focused on understanding how we can educate in ways that cultivate a sense of kinship with the more-than-human and a deep understanding of our inseparability with and in the natural world. Acknowledging how systems of schooling around the globe tend to separate humans from the more-than-human, perpetuate myths of human superiority over nature and ignore the richness of local contexts, I wonder: What is required to think beyond ecologically destructive beliefs that fuel climate crises? What is required to understand humankind's deep entwinement in a more-than-human world? In this volume, Julia Bentz and Jelena Ristić Trajković have collated a set of works that collectively offers answers to those questions. Importantly, the collection brings into focus two urgently needed but often neglected dimensions of socio-ecological change: imagination and emotional engagement.

My research has taught me that nothing will or can change without imagination and without strong emotional connections with the natural world. I understand imagination as the ability to envision the possible in all things; it is the soil in which all change, creativity, and innovation find their roots. To create a positive future, we need to be able to envision beyond what is, to the not yet and we need the strength, courage, and community to move in that direction. We need tools to cultivate our individual and collective imaginations. We need space and time to practice; cultivating imagination takes work. My learning journey has taught me the invaluable nature of Indigenous knowledges for this work. Re-imagining what is flourishes if we walk forward in learning with Indigenous knowledge holders and communities, learning about the interconnections of human and more-than-human worlds that have formed the heart of Indigenous cultures since time immemorial.

This work also requires that we honor and celebrate the affective dimensions of learning and living for a regenerative future. For meaningful and lasting change to occur, we urgently need emotional connections with the natural world. As activist Dr. Rob Hopkins suggests, we must long for that better future. The personal and affective threads connecting us with and in human and more-than-human communities will motivate and sustain the work required to regenerate, reimagine, and act.

With imagination, creativity, and hope as guiding lights, this volume offers a way forward for anyone seeking to think and feel differently about the future of our common world. This book amplifies the voices of researchers, practitioners, and community change-makers from around the globe. It offers provoking topics and theoretical questions along with practical activities. It is an action-oriented book for anyone, anywhere. Moreover, it is a collective work, creating a community of accomplices and space for shared hope and passion for a better future. As you read the work, you can't help but feel part of a space where unthinking and rethinking are not only invited but celebrated. Each chapter offers different ways of enacting and engaging with the more-than-human world, diverse ways of knowing and feeling forward to a new "what is." The book creates a space where accomplices engaged in the work of imagining and enacting a positive future can come together and find the support, strength and bravery to work toward what is possible.

If you are like me, you may find this book hard to put down. Each chapter intrigued me with its newness and imaginative perspective. Each chapter was a new morsel of a delicious meal, a new idea and clearly articulated practice to consider for my own work. This book should be required reading for educators (and educators-to-be) anywhere seeking support and community in cultivating a positive future. With this book in hand, educators of all kinds and in all contexts can unsettle, can dissonate, and can forge new pathways—pathways through anxious times to a hopeful, positive future in which the inseparable relationship between human thriving and more-than-human thriving is understood as the basis for right action.

Simon Fraser University, Faculty of Education
Surrey, BC, Canada

Gillian Judson

Acknowledgement

This publication is based upon work from COST Action CA21166, supported by COST (European Cooperation in Science and Technology).

COST (European Cooperation in Science and Technology) is a funding agency for research and innovation networks. Our Actions help connect research initiatives across Europe and enable scientists to grow their ideas by sharing them with their peers. This boosts their research, career and innovation.

www.cost.eu.

EUROPEAN COOPERATION IN SCIENCE & TECHNOLOGY

Funded by the European Union

Contents

Editors and Authors

About the Editors

Julia Bentz is a regeneration and transformation researcher at the University of Lisbon and a lecturer at Freie Universität Berlin. She has a background in interdisciplinary social sciences (MPhil Development Studies, PhD Economics) and conducts research on the many interactions between social and ecological systems in a variety of research fields and contexts including climate change mitigation, adaptation, and transformation. Her special interest has been community engagement and sustainability education. In her current projects Art for Adaptation and NBRACER she explores the transformative potential of art and story as an engagement practice for diverse communities and social groups. She is also the lead of one of the working groups of the Cost Action "Shift—Social Sciences and Humanities for Transformation and Climate Resilience."

Jelena Ristić Trajković is an Associate Professor and Vice Dean for Research and Innovation at the University of Belgrade—Faculty of Architecture. She also teaches at the University of Arts in Belgrade—Faculty of Applied Arts. With a PhD in Architecture and Urbanism, her primary research interests include design methodologies related to social transformation, sustainability and resilience, the relationship between space and health, and the application of environment-behavior theories in design. Her approach aims to bridge theoretical frameworks with practical applications, emphasizing critical reflection and design's role in fostering reconnection with nature, regenerative and healthier environments and communities. In the professional-artistic field, she acts through competitions, projects, policy briefs, and various initiatives aimed at contributing to the academic and broader community.

Authors

Fátima Alves Department of Social Sciences and Management, Universidade Aberta, Lisbon, Portugal

Centre for Functional Ecology—Science for People and the Planet, University of Coimbra, Coimbra, Portugal

Nagayamma Aragão Universidade Lusófona, Department of Architecture and Urban Planning & Research Centre for Architecture, Urban Planning and Design (University of Lisbon), https://urbanismo.ulusofona.pt/en/

Maria Bakatsaki School of Production Engineering and Management, Technical, University of Crete, Chania, Greece

Kitija Balcare Institute of Literature, Folklore and Art of the University of Latvia, Riga, Latvia

John Barry Centre for Sustainability, Equality and Climate Action, Queen's University Belfast, Belfast, UK

Julia Bentz Center for Ecology, Evolution and Environmental Change (CE3C), University of Lisbon, Lisbon, Portugal

Kristin Bodiford Community Strengths, Chattahoochee Hills, GA, USA

Paul Bolger Sustainability Institute, University College Cork, Cork, Ireland

Susanne Bosch Independent Artist, Berlin, Germany

Evan Boyle MaREI Centre, Environmental Research Institute, University College Cork, Cork, Ireland

Luciana Bragança Universidade Federal de Minas Gerais, Architecture School, Departament of Projects, https://naturezapolitica.arq.ufmg.br/

Miriam Mary Brgles Catholic University of Croatia, Zagreb, Croatia

Katarzyna Brożek Jan Kochanowski University of Kielce, Kielce, Poland

Serhat Burmaoglu Faculty of Economics and Administrative Sciences, Department of Health Management, Cigli, Izmir, Türkiye

Izmir Katip Celebi University, Faculty of Economics and Administrative Sciences, Department of Health Management, Cigli, Izmir, Türkiye

Marcus Bussey University of the Sunshine Coast, Sippy Downs, Australia

Edmond Byrne Process and Chemical Engineering, School of Engineering & Architecture, University College Cork, Cork, Ireland

Sustainability Institute, University College Cork, Cork, Ireland

Catarina Cadima Research Centre for Territory, Transports and Environment (CITTA), University of Porto, Porto, Portugal

Celiane Camargo-Borges Sustainability Transitions Research Group, Breda University of Applied Sciences, Breda, The Netherlands

Censu Caruana Centre for Environmental Education and Research, University of Malta, Msida, Malta

Eduardo Cassina University of Agder, Kristiansand, Norway

Ana Paula Claúdio University of Lisbon, Faculty of Sciences, LASIGE, Lisbon, Portugal

Carlos Smaniotto Costa Universidade Lusófona, Department of Architecture and Urban Planning & Research Centre for Architecture, Urban Planning and Design, (University of Lisbon), https://urbanismo.ulusofona.pt/en/

Elizabeth Creed Sustainability Institute, University College Cork, Cork, Ireland

Department of Sociology and Criminology, University College Cork, Cork, Ireland

Sarata Darboe Ca'Foscari University of Venice, Venice, Italy

Eglantina Dervishi Department of Psychology and Pedagogy, Faculty of Social Sciences, University of Tirana, Tirana, Albania

Niels Devisscher Sympoiesis.world, Glienicke/Nordbahn, Germany

Antje Disterheft CENSE – Center for Environmental and Sustainability Research & CHANGE – Global Change and Sustainability Institute, NOVA University Lisbon, Lisbon, Portugal

Instituto de Ciências Sociais, Universidade de Lisboa, Lisbon, Portugal

Justyna Doherty Institute of Art, Design + Technology, Dublin, Ireland

Zoran Đukanović University of Belgrade—Faculty of Architecture, Belgrade, Serbia

Niall Dunphy Cleaner Production Promotion Unit, Environmental Research Institute, University College Cork, Cork, Ireland

Sustainability Institute, University College Cork, Cork, Ireland

School of Engineering & Architecture, University College Cork, Cork, Ireland

Institute for Social Science in the 21st Century, University College Cork, Cork, Ireland

Mithat Ekici Ministry of Environment, Urbanization and Climate Change, Turkish State Meteorological Service, Ankara, Türkiye

Geraint Ellis School of Natural and Built Environment, Queen's University Belfast, Belfast, UK

Edy Fantinato Ca'Foscari University of Venice, Venice, Italy

Stephen Flood Irish Climate Analysis and Research Units, Department of Geography, Maynooth University, Maynooth, Ireland

Cristina Flora University of California Davis, Davis, CA, USA

Desiree Foerster Department of Cinema and Media Studies, University of Chicago, Chicago, IL, USA

Maria João Fonseca HTC – CFE – NOVA FCSH, Lisbon, Portugal

Ruth Förster Dr. Ruth Förster, Training & Counseling, Zürich, Switzerland

Martina R. Fröschl University of Applied Arts Vienna, Vienna, Austria

Brian Ó. Gallachóir MaREI Centre, Environmental Research Institute, University College Cork, Cork, Ireland

Sustainability Institute, University College Cork, Cork, Ireland

Augusta Gaspar Universidade Católica Portuguesa, Católica Research Centre for Psychological, Family and Social Wellbeing, Lisbon, Portugal

Gloria Gaviria Urban Planer, Urban Manager and Architect, Berlin, Germany

Curt Gervich Center for Earth and Environmental Science SUNY Plattsburgh, New York, NY, USA

Zoe Gilbertson Falmouth University, Cornwall, UK

Milena Grbić University of Novi Sad – Faculty of Civil Engineering, Subotica, Serbia

Catherine Hamel School of Architecture, Planning and Landscape, University of Calgary, Calgary, AB, Canada

Clodagh Harris Department of Government and Politics, University College Cork, Cork, Ireland

Sustainability Institute, University College Cork, Cork, Ireland

Institute for Social Science in the 21st Century, University College Cork, Cork, Ireland

Krisztina Jónás UiT The Arctic University of Norway, Norwegian College of Fishery Science, Faculty of Biosciences, Fisheries and Economics, Tromsø, Norway

Jasmin Jossin Department of Design, University of Applied Sciences, Potsdam, Germany

Predrag Jovanović University of Belgrade—Faculty of Architecture, Belgrade Serbia

Marina Kalashyan Department of Philosophy, Russian-Armenian University, Yerevan, Republic of Armenia

Cheryl Akner Koler Konstfack University of Arts, Crafts and Design, Stockholm, Sweden

Iuliia Kostynets Department of Marketing, Economics, Management and Administration, National Academy of Management, Kyiv, Ukraine

Valeriia Kostynets Department of International Economic Relations, National Academy of Management, Kyiv, Ukraine

Joanna Łabędzka Jan Kochanowski University of Kielce, Kielce, Poland

Neus Lagunas Neus Lagunas, CHAM Centro de Humanidades - Nova University of Lisbon, Lisboa, Portugal

Ksenija Lalović University of Belgrade – Faculty of Architecture, Belgrade, Serbia

Ieva Legzda Society "The House of Fairy-tales 'Undine'", "Omrači", Aizkraukle County, Latvia

Bem Le Hunte Transdisciplinary School, University of Technology Sydney, Ultimo, NSW, Australia

Stephanie Leite McGill University, Montréal, Canada

Marianna Marchesi Welsh School of Architecture, Cardiff University, Cardiff, UK

Rebeca Mateus HortaFCUL, University of Lisbon, Lisbon, Portugal

Scott Matter Transdisciplinary School, University of Technology Sydney, Ultimo, NSW, Australia

Connor McGookin Sustainable Energy Engineering, Simon Fraser University, Burnaby, BC, Canada

Cara Sophie Meffert EMJMD Adult Education for Social Change, University of Glasgow, Glasgow, United Kingdom

Marilyn Mehlmann Legacy17 Cooperative Association, Stockholm County, Sweden

Carolina Mello-Silva CENSE – Center for Environmental and Sustainability Research & CHANGE – Global Change and Sustainability Institute, NOVA School of Science and Technology, NOVA University Lisbon, Lisbon, Portugal

Daniel Metcalfe Technion- Israel Institute of Technology, Haifa, Israel

Myriel Milićević Department of Design, University of Applied Sciences, Potsdam, Germany

Beatriz Moriano Centre for English, Translation, and Anglo-Portuguese Studies (CETAPS)—Nova University of Lisbon, Lisbon, Portugal

Gerard Mullally Cleaner Production Promotion Unit, Environmental Research Institute, University College Cork, Cork, Ireland

Sustainability Institute, University College Cork, Cork, Ireland

Department of Sociology and Criminology, University College Cork, Cork, Ireland

Institute for Social Science in the 21st Century, University College Cork, Cork, Ireland

Kiat Ng Center for Ecology, Evolution and Environmental Changes (CE3C), University of Lisbon, Lisbon, Portugal

Dagmar Nolden Berghof Foundation Operations gGmbH, Tübingen, Germany

Barry O'Dwyer KPMG Ireland, Dublin, Ireland

Lígia Oliveira Independent Artist and Designer, Vila Real de Santo António, Portugal

Colăcel Onoriu University of Suceava, Suceava, Romania

Alina Östling RISE, Reseach Institutes of Sweden, Stockholm, Sweden

Leonie Paul LUT University Lappeenranta Finland & RMIT University, Melbourne, Australia

Anne Pender School of Architecture, Planning and Environmental Policy, University College Dublin, Dublin, Ireland

Ger Pepels Breda University of Applied Sciences, Breda, The Netherlands

Beata Poteralska Jan Kochanowski University of Kielce, Kielce, Poland

Lena Pozdnyakova Institute of Art History, Free University Berlin, Berlin, Germany

Julia Prakofjewa Ca'Foscari University of Venice, Venice, Italy

Susanne Pratt Transdisciplinary School, University of Technology Sydney, Ultimo, NSW, Australia

Baiba Prūse Athena Institute Vrije Universiteit Amsterdam, Amsterdam, The Netherlands

University College Cork, Cork, Ireland

Ca'Foscari University of Venice, Venice, Italy

Alexandra Revez MaREI Centre, Sustainability Institute, University College Cork, Cork, Ireland

Sustainability Institute, University College Cork, Cork, Ireland

Jelena Ristić Trajković University of Belgrade – Faculty of Architecture, Belgrade, Serbia

Luea Ritter Collective Transitions, Muttenz, Baselland, Switzerland

Fionn Rogan MaREI Centre, Sustainability Institute, University College Cork, Cork, Ireland

Pia Rox University of Tübingen, Tübingen, Baden-Württemberg, Germany

Kate Rudd University of York, York, UK

Florian Ruland Freie Universität Berlin, Berlin, Germany

IGB Berlin, Berlin, Germany

Náttúrustofa Vesturlands, Stykkishólmur, Iceland

Munir Moosa Sadruddin Sindh Madressatul Islam University, Karachi, Pakistan

Anaïs Sägesser scaling4good, Riehen, Switzerland

Cemre Sahinkaya i3mainz c/o Mainz University of Applied Sciences, Mainz, Germany

Sandra Sarfo Ca'Foscari University of Venice, Venice, Italy

Zerrin Savaşan Department of International Relations, Sub-Department of International Law, Faculty of Economics and Administrative Sciences, Selçuk University, Konya, Türkiye

Markus Schaffert i3mainz c/o Mainz University of Applied Sciences, Mainz, Germany

Annalena Schmidt-Faber Eberswalde University for Sustainable Development, Eberswalde, Germany

Sabrina Schork Aschaffenburg University of Applied Sciences, Faculty of Economy and Law, Aschaffenburg, Germany

Andra Simanova Society "Ecovillage", "Omrači", Aizkraukle County, Latvia

Renata Sõukand Ca'Foscari University of Venice, Venice, Italy

Ivan Spasojević University of Belgrade – Institute for Mutlidisciplinary Research, Belgrade, Serbia

Ben Stasny University of Colorado Boulder, Boulder, CO, USA

Karel Sterckx Field Research Leadership, Inari, Finland

Danica Stojiljković University of Belgrade – Institute for Mutlidisciplinary Research, Belgrade, Serbia

Andrea Troncoso Freelancer Consultant in Creativity and Communications, A mountain with wings, Heidelberg, Germany

Zoran Turza Catholic University of Croatia, Zagreb, Croatia

Elsa Kosmack Vaara RISE, Reseach Institutes of Sweden, Stockholm, Sweden

Diogo Guedes Vidal Department of Social Sciences and Management, Universidade Aberta, Lisbon, Portugal

Centre for Functional Ecology—Science for People and the Planet, University of Coimbra, Coimbra, Portugal

Līga Volkova A. Upitis Skriveri Secondary School, Skrīveri, Latvia

Ruttikorn Vuttikorn Club Creative Co., Ltd., Bangkok, Thailand

Justus Wachs McGill University, Montreal, QC, Canada

Robert Woodford Deep Time Walk C.I.C., Devon, UK

Samuel Yu Transdisciplinary School, University of Technology Sydney, Ultimo, NSW, Australia

Dārta Unda Zālīte Society "Ecovillage", "Omrači", Aizkraukle County, Latvia

Karla Santos Zambrano Department of Sociology and Criminology, School of Society, Politics and Ethics, University College Cork, Cork, Ireland

Nancy Zamierwoski Collective Transitions, Muttenz, Baselland, Switzerland Transformative Studies, California Institute of Integral Studies, San Francisco, CA, USA

Rūta Žemčugovaitė Sympoiesis.world, Glienicke/Nordbahn, Germany

Ulrike Zeshan University of Central Lancashire, Preston, UK

Ishara Press CIC, Lancaster, UK

Chapter 1
Co-Creating a Toolkit of Art-Science Approaches for Imagining, Designing and Teaching Regenerative Futures

Julia Bentz and Jelena Ristić Trajković

1.1 Introduction

Regenerative futures may sound like a daring idea. It is a bold and radical concept based on the realization that doing "less worse" is not good enough. As several planetary boundaries are being reached or even exceeded, calls for new ways of learning and engaging in and with nature, and with others, are clear and loud. The environmentalist Paul Hawken writes that "regeneration is not only about bringing the world back to life; it is about bringing each of us back to life" (2021, p. 9). It emphasizes solutions and approaches in which humans are participating as an integral part of nature and *as* nature and in which they have the capacity to restore, support, and regenerate living systems (Reed 2007). It addresses the challenging topic of climate change via new channels that inspire its learners and underline the role of each individual with their specific talents and world views.

While sustainability often focuses on reducing harm—using resources wisely, minimizing negative impact, and ensuring future generations' needs are met (Brundtland Report 1987)—regeneration focuses not just on maintaining the status quo but actively restoring, healing, and strengthening ecosystems and communities—helping systems thrive, not merely survive. Sustainability science has also evolved, incorporating more interdisciplinary and transdisciplinary approaches (Clark and Dickson 2003; Komiyama and Takeuchi 2006). It has grown into an

J. Bentz (✉)
Center for Ecology, Evolution and Environmental Changes (CE3C), University of Lisbon, Lisbon, Portugal
e-mail: jhbentz@fc.ul.pf

J. Ristić Trajković
University of Belgrade – Faculty of Architecture, Belgrade, Serbia
e-mail: jelena.ristic@arh.bg.ac.rs

J. Bentz, J. Ristić Trajković (eds.), *Imagining, Designing and Teaching Regenerative Futures: Art-Science Approaches and Inspirations From Around the World*, Science for Sustainable Societies,
https://doi.org/10.1007/978-981-96-9029-9_1

innovative, solution-oriented field that seeks to advance the understanding of the complex interactions between human and environmental systems and to contribute actionable knowledge for societal transformation (Kates et al. 2001; Bettencourt and Kaur 2011). Regeneration is an important perspective on current social–ecological challenges and deepens the goals of sustainability science, while moving toward conditions for renewal and flourishing (Hawken 2021).

Regeneration means change across scale and presenting alternatives to the structures that dichotomize and distance humans from nature. It implies going to the foundations, to our relationships, cultural values, and moral principles. It requires profound and systemic shifts and resetting the dichotomous relationship with nature perpetuated by Western epistemologies (Castree 2005; Welden et al. 2021). Emphasizing the deep human–nature connections, regeneration implies an interdependent process within which the healthy becoming of every life form is inseparably linked with the healthy unfolding of other living beings and spaces. Consequently, human well-being is tightly linked to the well-being of the environment as a whole. Fostering and evolving an emerging regenerative view of well-being demands a transformative journey in order to radically shift the way we understand, imagine, and design our place within the future world (Walsh et al. 2021; Ristić Trajković et al. 2021; Ristić Trajković et al. 2024; Nikezić et al. 2021).

This book aligns with a growing recognition that addressing today's complex challenges requires moving beyond maintaining the status quo toward actively regenerating social–ecological systems. In this sense, regenerative approaches deepen and extend the transformative aspirations of sustainability science (Berkes and Folke 1998; Wahl 2016). By integrating regenerative design, creative imagination, and experiential learning into educational practices, the 65 methods presented in this book aim to actively nurture thriving, equitable, and resilient futures, in alignment with the evolving frontiers of sustainability science. They foster holistic perspectives, support collective knowledge production, and strengthen transformative capacities such as imagination, empathy, critical reflection, and collaboration, which are crucial for enabling societal transformation toward equitable and regenerative futures.

1.2 Transformative Capacities of Creativity, Imagination, and Hope

In times like today that are shaped by multiple crises and complex global issues, there is a growing recognition that education needs to shift in order to prepare learners for these challenges. With this book, we propose creativity, imagination, and hope as orienting principles and helpful guides in the process of finding alternatives to mainstream approaches to education about global challenges. Their ability to challenge assumptions and invite new perspectives holds transformative potential and supports learning about regenerative futures.

Creativity has many superpowers. The writer Elisabeth Gilbert claims that "creativity is a force of enchantment" (2016, p. 17). Others have depicted it as a fundamental and intrinsic human characteristic that has brought forth new ideas and fostered innovation (Florida 2003). Formerly, creativity was attributed to some individuals who had a rare and innate talent and conceived brilliant ideas. Newer interpretations of creativity suggest it's a skill that anyone can acquire. John Cleese, co-founder of Monty Python, for instance, writes that "creativity is not a talent. It is a way of operating" that anyone can train (2020, p. 22). Everyday creative practices proliferate in the ordinary spaces we live in, and collective approaches to creativity, in which people exercise their creative thinking together and come up with innovative ideas, have also provided novel interpretations of the generation of creativity (Catmull 2008; Hawkins 2016; Montuori 2011). The physicist David Bohm provided the perspective on creativity as a process of unfolding meaning that occurred through the interaction or dialogue between the (individual) parts and the whole, which permitted the emergence of novel insights and perspectives (Bohm 2004).

Focusing on the processes, whether individual or collective, that support creativity can invite new imaginaries of regeneration and well-being. Imagination is usually seen as an integral part of creativity in its capacity to generate new ideas and images and consider new possibilities to solve problems (Sawyer 2011). Mangabeira (2007) views imagination as the aspect of the human mind that not only perceives reality but also explores new combinations of meaning. In this sense, imagination involves discovering new pathways for thinking and acting by challenging and freeing ourselves from established assumptions and mental categories (Wapner and Elver 2016). For John Dewey, imagination is "the ability to look at things as if they could be otherwise" (1980). More and more researchers have pointed out the potential of imagination in addressing sustainability challenges and societal issues, as it is an important prerequisite to change the reality of any specific society. Imagination has this potential because it gives the necessary motivation or impulse to change reality based on its ability to facilitate escape from reality by mentally generating alternatives (Moore and Milkoreit 2020).

The origins of fundamental change are mostly hard to track because they first take place in our minds. Visible, large-scale change arises from invisible, wild ideas and imaginings. According to Mohsin Hamid, "engaging with the future, imagining that we are willing to author it together as opposed to having it given to us, is important. And 'radical' is to say that we should be ambitious and slightly crazy and idiosyncratic about how we think that future could be" (Hamid 2017 in Hopkins 2019). This willingness to author the future and to ask, "what if?" is central to envisioning and enacting better lives. Collectively envisioning desirable futures provides the motivation and guidance for change, or as the activist Rob Hopkins has put it: "We need to create stories where the kind of future we want to see becomes commonplace every day. We need to tell stories with an underlying sense that the mere telling of them can create a degree of inevitability about their becoming reality and a sense that speaking them out loud is also a great benefit to our own mind, a powerful antidote to despondency and trauma" (2019, p. 127).

Looking at today's challenges and those ahead through this angle may feel unfamiliar or naive. Yet, a continually warming future is not inevitable, and many climate change risks are still avoidable. The differences between low-end and high-end scenarios are profound, and the future will depend, to a large degree, on the choices and actions we take today. This gives us space and reason for hope. Relatedly, activist and author Rebecca Solnit writes: "Hope locates itself in the premises that we don't know what will happen and that in the spaciousness of uncertainty is room to act. When you recognize uncertainty, you recognize that you may be able to influence the outcomes – you alone or you in concert with a few dozen or several million others. Hope is an embrace of the unknown and the unknowable and alternative to the certainty of both optimists and pessimists. Optimists think it will all be fine without our involvement; pessimists take the opposite position; both excuse themselves from acting. It's the belief that what we do matters, even though how and when it may matter, who and what it may impact, are not things you may know beforehand" (2016, p. xii).

Moving beyond optimistic and pessimistic future scenarios, hope can help us imagine a shared vision of an alternative future that catalyzes regeneration. Hope can be described as an embodied vision that activates agency and constitutes a starting point for imagining and working toward a different world. Rather than a substitute for action, it constitutes the basis for it. The action strengthens feelings of hope, creating a positive feedback loop (Ojala 2012). This can create spaces where alternative, regenerative futures can emerge.

1.3 Bringing Regenerative Futures into Educational Spaces: Imagining, Designing, and Teaching Across Disciplines, Cultures, and Practices

The complexities of sustainability transformation and regeneration require new and possibly different competencies on the part of the younger generation and demand that teachers, practitioners, and lecturers constantly adapt their teaching methods and content to a changing world. Bringing Hawken's idea of regeneration into a learning setting may mean a totally new way of approaching global challenges. In most education contexts, learning about global challenges such as climate change and biodiversity loss mainly focuses on understanding the problem, i.e., explaining the greenhouse effect and reasons for habitat destruction as well as discussing the potential consequences of rising temperatures and changing precipitation patterns (Monroe et al. 2019; Stevenson et al. 2017). Communicating mainly messages of fear rather than showing real examples for active engagement, this approach has been criticized for contributing to feelings of denial, numbing, and apathy (Stoknes 2015). Not surprisingly, young people feel increasingly hopeless, pessimistic, and powerless about climate change and the future in general (Hickman et al. 2021). Yet offering alternatives to mainstream education is challenging because it is far from obvious what is needed to imagine and design regenerative futures in a learning context. Many educators may ask themselves what kind of approaches are conducive to creating hope and empowerment among learners and which tools support imaginaries of regenerative futures and help learners see a pathway for realizing them.

By introducing art–science approaches such as artistic experimentation, creative community-building practices, and transdisciplinary pedagogies, this book aims to equip educators and change-makers with tools for fostering critical competencies among their learners, including systemic understanding, anticipatory capabilities, strategic skills, and interpersonal competencies (Wiek et al. 2011). Art–science approaches are increasingly seen as a powerful way to engage in global challenges and explore the idea of sustainable and regenerative futures (Bentz 2020; Bentz and O'Brien 2019). Involving creativity, imagination, and art can help us expand our visions of the future, opening our minds to new scenarios and more-than-human perspectives. It can create spaces for people to engage with societal transformation and regenerative futures on a personal and emotional level, linking the issue to their own lives and communities (Bentz 2023; Bentz et al. 2022). Creative and imaginative forms of learning can create spaces for meaning-making where relationships to self, others, and nature can be questioned and redefined. By situating these methods within the framework of sustainability science, this book serves as a bridge between theoretical foundations and practical application, equipping educators to inspire and empower learners to envision and enact regenerative futures. In doing so, this book invites readers to imagine and design sustainable and regenerative futures—not as distant ideals, but as tangible possibilities nurtured through creative, integrative, and participatory educational practices grounded within the evolving field of sustainability science.

Many educational contexts, including universities, community centers, and schools, can offer environments for art–science engagement and transdisciplinary learning approaches. Art–science methods can be integrated into (almost) all disciplines, providing spaces for creation, imagination, experimentation, hope, and perspective-taking (Bentz 2020). Integrating art–science tools in education and engagement about regenerative futures can serve as a means of expanding people's imagination and empowering them to co-create new imaginaries and transformative change (Đorđević et al. 2022). Notwithstanding the increasingly recognized potential of creative approaches to support learning about social-environmental challenges and regenerative futures, wide application across disciplines and geographical contexts is still lagging behind.

This book was developed within the transdisciplinary environment of the COST Action SHiFT—*Social Sciences and Humanities for Social Transformation and Climate Resilience (https://shift-cost.eu/)*. COST (European Cooperation in Science & Technology) Actions are interdisciplinary research networks that bring together researchers and innovators from academia, public institutions, and other relevant organizations or interested parties. They are open to all scientific and technological fields, focused on excellence, and inclusive. SHiFT is a COST action that addresses existing social–ecological challenges through transformative and transdisciplinary ways of working across different social, political, economic, environmental, and technological contexts. The SHiFT Action network, within which this collaborative book project emerged, is organized into four interdisciplinary Working Groups that reflect different facets of advancing sustainability and regenerative futures: Working Group 1: Reflecting on critical practice interventions; Working Group 2: Action-based and transdisciplinary tools for engagement; Working Group 3: Creative practices, arts, and outreach; and Working Group 4: Sharing, disseminating and exchanging knowledge. This interdisciplinary network fosters collaborations between arts, technology, science, and humanities for creative solutions and impactful outreach. This is reflected in the book's methods that include different kinds of knowledge and in the diversity of the book's contributors, who bring perspectives from over 25 disciplinary fields (Fig. 1.1). There is a strong representation of social

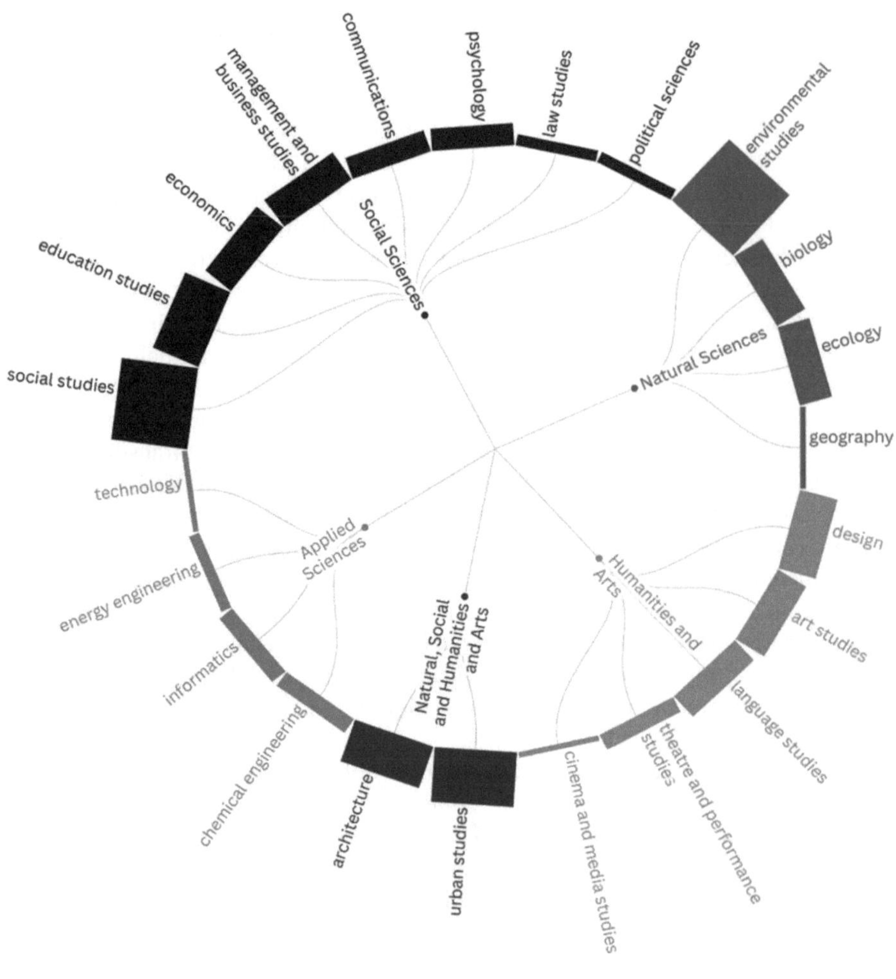

Fig. 1.1 Representation of disciplinary fields

sciences, architecture, urban studies, environmental sciences, as well as creative and performing arts, but also informatics, engineering, economics, psychology, and law.

This book was initiated, developed, and coordinated within Working Group 3 which focuses on developing creative capacity-building opportunities for early-career, mid-career, and senior researchers by creating and offering transdisciplinary learning spaces (Bentz and Ristić Trajković 2024). Moreover, this book also draws upon a diverse and complementary mix of knowledge and expertise of the members of all four Working Groups. This collaborative book emerged as the result of nearly 2 years of concerted effort and joint activities, including workshops, meetings, and knowledge exchanges to align ideas into a meaningful, user-friendly, and inspiring resource. It builds upon a wide range of experiences, practices, and methodologies that were tested and implemented across diverse disciplines, cultures, and contexts worldwide.

This collaborative book project aims to support educators in exploring and applying new ways of teaching and learning about regenerative futures. It provides guidance on how art–science approaches can support change makers and educators in teaching about regenerative futures and which methods can be used for designing and imagining a thriving world in which humans are participating as an integral part of nature. This book compiles creative methods and tools with the aim of facilitating educators to apply them in their educational contexts. It comprises a selection of 65 art-science methods brought together in a collaborative effort by 120 authors from 30 countries from 5 continents (Fig. 1.2), reflecting the global, pluralistic, and

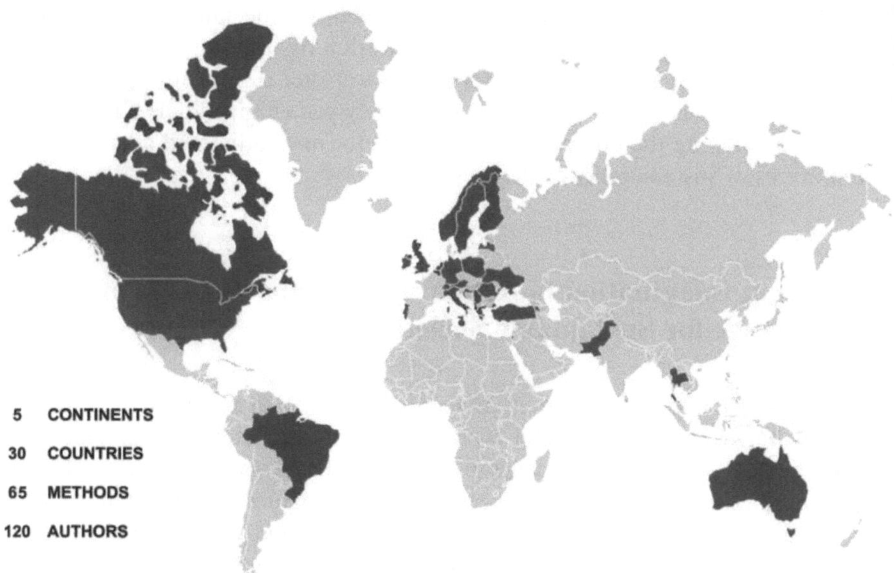

Fig. 1.2 Contributors overview. (The map has been generated using the online and free tool, available at https://visitedplaces.com/. Accessed 15 Dec 2024)

integrative spirit that sustainability science encourages. Each of its short contributions briefly introduces a method, identifies its regenerative or transformative potential, and describes how it can be used in a learning or community engagement setting, e.g., through the provision of a workshop design, providing instructions on required time and material and/or through illustrating the method's application through an example or actual project.

The enthusiasm for contributing to this book from authors coming from many parts of the world highlights the growing relevance of more generative, imaginative, and hopeful approaches to sustainability education across diverse fields and global contexts. The broad interest also significantly enriches the diversity of the book's content, bringing together a wide range of methods, disciplinary perspectives, and practical experiences as well as cultural and geographical backgrounds The high level of motivation in contributing to this project also resulted in the creation of an additional resource: an open-source Creative Methods Toolkit (Ristić Trajković and Bentz 2024). The Toolkit presents an open-access preview of all methods in a concise, user-friendly format, complemented by additional graphic material to support implementation. To facilitate finding suitable methods for different contexts and target audiences, the Toolkit uses a system of thematic and format-based tags, helping educators, practitioners, and researchers quickly identify appropriate tools and methods. It is intended as complementary material to this book, supporting the practical application of creative and transdisciplinary methods for teaching and learning about regenerative futures.

The methods presented in this book illuminate the many possible intersections and synergies of the arts, natural and social sciences, technologies, and humanities—illustrating their power of collaboration to foster meaningful change. They align with the core tenets of sustainability education by offering a diverse array of creative, integrative, and participatory methods to foster regenerative thinking and action. The methods presented here are not merely pedagogical tools but also catalysts for developing the competencies essential for navigating and addressing the complexities of sustainability challenges.

1.4 From Imagination to Action: Embracing Transformative Change for Sustainable Societies

Imagination and hope are important prerequisites for action as they activate agency and constitute a starting point for working toward a different world. This book invites readers to envision regenerative futures while transcending traditional boundaries between disciplines, between humans and nature, and between body and mind. It draws attention to the creativity, imagination, and wisdom in each one of us, and to our potential to contribute to transformative change. It invites educators to collaborate with people coming from different disciplines, contexts, and mindsets. In an increasingly polarized world, it is more important than ever to cultivate open dialogues, encourage diverse perspectives, and nurture an environment where new ideas and hopeful voices can be heard. By adopting imaginative and creative art–science approaches, we may unlock new pathways to regeneration and resilience that go beyond contexts of education and learning but affect our societies as a whole.

In alignment with the aims of the *Science for Sustainable Societies* series, our aim is to deepen and expand applied research in sustainability science by providing innovative and creative educational approaches that bridge disciplinary divides and foster systemic change. In our work as sustainability transformation researchers and educators, we have increasingly seen the importance of creating spaces for exploring hope and imagination. Working in diverse learning contexts has shown us that involving creativity, art, and imagination can catalyze transformative learning and alternative visions of the future, opening our minds to more-than-human perspectives and new avenues for collective action. Through the emphasis on transdisciplinary knowledge co-production, participatory learning, and regenerative thinking, this book contributes to critical perspectives on sustainable development, such as the disconnection between human and natural systems, and the inertia of traditional education systems. By equipping educators and change-makers with tools to imagine, design, and implement more sustainable ways of living, the book contributes to narrowing the knowledge-action gap in sustainability science (Kajikawa 2008).

The methods and approaches within the book are structured according to seven overarching themes and sections: (1) Building communities: collaborative decision-making and participatory learning, (2) Artistic expression and experimentation for environmental change, (3) Envisioning tomorrow: scenarios and possibilities, (4) Enhancing well-being: experiential learning and nature connection, (5) Playful learning and interactive digital tool, (6) Complexity and integrative thinking, and (7) Innovative pedagogies for regeneration and societal transformation. The sections are intended to help readers navigate the book.

The compiled methods for transformative education show how art–science approaches, systems thinking, and regenerative design can serve as powerful catalysts for collectively designing societies that live in harmony with nature, and cultivate resilient, inclusive communities. It responds to the widely recognized need for more holistic, collaborative forms of learning that bridge the "human gap" between knowledge and action and catalyze cultural change (UNESCO 2017; Club of Rome et al. 2025). Nurturing imagination, creativity, and hope in learning contexts can support us to collectively envision desirable futures and contribute to a more just, resilient, and thriving world.

References

Bentz J (2020) Learning about climate change in, with and through art. Clim Chang 162:1595–1612. https://doi.org/10.1007/s10584-020-02804-4

Bentz J (2023) Creative approaches to climate and peace education: an educator's guide to using storytelling and art. Leibniz Institute for Educational Media | Georg Eckert Institute, Braunschweig

Bentz J, O'Brien K (2019) ART FOR CHANGE: transformative learning and youth empowerment in a changing climate. Elementa Sci Anth 7:52. https://doi.org/10.1525/elementa.390
Bentz J et al (2022) Creative, embodied practices, and the potentialities for sustainability transformations. Sustain Sci 17(2):687–699. https://doi.org/10.1007/s11625-021-01000-2
Bentz J, Ristić Trajković J (2024) Creative capacity building for early, mid-career and senior researchers: a portfolio of WG3-facilitated activities. Nova University Lisbon, Lisbon
Berkes F, Folke C (1998) Linking social and ecological systems: management practices and social mechanisms for building resilience. Cambridge University Press, New York
Bettencourt LMA, Kaur J (2011) Evolution and structure of sustainability science. Proc Natl Acad Sci 108(49):19540–19545. https://doi.org/10.1073/pnas.1102712108
Bohm D (2004) On creativity. Routledge Classics, London
Castree N (2005) Nature. Routledge, London
Catmull E (2008) How Pixar fosters collective creativity. Harv Bus Rev 86(9):64–72+134
Clark WC, Dickson NM (2003) Sustainability science: the emerging research program. Proc Natl Acad Sci 100(14):8059–8061. https://doi.org/10.1073/pnas.1231333100
Cleese J (2020) Creativity: a short and cheerful guide. Crown Publications, New York
Club of Rome, World Environmental Education Congress Network and the Fifth Element (2025) No limits to hope – transforming learning for better futures, call for contribution and concept note. https://www.clubofrome.org/wp-content/uploads/2025/03/NLTH_Call-and-Concept-Note-1.pdf
Dewey J (1980) Art as experience (Revised edition, edited by Joan L Seigfried). Perigee Books, New York
Đorđević A, Milovanović A, Milojević MP, Zorić A, Pešić M, Ristić Trajković J, Nikezić A, Đokić V (2022) Developing methodological framework for addressing sustainability and heritage in architectural higher education—insights from HERSUS project. Sustainability 14(8):4597. https://doi.org/10.3390/su14084597
Florida R (2003) Cities and the creative class. City Community 2(1):17
Gilbert E (2016) Big magic—how to live a creative life, and let go of your fear. Bloomsbury Academic, London
Hamid M (2017) Imagination taking power (Interview by Rob Hopkins). https://www.robhopkins.net/2017/05/15/mohsin-hamid-imagination-allows-us-to-imagine-futures-not-bound-by-the-tyranny-of-the-past-and-the-present/. Accessed 28 Jan 2025
Hawken P (2021) Regeneration: ending the climate crisis in one generation. Penguin Books, New York
Hawkins H (2016) Creativity, 1st edn. Routledge, London
Hickman C et al (2021) Climate anxiety in children and young people and their beliefs about government responses to climate change: a global survey. Lancet Planet Health 5(12):863–873. https://doi.org/10.1016/S2542-5196(21)00278-3
Hopkins R (2019) From what is to what if: unleashing the power of imagination to create the future we want. Chelsea Green Publishing, White River Junction, VT
Kajikawa Y (2008) Research core and framework of sustainability science. Sustain Sci 3(2):215–239. https://doi.org/10.1007/s11625-008-0053-1
Kates RW et al (2001) Sustainability science. Science 292(5517):641–642. https://doi.org/10.1126/science.1059386
Komiyama H, Takeuchi K (2006) Sustainability science: building a new discipline. Sustain Sci 1:1–6. https://doi.org/10.1007/s11625-006-0007-4
Mangabeira RU (2007) The self awakened: pragmatism unbound. Harvard University Press, Cambridge, MA
Monroe MC et al (2019) Identifying effective climate change education strategies: a systematic review of the research. Environ Educ Res 25:791–812. https://doi.org/10.1080/13504622.2017.1360842
Montuori A (2011) Beyond postmodern times: the future of creativity and the creativity of the future. Futures 43:221–227
Moore ML, Milkoreit M (2020) Imagination and transformations to sustainable and just futures. Elementa Sci Anth 8(1):081. https://doi.org/10.1525/elementa.2020.081

Nikezić A, Ristić Trajković J, Milovanović A (2021) Future housing identities: designing in line with the contemporary sustainable urban lifestyle. Buildings 11(1):18. https://doi.org/10.3390/buildings11010018

Ojala M (2012) Hope and climate change: the importance of hope for environmental engagement among young people. Environ Educ Res 18(5):625–642. https://doi.org/10.1080/13504622.2011.637157

Reed B (2007) Shifting from 'sustainability' to regeneration. Build Res Inf. 35(6):674–680. https://doi.org/10.1080/09613210701475753

Ristić Trajković J, Bentz J (eds) (2024) Creative methods toolkit for imagining, designing and teaching regenerative futures. NOVA University of Lisbon—FCSH, Lisbon. https://doi.org/10.5281/zenodo.14000111

Ristić Trajković J, Krstić V, Milovanović A, Calheiros CSC, Ćujić M, Karanac M, Kazak JK, Di Lonardo S, Pineda-Martos R, Garcia Mateo MC et al (2024) Moving towards a holistic approach to circular cities: obstacles and perspectives for implementation of nature-based solutions in Europe. Sustainability 16(16):7085. https://doi.org/10.3390/su16167085

Ristić Trajković J, Milovanović A, Nikezić A (2021) Reprogramming modernist heritage: enhancing social wellbeing by value-based programming approach in architectural design. Sustainability 13(20):11111. https://doi.org/10.3390/su132011111

Sawyer RK (2011) Explaining creativity: the science of human innovation. Oxford University Press, New York

Solnit R (2016) Hope in the dark, 3rd edn. Canongate Books, New York

Stevenson RB, Nicholls J, Whitehouse H (2017) What is climate change education? Curric Perspect 37(1):67–71. https://doi.org/10.1007/s41297-017-0015-9

Stoknes PE (2015) What we think about when we try not to think about global warming. Chelsea Green Publishing, White River Junction, VT

UNESCO (2017) Guidelines on sustainability science in research and education

Wahl DC (2016) Designing regenerative cultures. Triarchy Press, Axminster

Walsh Z, Böhme J, Wamsler C (2021) Towards a relational paradigm in sustainability research, practice, and education. Ambio 50(1):74–84. https://doi.org/10.1007/s13280-020-01322-y

Wapner P, Elver H (2016) Reimagining climate change. Routledge, Oxon/New York

Welden EA, Chausson A, Melanidis MS (2021) Leveraging nature-based solutions for transformation: reconnecting people and nature. People Nat 3(5):966–977. https://doi.org/10.1002/pan3.10212

Wiek A, Withycombe L, Redman CL (2011) Key competencies in sustainability: a reference framework for academic program development. Sustain Sci 6:203–218. https://doi.org/10.1007/s11625-011-0132-6

World Commission on Environment and Development (1987) Our common future (The Brundtland Report). Oxford University Press, Oxford

Part I
Building Communities: Collaborative Decision-Making and Participatory Learning

At the heart of thriving communities often lies collaborative decision-making and participatory learning. These processes empower individuals to co-create solutions and share knowledge. Fostering open dialogue and inclusive participation are ways to dismantle hierarchies, ensuring that diverse voices are heard and different needs are met. Collaborative decision-making harnesses collective intelligence, supporting groups to navigate complex challenges through consensus-building and shared accountability. Meanwhile, participatory learning cultivates a culture of continuous growth, where people can exchange skills, perspectives, and experiences. It can strengthen social bonds and problem-solving capacity, and contribute to resilient communities. This part brings forward methods for commitment to inclusive, place-based engagement and collective agency. In times of fragmentation and disconnection, they offer not only tools for decision-making and participatory learning but also practices for building trust, belonging, and mutual care—essential foundations for regenerative and sustainable societies.

Chapter 2
Deliberative Futures Workshops for Transformative Sustainability

Alexandra Revez, Clodagh Harris, Niall Dunphy, Brian Ó. Gallachóir, Edmond Byrne, Evan Boyle, Connor McGookin, John Barry, Paul Bolger, Fionn Rogan, Geraint Ellis, Barry O'Dwyer, Stephen Flood, Elizabeth Creed, and Gerard Mullally

2.1 Introduction

The deliberative futures workshop is both a method and a model of engagement that is designed to support transformative change, thinking ahead, and getting communities involved in shaping their future. It promotes collaboration with experts, stakeholders, and communities in a co-productive process that aligns with deliberative action research. The process seeks to be inclusive of everyone with a stake in the future by facilitating the development of multiple tools for their involvement in the planning and decision-making processes, following the deliberative principles of equality, inclusion, and reasoned argument. The deliberative futures workshop

A. Revez (✉) · B. Ó. Gallachóir
MaREI Centre, Sustainability Institute, University College Cork, Cork, Ireland

Sustainability Institute, University College Cork, Cork, Ireland
e-mail: alexandra.revez@ucc.ie; b.ogallachoir@ucc.ie

C. Harris
Department of Government and Politics, University College Cork, Cork, Ireland

Sustainability Institute, University College Cork, Cork, Ireland

Institute for Social Science in the 21st Century, University College Cork, Cork, Ireland
e-mail: clodagh.harris@ucc.ie

N. Dunphy
Cleaner Production Promotion Unit, Sustainability Institute, University College Cork, Cork, Ireland

Sustainability Institute, University College Cork, Cork, Ireland

School of Engineering & Architecture, University College Cork, Cork, Ireland

Institute for Social Science in the 21st Century, University College Cork, Cork, Ireland
e-mail: n.dunphy@ucc.ie

J. Bentz, J. Ristić Trajković (eds.), *Imagining, Designing and Teaching Regenerative Futures: Art-Science Approaches and Inspirations From Around the World*, Science for Sustainable Societies,
https://doi.org/10.1007/978-981-96-9029-9_2

embraces a diversity of communication and practices that recognize the systemic turn in deliberative democratic theory. The various techniques proposed draw upon multiple experiences, voices, and perspectives with a view to increasing and enhancing the number of "deliberative" moments within the process.

The deliberative futures workshop is more than a typical discussion group held at a given time and place. It consists of a staged and adaptable engagement process

E. Byrne
Process and Chemical Engineering, School of Engineering & Architecture, University College Cork, Cork, Ireland

Sustainability Institute, University College Cork, Cork, Ireland
e-mail: e.byrne@ucc.ie

E. Boyle · F. Rogan
MaREI Centre, Sustainability Institute, University College Cork, Cork, Ireland
e-mail: evan.boyle@ucc.ie; f.rogan@ucc.ie

C. McGookin
Sustainable Energy Engineering, Simon Fraser University, Burnaby, BC, Canada
e-mail: connor_mcgookin@sfu.ca

J. Barry
Centre for Sustainability, Equality and Climate Action, Queen's University Belfast, Belfast, UK
e-mail: j.barry@qub.ac.uk

P. Bolger
Sustainability Institute, University College Cork, Cork, Ireland
e-mail: p.bolger@ucc.ie

G. Ellis
School of Natural and Built Environment, Queen's University Belfast, Belfast, UK
e-mail: g.ellis@qub.ac.uk

B. O'Dwyer
KPMG Ireland, Dublin, Ireland
e-mail: barry.odwyer@kpmg.ie

S. Flood
Irish Climate Analysis and Research Units, Department of Geography, Maynooth University, Maynooth, Ireland
e-mail: Stephen.Flood@mu.ie

E. Creed
Sustainability Institute, University College Cork, Cork, Ireland

Department of Sociology and Criminology, University College Cork, Cork, Ireland
e-mail: elisabeth.creed@ucc.ie

G. Mullally
Cleaner Production Promotion Unit, Sustainability Institute, University College Cork, Cork, Ireland

Sustainability Institute, University College Cork, Cork, Ireland

Department of Sociology and Criminology, University College Cork, Cork, Ireland

Institute for Social Science in the 21st Century, University College Cork, Cork, Ireland
e-mail: G.Mullally@ucc.ie

that urges for an integrated approach to deliberation. The model is structured to encourage long-term dialogue by linking locally based deliberative workshops to wider processes and stakeholders across scales, using interactive and future-focused visions and scenarios to help define systemic pathways for transformation (Flood et al. 2023). Advancing the value of partnership within our complex societal systems, the staged model is proposed as a contribution to engagement, which must be designed within a recognized wider network of other exchanges and engagements.

Deliberative democracy has been operationalized through various deliberative forums ranging from futures workshops to citizens' assemblies and juries. Such fora can provide spaces for deeper public discussion and examination on an issue as they invite ordinary citizens to consider the arguments of differently situated and opinionated others, present reasons for their own preferences, weigh up the arguments, and be open to changing their minds as a result (Böker and Elstub 2015). They have much to offer in addressing the climate emergency by bringing different perspectives and experiences to bear on an issue, moving beyond vested interests and electoral cycles. These forums offer a mix of scientific evidence with public opinions and values and help citizens imagine new ways of living.

2.2 Transformative Potential

In the context of climate change, processes such as Citizens' Assemblies are seen as critical and transformative practices that can help turn countries across Europe from climate action "laggards" to climate action "leaders" (Torney and O'Gorman 2019). In an increasingly polarized society grappling with multiple political and ecological crises, deliberative processes offer the opportunity to bring together different perspectives in a way that helps create a more coherent and thoughtful integration of multiple views. It does so by creating meaningful ways to share knowledge, express public opinion, and widen the range of choices available to engage in climate action.

Deliberative tools and approaches have expanded considerably. The impact of these tools in facilitating change in contentious areas such as the Irish abortion laws has given rise to optimism in its ability to offer breakthroughs and solutions in situations of deadlock. However, two common assumptions about deliberative democracy require further scrutiny. First, that the main purpose of deliberative tools is to generate consensus. Second, that deliberative processes will "transform" our systems by producing a mobilized and agreeable "public." Recent developments in deliberative democratic theory recognize that seeking consensus shouldn't be the main goal in real-world decision-making. Emphasizing consensus may limit inclusivity, especially with regard to minority perspectives, and constrain open discussion. Furthermore, developing deliberative spaces beyond institutional settings can deepen democracy, not as a disciplining action and a way to mobilize "agreeable publics," but as a means to avoid manipulation and arbitrary decision-making and produce robust outcomes through "reason-recognizing" procedures (Estlund 1997).

Reforms suggested by the deliberative approach are thus seen as a catalyst for re-imagining more inclusive, equal, and reasonable engagement and interaction with citizens. Multiple tools and approaches are emerging in this space, reinforcing the idea that while democratic innovations and reforms can be positive, there is no single solution, or rather a one-size-fits-all approach to help connect decision-making with local communities (Revez et al. 2022). Furthermore, while the deliberative forums can be used to enhance "deliberation-making" sites, their critical dimension hinges on the existence of a much wider and more dynamic system of multiple interlocking instances of democratic emancipation.

To start, consider how much time and resources you have to hold the workshops. There are many exciting ways to bring citizens and communities together in deliberative processes to tackle real-world challenges. While deliberative processes might vary in how they recruit participants, their size, and how long they last, they all have some important things in common. They typically include gathering information from various sources, like presentations by academic experts, stakeholders, or witnesses. Plus, they usually involve small group discussions led by a moderator, which help to shape a shared opinion or recommendation that can be communicated with a broader public.

We propose three main principles to guide the use of the Deliberative Futures Workshops: inclusion, equal voice, and thoughtful decision-making. Inclusion means ensuring that different kinds of people are represented and that everyone's views are considered. Equal voice means ensuring everyone has an equal chance to speak and be heard in discussions. Considered judgment means engaging in informed, reasoned, and respectful conversations and considering the needs of others. Participants should be encouraged to justify their preferences, consider differing opinions, and be open to changing their views based on new information. Sharing lived experiences and emotions can also enrich the discussions.

The best way to run the workshops is to follow a simple, staged process that keeps things organized and engaging! While there's much flexibility in how you can approach it, Deliberative Futures Workshops usually focus on three main stages over two weekend-long events:

Critical Analysis: This is where you collaboratively identify the target problems.

Visioning: Here, you brainstorm and imagine creative solutions and exciting future possibilities.

Implementing: Finally, you collaboratively develop a concrete action plan to put ideas into motion and share these widely across other forums and with multiple stakeholders.

2.3 Using Creative Tools

The deliberative futures workshops, involve the deployment of multiple methods as illustrated in Table 2.2 are designed to be engaging and approachable as seen in the method overview presented as Table 2.3. They start with short, easy-to-understand presentations from experts, which set the stage for thoughtful discussion. Participants then discuss what climate change means for them and their communities. One of the exciting aspects of these workshops is the use of creative tools. These tools help enhance our ability to imagine and plan different futures, especially when dealing with climate change. The aim is to create a welcoming space where everyone can connect with information and each other in a friendly and interactive way. Together, participants can dream up a shared future and develop practical climate action plans. We also embrace methods like storytelling, empathy mapping, creating visual materials that are easy to share, and working together to develop different scenarios. These techniques help bridge the gap between expert insights and community voices, ensuring everyone's ideas are valued. Our goal is to move beyond the usual political discussions by opening the door to new ways of thinking about society and culture. We focus on fostering collaboration, sharing knowledge among various stakeholders, and encouraging creative expressions beyond traditional formats like books and reports. It's all about making the conversation more inclusive and inspiring for everyone involved.

One such tool is empathy mapping. An empathy map is a fantastic tool that helps us step outside our perspectives and connect with others. It's versatile and can be used in many areas, from building awareness to developing targeted strategies for change. The benefits of using this tool include uncovering surprising information about your audience, community, or users that you hadn't thought of before, resolving conflicts by building empathy toward others, and identifying knowledge gaps. Empathy mapping is all about connecting and understanding. It's great to include an empathy map exercise in a multiple-group setting! Start by identifying different "citizens" based on various factors like gender, age, job, location, disability, and economic status. Then, discuss who might be most impacted by climate change. It can be helpful to create doodles of these different "citizens" at this stage. Next, ask each group to draw a larger outline of one pre-selected citizen. Each group can then choose a citizen to map out in more detail. They'll create a character profile that highlights the citizen's hopes, fears, priorities, behaviors, influences, agency, and climate concerns. Lastly, each group should present their citizen to everyone else. Encourage discussion and sharing of insights that came up during this activity! This way, everyone can learn and contribute to a better understanding together.

Community mapping is also a valuable creative tool; it's an easy and quick way to create climate transformation strategies rooted in local experiences. This approach allows communities to create maps visually showcasing what they love about their area and the key features that matter to them. When it comes to planning for climate action, these maps can be powerful as they let community members share their thoughts and experiences regarding the impacts of climate change—both what they've faced in the past and what they anticipate in the future. They are tools that can be used to connect scientific information with community knowledge. For example, community mapping is a great way to develop strategies for adapting to climate change! To get started, you could look at recent weather events and any climate changes that affected the local area. Next, use a base map to highlight the regions that are most vulnerable, along with the social, economic, and environmental impacts. Don't forget to consider any existing preventative measures, like flood defenses, and evaluate how well they are working. It's also important to determine what climate changes are projected in the future and how they might affect your community. Keep an eye on upcoming social, economic, and environmental plans, and think about factors like an aging population that could play a role. Make sure to identify any other areas that might be at risk from climate change and note any key social, economic, or environmental assets that could be impacted. Lastly, take some time to evaluate how current and planned measures stack up. Think about any new hazards or opportunities that might arise from climate change, and suggest some actionable ideas for adaptation and mitigation, marking the locations on your map (refer to Imagining2050 toolkit for more on creative tools) (Tables 2.1 and 2.2).

Table 2.1 Deliberative creative tools for transformation. You can refer to Imagining2050 Toolkit for further details on co-creative deliberative tools

VISIONING
Sense-making A tool for exploring meaning formation in the context of uncertainty. Used to explore how people come to define complex and unfamiliar issues, to find key anchors for sustaining meaning, and their role as empowering or disempowering drivers for action, and to note changes of meaning over time.
Empathy mapping A valuable human-centered tool to draw out unexpected insights about your audience or users, to broaden individual perspectives, shape and transform collective thinking, and to discover gaps in current knowledge. It helps better understand and empathize with others.
Story boarding Tool used to break down a vision into smaller and more detailed elements, to explore user–system interactions, and to pilot the development of new technologies and practices. It gives the opportunity to sequence action and consider the passage of time within change processes.
RECOMMENDATIONS
Community mapping This is a key tool for "localizing" action and considering the siting of new actions. It is valuable for engaging communities in planning, developing a common understanding of emerging risks, and providing a platform to explore potential solutions drawing on local knowledge.
Audience polls A tool used to understand the implications of individual decisions on collective goals, to develop scenarios based on specific group trends, and to establish levels of consensus or divergence of opinion around specific issues. Live polling audiences can also be used to reveal trends and expectations around the future.
Ballot Citizen engagement processes can use various methods to decide on their final recommendations, reports, action plans, etc. future workshops may culminate in the development of an action plan based on establishing the consensus/divergence rate of the various recommendations. An anonymous ballot process is ideal for developing final recommendations, each defined by varying degrees of consensus.

Table 2.2 Method overview

Main purpose
New examples and models of collective deliberative decision-making that educators and groups can aspire to replicate in their work to help deliver fair and equitable climate action.
Gained competences
Enriched knowledge base on participatory and deliberative practices with a future-oriented focus. Creating a staged strategy to link local deliberations to wider processes with various stakeholders.
Educational setting
Informal workshop format guided by deliberative democracy principles.
Space requirements/restrictions
Indoor activities.
Resources and necessary materials
Expert presentations based on locally defined thematic needs and testimonials from local community representatives.
Number of participants
Group work with 8–12 participants.
Facilitator competences and skills
Group work competence and skills include facilitating inclusive dynamics, stimulating discussion, conflict resolution, and moderation of debate.
Participants' skills/age/competences
Willingness to engage in respectful discussions that focus on facts, the future, and the consideration of the needs of others in decision-making.
Duration
A deliberative process over two weekends is linked to a longer-term strategy to ensure the development of actionable partnerships.

References

Böker M, Elstub S (2015) The possibility of critical mini-publics: realpolitik and normative cycles in democratic theory. Representation 51(1):125–144

Estlund D (1997) Beyond fairness and deliberation: the epistemic dimension of democratic authority. In: Bohman J, Rehg W (eds) Deliberative democracy: essays on reason and politics. MIT Press, Cambridge, MA, pp 173–204

Flood S et al (2023) Imagining climate resilient futures: a layered Delphi panel approach. Futures 147:103100

Revez A et al (2022) Mapping emergent public engagement in societal transitions: a scoping review. Energy Sustain Soc 12:1–18

Torney D, O'Gorman R (2019) A laggard in good times and bad? The limited impact of EU membership on Ireland's climate change and environmental policy. Ir Polit Stud 34(4):575–594

Chapter 3
Collaborative Maps of Curiosity

Catherine Hamel

3.1 Introduction

Collaborative Maps of Curiosity is an invitation to visualize a city to explore its space toward creating alternative experiences for people adapting and integrating in environments. The method has been applied in a range of contexts, including communities of resettled refugees, explorations of public safety after addiction recovery, and the interplay of visual and olfactory experiences with participants living with Alzheimer's disease. The proposed approach to collaborative map-making is a three-layered, adaptable framework, with each layer corresponding to a different scale of observation explored in a different medium. Beginning with the personal and expanding to the communal, mapping becomes a form of expression. Oscillating between clarity and disorder, the most meaningful aspect of the process lies in the exchange between participants. Initial individual images in one layer respond to neighboring depictions in the next, gradually developing and merging into a final experiential map of the participants' new city—one that invites them to engage with unfamiliar places and embrace new experiences.

Collaborative mapping is a visual discussion where a small community of people are brought together to converse though material expression. Using one large surface and multiple forms of mark-making, participants contribute to a collective composition. Crucially, this practice moves beyond the expectation of artistic expression as the sole outcome. It is not about creating art, but rather about surfacing thought through observation and reimagining the built environment as a space for engagement. It is not a soliloquy. It is a process where the material marks of others can be viewed, responded to, and respectfully altered toward one large

C. Hamel (✉)
School of Architecture, Planning and Landscape, University of Calgary, Calgary, AB, Canada
e-mail: chamel@ucalgary.ca

J. Bentz, J. Ristić Trajković (eds.), *Imagining, Designing and Teaching Regenerative Futures: Art-Science Approaches and Inspirations From Around the World*, Science for Sustainable Societies,
https://doi.org/10.1007/978-981-96-9029-9_3

composition. A fundamental goal of the practice is the making of creative spaces and community environments for further vitality and healing. It is a form of engagement that visualizes the space of the city by supporting and inviting individuals to engage in its public realm.

Creating collaborative drawings and diagrams is not a new methodology. Traditionally applied in contexts of community engagement, the focus of the process is the single large, layered experiential map. The impact of this intimate horizontal and vertical cooperation lies in the observations, exchanges, and lessons learned in the making. At a time when many are questioning what happens when traditionally marginal considerations in city design are brought to the center, even temporarily, this practice offers an important step toward adapting to current civic realities. As a humble step, this method, before even giving voice, sets a tone. The practice calibrates an attitude. It inspires and grounds participants in being part of something as a way to organize thought and disorganize habits of imposed patterns. It provides a platform to give expression to members of a community who temporarily give each other attention. It has the potential to become a lingering conversation that accompanies each as they navigate their environments endowed with new forms of curiosity.

3.2 Transformative Potential

In a time where many are experimenting with which methodologies are most effective in exposing and unsettling socio-spatial injustices in order to counter them, the aim is to establish collaborations serving the needs of different communities using the creative arts and design. The intent is to infuse the process with new ideas toward different outcomes. Ecology differs from other sciences in that it defines a scope and an attitude of study, rather than a discrete field. Once unleashed, ecological thought permeates all areas of life, aligning itself with social practice to shift public awareness toward less harmful, more sustainable ways of relating to the world. The term derives from the Greek *ecos*—meaning house, or environment, and *hora*, dwelling place. It is not merely where we find ourselves, but how everything that surrounds us allows us to live (Bridle 2022). When what matters resides in relationships rather than in things, the transformational and generative potential of these maps of curiosity is the nature of the collaboration. The heart of the practice lies in temporarily and periodically inhabiting a space where we reimagine our relationships—with one another and with the world around us. It is not about my art or my creation, but about our process—our commitment to showing up, to engaging in shared reflection and reconsideration. It is a way of re-forming a world for the many that have been cast aside. Though sometimes stranded, they are seldom defeated. In this work, the rigid and imposing categories we inherit can begin to crack. And in those cracks lies a space of unknowing—where stories falter and the gap between paradigms opens. It is here that creativity speaks, not to build a wall, but to say: there is more—open the door.

The social is entangled in the interactions and tensions between the imposed space of dominant orders and the survival of those who do not fit within its parameters. In the challenge of generating actionable pathways by engaging with the critical practice dimensions of transformation, the intent is to dilute the dismissive attitudes and reorganize the barriers for some or dissolve them for others. The transient experiences of the forcibly displaced adapting to new environments silently escape and elude city policies and procedures that are more often grounded in stable and local data. Collaborative mapping can be used to survey these transitional territories to make issues that are inaccessible in space be relocated, re-dressed, re-addressed, and made visible and responded to in city design. Mapping and explorations are coordinated by both the participants and the facilitators. Collaborative imagining and mapping reshape interactions toward new attentions and interpretations. From the longed-for past and unfamiliar places, attention is directed to new forms of distributed curiosity in the explorations of the participants' new environments. The real and the not yet real defy exclusion through the integration of imaginary mapping. The relationship between the visual world created by such methods and the actuality of the lived world gets reorganized through the role of creative explorations. The imagination restored opens the ability to engage a city to imagine acceptance in its space. Resilience comes out of imagination and speculation. Having been cast out, difficult histories of expulsion and rejection are reshaped through fluid visualization. By creating new modes of communication, this work preserves not just the problematic present, but a malleable future—one continuously shaped and reshaped with others. For people to truly belong, spaces must exist where they feel free to imagine something beyond their current realities. To keep, and to be kept—like a field turning its secrets into poppies, flourishing in the disturbed earth (Vuong 2016).

3.3 Application

3.3.1 Participants

12–15 participants working toward engagement in a new city environment.

3.3.2 Timeframe

Three 90-min sessions with a minimum of 48 h in between for materials to dry.

3.3.3 Location

Informal classrooms, community centers, or studio spaces that welcome creative production with a central space open enough for people to move around the drawing surface.

3.3.4 Facilitators

1 or 2 facilitators who are comfortable and skilled in working with visual creativity. The accountable facilitator needs to act within a generous and expansive perspective to create social arrangements, to recognize the forces of change, and create the option, where the unimaginable becomes ordinary. The aim is not to direct the work aesthetically but to guide the process. This includes creating a safe space for the group, framing an appropriate theme in response to the specific group invited to participate. This focused response helps each group's response focus on a relevant aspect of their experience and interaction in the city (e.g., safety and security; spaces of play, etc.). In addition to having the role of provokers of possibilities to move the participants toward new wonders and glories in imagining new ways to explore their new environments, it is crucial to inspire participants beyond rigid art making. In addition, the facilitators engage in the material collaboration by developing the work between sessions to help shift the attention of scale and materials between layers (Fig. 3.1).

3.3.5 Ingredients

One large surface (4 × 10 ft. medium thickness watercolor paper). Each of the three layers focuses on a different scale of attention explored with a different medium.

3.4 LAYER 01: Individual Mark Making

Medium/tools: Minimum 24 color watercolor pencils, sharpeners, cups, paper towels, and brushes.

Objective: Exploring a defined theme with a focused scale of attention (e.g., the material textures of an appealing space; the colors of a safe space; details that invite inclusivity).

Process: A brief visual presentation (or printed examples) of examples and city scenes can be distributed to invite a gentler interaction with less intimidation toward color, texture, and proportions, rather than a realistic representation of space.

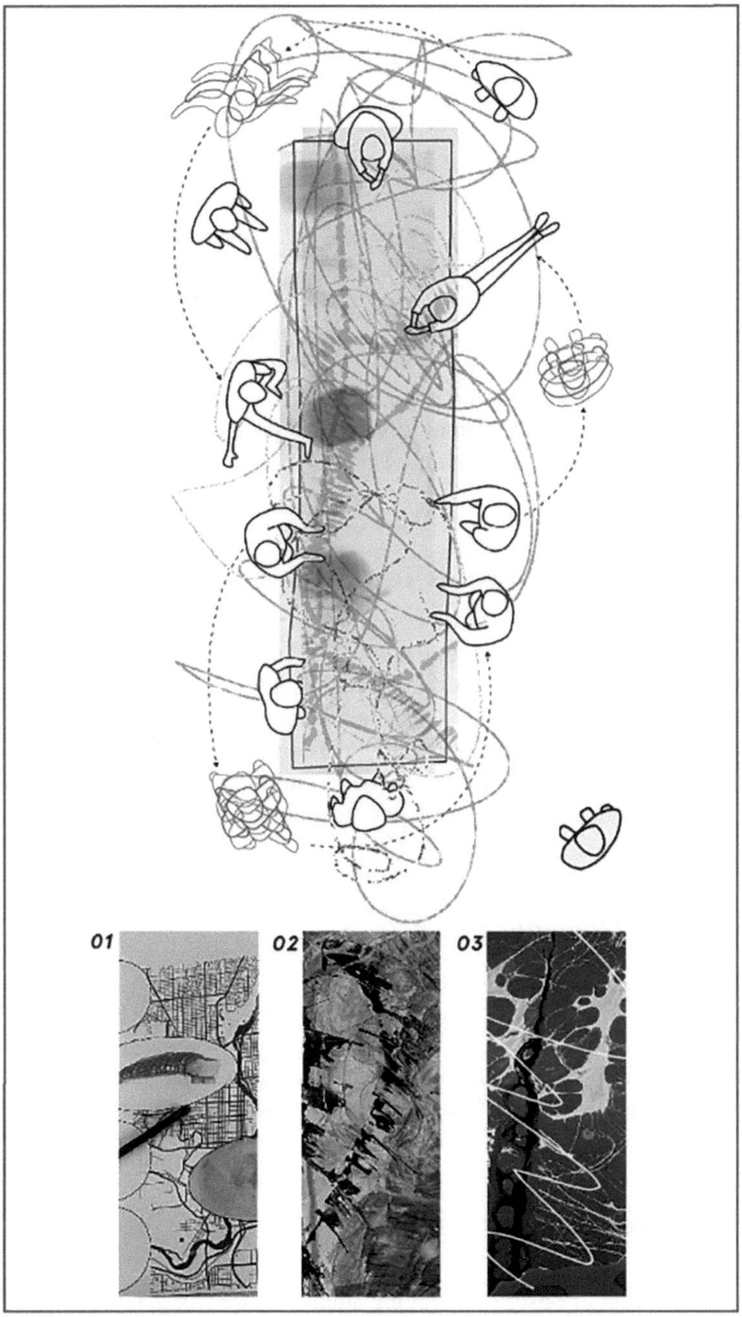

Fig. 3.1 Graphics. Kyra Richter. (Images C. Hamel, from *Maps of Curiosity: Adaptation*, visualizations of the city to explore its spaces (workshop with displaced Ukrainian teens). Canada, 2022)

Shift: By externalizing selective memories, individuals pay attention within themselves and raise awareness to others of qualities in experienced positive spaces.

3.5 LAYER 02: From the Individual Toward the Collective

Medium/tools: Acrylic paint, fat markers, larger brushes, and spatulas.

Objective: Translating individual interpretations toward a collaborative landscape on paper.

Process: Individual and more careful images get incorporated into the larger and more collective layer as the scale of attention shifts, using existing drawings from layer 01 as visual notes to expand upon.

Shift: Looser medium becomes connective tissue between initial individual images as a single large image emerges.

3.6 LAYER 03: Sharpness Toward Refined Organization

Medium/tools: Markers, acrylic paint markers to consolidate the previous two layers.

Objective: Regain attention to smaller details and consolidate a single map.

Process: Reflecting on the two previous layers, cooperatively reflecting on the existing marks and refining them toward a cohesive single image.

Shift: Collaborate toward a final visual intention, reflect, receive, and interpret the suggestions from the map as guidelines to experience the city with new attention.

3.7 Further Considerations

3.7.1 Before Sessions

A visual brochure/presentation that frames the process as more than just an art class—awakening participants' belief in their capacity to contribute.

3.7.2 Between Sessions

Encouraging the act of observing: paying attention, sketching, and looking for … in the city.

Table 3.1 Method overview

Main purpose
Awareness building/decision-making/behavior change.
Gained competences
Competences acquired through the method.
Educational setting
Informal (format—Collaborative workshop).
Space requirements/restrictions
Indoor activities/water source for material cleanup/minor spills possible for cleanup.
Resources and necessary materials
Mark making materials (range in scale/water color pencils/acrylic paint/markers/brushes/cups)—Projectors optional.
Number of participants
Collaborative work group of 10–12.
Facilitator competences and skills
Ease of working with visual creativity.
Participants' skills/age/competences
Community/researchers/designers/social workers/rehabilitation coaches/expressive arts coaches
Duration
Three 90-min sessions with a minimum of 48 h in between for materials to dry.

3.7.3 *After Sessions*

A postcard derived from the produced map, to serve as a reminder to continue looking for … in the city (Table 3.1).

References

Bridle J (2022) Ways of being, animals, plants, machines: the search for planetary intelligence. Picador, New York

Vuong O (2016) Night sky with exit wounds. Copper Canyon Press, Port Townsend

Chapter 4
Future and Nature Stakeholder Integration in Climate Deliberation

Fátima Alves and Diogo Guedes Vidal

4.1 Introduction

Traditional participatory approaches are insufficient to handle the complexity of growing climate difficulties and threats to socioecological balance (Darwish et al., 2023). There is an increasing need to broaden climate discourse by including vulnerable and excluded social groups and conceptualizing how to merge nature representatives with future generations. Inclusive discussion increases the effectiveness of solving climate issues by relying on a diverse variety of information, experiences, and expertise, as well as embracing the interests and concerns of all impacted parties, including those who are not currently present (Vidal & Alves, 2024).

We suggest including future generations and nature as stakeholders in climate discussions. This approach acknowledges the agency of these actors and attempts to guarantee that their interests are considered in decision-making processes (Ekeli, 2005). By giving nature and future generations legal standing and a voice, we hope to reshape our understanding of life and rebuild our existence to promote regenerative practices and socioecological harmony.

The participation of nature representatives and future generations as stakeholders is a novel method that has the potential to transform environmental preservation and intergenerational equity. This method entails extending legal rights to nature and evaluating the long-term effects of present policies on future generations. Combining these approaches can promote transdisciplinary conversation and envision socio-ecological futures that address unexpected difficulties (Pigott, 2018).

F. Alves (✉) · D. G. Vidal
Department of Social Sciences and Management, Universidade Aberta, Lisbon, Portugal

Centre for Functional Ecology—Science for People and the Planet, University of Coimbra, Coimbra, Portugal
e-mail: fatimaa@uab.pt; diogo.vidal@uc.pt

J. Bentz, J. Ristić Trajković (eds.), *Imagining, Designing and Teaching Regenerative Futures: Art-Science Approaches and Inspirations From Around the World*, Science for Sustainable Societies,
https://doi.org/10.1007/978-981-96-9029-9_4

4.2 Transformative Potential

The participation of Nature Representatives and Future Generations as stakeholders has emerged as a game-changing method for promoting regenerative change for a flourishing planet. This approach entails giving nature and future generations legal stature and a say in decision-making processes, altering our view of life, and reconstructing human existence (Boyd, 2017). The case of Nature Representatives is inspired by the Rights of Nature framework, which recognizes entities other than humans as rights holders. This recognition of rights implies an understanding that nature has intrinsic value and should be handled as more than mere property or a resource for human use (Alves et al., 2023). Future Generations' strategy is built on intergenerational equity and justice. It acknowledges that our actions now will influence the planet that future generations will inherit. Recognizing intergenerational fairness in socioecological debate entails ensuring that the needs and interests of future generations are effectively represented (Byskov & Hyams, 2022).

Nature Representatives and Future Generations contribute to cultural narratives emphasizing all life forms' interdependence, reinforcing the ethical requirement for conservation and regeneration (Braun, 2015). Their inclusion in climate resilience contributes to a broader opportunity for transdisciplinary conversation, encouraging envisioning socioecological futures that can accept unexpected problems. Thus, incorporating Nature Representatives and Future Generations as stakeholders provides a regenerative approach that addresses climate change from several linked perspectives: promoting regenerative change by including ecological viewpoints, ethical considerations, cultural narratives, and transdisciplinary collaboration, so broadening the circle of climate discourse for a more sustainable and inclusive future.

4.3 Application

Including Nature Representatives and Future Generations as stakeholders in the climate discourse is a viable strategy to address environmental concerns' complexities and promote regenerative change. Each step in this process (Table 4.1) is considered critical for ensuring broad representation, improving communication, and encouraging collaboration among varied stakeholders. The approach derives from our experience in the project "Phoenix" (GA: 101037328), which brings together partners from seven different European countries to investigate novel approaches to involvement for an inclusive, sustainable ecological transition.

To ensure a variety of viewpoints, we first identified stakeholders by speaking with youth organizations, environmental organizations, and local communities while ensuring that marginalized groups were adequately represented. Through

Table 4.1 Steps and considerations for implementing the method of integrating. Nature Representatives and Future Generations as stakeholders in climate deliberation

Step	Description	Considerations for implementation
1	Identifying stakeholders	Consult with local communities, environmental organizations, and youth groups. Ensure comprehensive representation. Consider the diversity of perspectives and interests
2	Establishing legal standing	Advocate for policy changes, support legal initiatives, and raise awareness about the importance of recognizing rights
3	Facilitating communication	Develop innovative approaches. Use socioecological indicators; involve local communities in identifying nature's representatives. Include non-human communication
4	Incorporating future generations	Create platforms and forums for intergenerational dialogue. Organize youth-led initiatives, workshops, and educational programs. Empower young activists through mentorship programs, networking opportunities, online and offline platforms, creating safe spaces for dialogue, and policy engagement training
5	Building transdisciplinary collaboration	Foster collaboration among diverse stakeholder groups. Convene multi-stakeholder workshops. Conduct scenario planning exercises. Integrate diverse perspectives into decision-making processes

legislative modifications, these delegates were given legal standing while spreading awareness of how crucial it is to respect their rights. In this process, it is critical to realize the inherent difficulties in interacting with non-human stakeholders and to devise inclusive and accessible communication approaches, moving beyond human forms and introducing alternative forms to communicate, such as art, emotions, or even experiences. By setting up forums for intergenerational conversation, planning youth-led projects, and providing young activists with training and mentoring opportunities, we engaged future generations. Through encouraging partnerships among stakeholders, hosting multi-stakeholder workshops, running scenario planning exercises, and incorporating various viewpoints into decision-making processes, we developed transdisciplinary collaboration.

Thus, including Nature Representatives and Future Generations as stakeholders in climate discourses provides a promising approach to tackling environmental concerns and promoting regenerative change. By adopting this method, stakeholders can effectively traverse the intricacies of climate discourse and strive toward a more regenerative, emphatic, and inclusive future (Table 4.2).

Table 4.2 Method overview

Main purpose
This method aims to enhance the inclusivity and comprehensiveness of decision-making processes by incorporating the perspectives and interests of both nature and future generations, ultimately leading to more sustainable and regenerative outcomes.
Gained competences
The method fosters systems thinking, interdisciplinary collaboration, and ethical reasoning. It enhances communication skills, long-term planning, and cultural competence, empowering diverse stakeholders and promoting sustainable decision-making.
Educational setting
The educational setting for this method can be both formal (higher education, undergraduate and graduate courses in environmental science, sustainability, social sciences, arts, humanities, law, and policy) and informal (workshops, seminars, community forums, and youth engagement programs).
Space requirements/restrictions
Classrooms or lecture halls for formal education; meeting rooms for workshops, seminars, and stakeholder dialogues; field sites for ecological observation and participatory environmental assessments; community spaces for engaging with local stakeholders and conducting participatory planning sessions.
Resources and necessary materials
Educational materials such as textbooks, articles, and case studies; computers and projectors for presentations and data analysis; recording equipment for capturing stakeholder dialogues and feedback; communication tools like video conferencing software for remote collaboration.
Number of participants
This should be adapted to the local context specificities, but the ideal group size would be 10–20 participants per group to ensure effective discussion and engagement; multiple groups can be formed depending on the scale of the project and the number of stakeholders involved. The method should accommodate a diverse range of participants, including local community members, youth representatives, environmental scientists, policymakers, and legal experts.
Facilitator competences and skills
Background in social sciences Experience in stakeholder engagement and mediation Strong communication, analytical, and interpersonal skills, coupled with cultural competence and advocacy abilities
Participants' skills/age/competences
Diverse ages and skills, including traditional knowledge and advocacy. Varied educational backgrounds, adept in research and critical thinking. Experienced in governance and decision-making processes. Proficient in scientific inquiry, creative problem-solving, and project implementation. Nature representatives. Future generations representatives.
Duration
This is dependent on the local context specificities, but the workshops or seminars may range from a few hours to several days, depending on the depth of engagement and the number of participants; advocacy efforts and policy development may extend over weeks or months, requiring ongoing collaboration and negotiation; long-term projects, such as community-based initiatives or research studies, may span months or even years, requiring sustained effort and commitment.

References

Alves F, Costa PM, Novelli L, Vidal DG (2023) The rights of nature and the human right to nature: an overview of the European legal system and challenges for the ecological transition. Front Environ Sci 11(1175143):1–10. https://doi.org/10.3389/fenvs.2023.1175143

Boyd DR (2017) The rights of nature: a legal revolution that could save the world. ECW Press, Toronto, ON

Braun B (2015) Futures: imagining socioecological transformation—an introduction. Ann Assoc Am Geogr 105(2):239–243. https://doi.org/10.1080/00045608.2014.1000893

Byskov MF, Hyams K (2022) Who should represent future generations in climate planning? Ethics Int Aff 36(2):199–214. https://doi.org/10.1017/S0892679422000168

Darwish S, Allen RR, Lempke M (2023) Necessary complexity in the Anthropocene: new approaches in socioecological systems thinking, Do No Harm, and fragility integration. Dev Pract 33(5):534–547. https://doi.org/10.1080/09614524.2023.2219042

Ekeli KS (2005) Giving a voice to posterity – deliberative democracy and representation of future people. J Agric Environ Ethics 18(5):429–450. https://doi.org/10.1007/s10806-005-7048-z

Pigott A (2018) Imagining socioecological transformation: an analysis of the Welsh Government's policy innovations and orientations to the future. Elementa 6:60. https://doi.org/10.1525/elementa.315

Vidal DG, Alves F (2024) Voices of the absent: the agency of Nature and Future in climate regeneration. PLOS Clim 3(6):e0000420. https://doi.org/10.1371/journal.pclm.0000420

References

Chapter 5
Building Communities in Action

Celiane Camargo-Borges and Kristin Bodiford

5.1 Introduction

Building Communities in Action is a relational framework that supports communities in coming together around shared challenges to imagine and co-create meaningful change. It emphasizes trust, adaptability, and collective creativity, placing relationships at the heart of the design process.

By fostering understanding and creating space for multiple perspectives, *Building Communities in Action* cultivates inclusive and generative spaces for shared learning and innovation. It moves beyond reliance on external "experts" and encourages collaborative processes rooted in lived experience, collective agency, and mutual responsibility. It supports transformative, community-led action that honors and embraces diverse knowledge systems (Leibowitz 2017).

This framework is especially valuable for addressing complex, systemic challenges, such as, climate change, social inequities, and fragmentation, where technical solutions alone are not sufficient. *Building Communities in Action* offers a pathway for educators, students, and communities to co-create regenerative responses and contribute to a more just and flourishing future (Camargo-Borges and Gergen 2022; Bodiford and Whitehouse 2020).

C. Camargo-Borges (✉)
Sustainability Transitions Research Group, Breda University of Applied Sciences, Breda, The Netherlands
e-mail: camargo-borges.c@buas.nl

K. Bodiford
Community Strengths, Chattahoochee Hills, GA, USA
e-mail: bodiford@communitystrengths.org

J. Bentz, J. Ristić Trajković (eds.), *Imagining, Designing and Teaching Regenerative Futures: Art-Science Approaches and Inspirations From Around the World*, Science for Sustainable Societies,
https://doi.org/10.1007/978-981-96-9029-9_5

5.2 Transformative Potential

The future is ours—together—to create – Gergen Kenneth (2023)

The transformative power of *Building Communities in Action* lies in the interplay between *Know-Being* principles and *Know-How* practices:

- *Know-Being* principles foster relational awareness and enable engagement with real-world complexities while holding space for multiple perspectives throughout the process.
- *Know-How* practices provide adaptable tools to activate these principles, creating spaces for dialogue and shared journeys of co-creation.

Together, these principles and practices encourage the application of relational values with practical action to collectively address challenges and create pathways forward. They invite us to consistently ask during collaborative processes: "How are we relating?" and "What are we creating?"

5.2.1 *Know-Being Principles*

Each *Know-Being* principle serves as a resource for cultivating relational ways of working–fostering deeper connections, adaptability, and collective creativity (Bodiford and Camargo-Borges 2014).

- *Centering Collaborative and Relational Processes*:
- Focuses on collective intelligence and the power of collaboration to generate innovative ideas. Fostering meaningful connections and encouraging active engagement helps create opportunities for thriving through sharing commonalities, celebrating differences, and exploring new possibilities.
- *Engaging Organically and Dynamically*:
- Values the dynamic, ever-evolving nature of communities and teams as living systems. Balancing planning with flexibility allows participants to respond to what unfolds in the moment, nurturing creativity and openness to unexpected insights.
- *Recognizing Complexity and Multiplicity*:
- Acknowledges the importance of embracing differences and valuing diverse perspectives. Creating space for diverse voices and stories strengthens the group's capacity to approach challenges from multiple angles.
- *Focusing on What Is Useful*:
- Aligns efforts with practical needs and local realities, ensuring solutions are meaningful and actionable. Emphasizing relevance and context-sensitive impact helps build more sustainable responses.

5.3 Application of *Know-Being* Principles and *Know-How* Practices

Know-Being principles are made actionable through *Know-How* practices with design thinking phases of *Framing, Designing, Enacting*, and *Reflecting*. These phases guide teams and communities through a collaborative process to address challenges. Unlike linear approaches, this framework evolves based on the aspirations of participants, allowing for choice, adaptability, and emergence.

5.3.1 Framing

This phase sets the intention and aligns values for collaboration around a specific topic that is relevant and meaningful for the team or community. Framing invites curiosity and openness with methods like Appreciative Inquiry, which emphasizes possibilities and strengths. It involves small-group conversations to explore collective hopes, values, and areas of common ground. This means focusing on how people engage with one another and their environment.

5.3.1.1 Steps in Framing

- Frame powerful and affirmative questions and intentions to support generative dialogue that helps to explore new meaning and possibilities.
- Focus on collective hopes and values to create alignment.
- Explore diverse perspectives to build a deeper understanding.

5.3.1.2 Tips and Tools

- Form small groups (3–5 people) to ensure all voices are heard.
- Mix participants across ages, contexts, or roles for diversity.
- Use circles to encourage equity, openness, and deep and active listening.
- Focus on what connects in order to create common ground as well as what surprises or changes perspectives.
- Capture and share hopes and themes to guide the next phase.

5.3.1.3 Example Activity

In small groups, participants share something they are proud of and their hopes for collaboration. Groups record and share common and unique or surprising hopes, identifying themes to guide the process forward.

5.3.2 Designing

This phase involves generating possibilities based on the themes from Framing. Teams focus on strengths, challenges, and co-design opportunities that they can put into action together.

5.3.2.1 Steps in Designing

- Build on strengths and collective insights.
- Generate a range of possibilities for addressing challenges.

5.3.2.2 Tips and Tools

- Encourage strengths-based dialogue and collaborative support.
- Emphasize shared understandings of potential ways forward.
- Use diverse or similar groupings based on the activity's goals.
- Use visual aids like sticky notes, whiteboards, and natural elements to help inspire creativity.

5.3.2.3 Example Activity

Teams discuss their strengths, challenges, opportunities, and potential ideas. Each group presents their discussions, noting common themes and actionable ideas.

5.3.3 Enacting

This phase transforms insights from Designing into prototyping and action. It focuses on connecting ideas to tangible steps that can be implemented collaboratively.

5.3.3.1 Steps in Enacting

- Translate ideas into concrete actions.
- Emphasize flexibility and responsiveness to emerging dynamics.

5.3.3.2 Tips and Tools

- Guide participants to align existing efforts with new ideas.
- Use creative tools like storytelling or prototyping to bring abstract ideas to life. This helps to also foster collaboration and shared ownership.
- Discuss commitments, collaboration, and mutual support.

5.3.3.3 Example Activity

Teams identify how new ideas connect to current efforts. They outline specific actions to test these ideas, ensuring feasibility and positive impact.

5.3.4 Reflecting

Reflection creates regular and intentional pauses in the process to deepen understanding, refine approaches, and adapt to new insights. It is woven throughout the process to guide learning and alignment.

5.3.4.1 Steps in Reflecting

- Evaluate progress and refine plans based on emerging learning.
- Use reflections to align priorities, make shifts where needed, and deepen understanding.

5.3.4.2 Tips and Tools

- Create spaces for participants to share experiences and learnings.
- Active listening, paraphrasing, and reflective pauses ensure all voices are heard and allow ideas to deepen.
- Stay curious about differing perspectives and explore tensions.

5.3.4.3 Example Activity

At the beginning or end of each phase, teams reflect on their experiences, learning, and what adaptations are needed. Key questions include:

- What are we experiencing?
- What are we learning?
- What needs to shift or adapt?

5.4 Summary

Building Communities in Action is a dynamic, relational approach that integrates *Know-Being* principles and *Know-How* practices that center on collaboration, adaptability, and collective creativity. By fostering trust and embracing diverse perspectives, this framework empowers communities to co-create actionable, context-sensitive responses for meaningful and sustainable change (Table 5.1).

Table 5.1 Method overview

Main purpose
The intention for building communities in action is to: Create different levels of connections among people and their systems Integrate diverse skills, knowledge, interests, and experiences Co-design solutions that matter and that match the local challenges, context, and culture
Gained competences
Building communities in action invites us to focus more on processes and relations/interactions rather than content or final solutions. By doing so we strengthen our abilities to engage in: Co-creation: Integrating diverse skills, knowledge, interests, and experiences to develop meaningful solutions Collaborative reflexivity: Engaging reflexively and collaboratively across differences to generate new understandings and possibilities Curiosity positions: Welcoming and exploring multiple perspectives with a curious and open stance
Educational setting
Resources can be shared in formal educational settings and workshops. The beauty and the challenge is that there is no such thing as the right technique to follow. This methodology is action-oriented, experiential, and relational. That means it is process-focused and context-related. Each system is unique and the community is created in action. Aligned *Know-How* practices include methods like design thinking, appreciative inquiry, community-based participatory research, liberating structures, solutions/progress-focused methods, and other intentionally generative approaches. To learn more, facilitators may find Design Thinking and Social Construction: A Practical Guide to Innovation in Research by Celiane Camargo-Borges and Sheila McNamee useful. Also, for resources to deepen your understanding and support meaningful action, visit our Social Constructionist community: www.taosinstitute.net.
Space requirements/restrictions
Building Communities in Action invites us to come together in a space that has meaning and relevance to the people involved and the cultural context. We recommend space to move, to create groups, and to connect to the local environment.

(continued)

Table 5.1 (continued)

Resources and necessary materials
Suggested materials to support meaningful engagement and shared learning: Name tags—Help participants connect by using and practicing each other's preferred names Flip charts, sticky notes, markers—Use visual tools to organize, share, and connect ideas Food!—We create community around food. Consider especially sharing food from the local community Accessibility considerations—Ensure the space is welcoming and usable for all ages and abilities Time—Give time for conversations to unfold; give space for people to share their perspectives and what is important to them
Number of participants
The number of participants will depend on the local situation. What is most important to consider is the diversity of people involved to bring many perspectives to the process.
Facilitator competences and skills
The principles and practices we suggest for building communities in action invite us to all learn and unlearn ways of being together. Be mindful of inconsistencies. Align practices with principles Pay attention to our discomfort in trying new ways of co-creating solutions to challenges. Embrace it. It is often in these spaces of discomfort that we have our greatest learnings Avoid taking anything for granted: be open to new and different perspectives Craft questions before the meetings: curious questions, reflexive questions, and imaginative questions in order to expand meanings around the topic and to highlight different perspectives Hold your plans lightly. Make plans, but stay present and adapt to the needs of the moment and the people around you
Participants' skills/Age/Competences
There are no requirements for participants' skills, age, or competencies. Consider diversity of age, gender, role in the community, and cultural background, in order to expand understanding and create inclusion.
Duration
Building communities in action takes time and intentionality of developing relationships, trust, and commitment. The duration of time will depend on the history of relationships in the community and the intention for implementing co-created solutions.

References

Bodiford K, Camargo-Borges C (2014) Bridging research and practice. Designing research in daily practice. AI Practit J 16(3):4–8

Bodiford K, Whitehouse P (2020) Intergenerative community building: intergenerational relationships for co-creating flourishing futures. In: The sage handbook of social constructionist practice. Sage Publications, Thousand Oaks, CA, pp 618–629

Camargo-Borges C, Gergen K (2022) Social construction and the forming of futures. In: Montuori A, Donnelly G (eds) The handbook for creative futures. Taylor & Francis/Routledge, Oxon/New York, pp 52–60

Gergen Kenneth J (2023) An invitation to social construction: co-creating the future fourth edition, 4th edn. Sage Publications, London

Leibowitz B (2017) Cognitive justice and the higher education curriculum. J Educ 68:93–112

Chapter 6
Community Creativity Booster Method

Predrag Jovanović, Ksenija Lalović, and Zoran Đukanović

6.1 Introduction

The open public space within one neighborhood represents a critical space for socialization. However, local governance participatory mechanisms need to be more sophisticated and developed in the context of a post-socialist society in transition, such as Serbia or other Eastern European countries, to achieve this quality. In such systems, children and youth usually do not participate in decision-making processes regarding the enhancement of these spaces, so their needs are neither included nor considered. The usual practice is that the open public space maintenance or renewal projects are top-down decided, and with limited funds, so often design solutions are replicated, not fitting the contemporary real needs of the end users, without relation to the climate change policies and measures. Therefore, this method focuses on including marginalized groups and promoting the Nature-Based Solutions co-creation on a community level.

The starting assumption is that children and youth, unburdened with systems, rules, and procedures, are more creative than adults and usually have deeper emotional relations toward nature. Through co-creation initiated by children and youth, strengthened by their parents or guardians, the decision-makers and other stakeholders are challenged with new perspectives and sustainable Nature-Based Solutions ideas (Bauduceau et al. 2015). Creating an unordinary, bottom-up, inspiring initiative and communication forms a niche for innovation (Mitić-Radulović and Lalović 2021), where decision-makers adopt new approaches. The innovation of this method is that it stimulates the optimization of the decision-making process according to the community's real needs, provides "on-the-job" training and

P. Jovanović (✉) · K. Lalović · Z. Đukanović
University of Belgrade—Faculty of Architecture, Belgrade, Serbia
e-mail: predrag.jovanovic@arh.bg.ac.rs; ksenija.lalovic@arh.bg.ac.rs; duke@arh.bg.ac.rs

J. Bentz, J. Ristić Trajković (eds.), *Imagining, Designing and Teaching Regenerative Futures: Art-Science Approaches and Inspirations From Around the World*, Science for Sustainable Societies,
https://doi.org/10.1007/978-981-96-9029-9_6

education for all participants (Živković and Lalović, 2018; Živković et al. 2019), and inspires and stimulates establishing trustful relationships between all stakeholders. In the Community Creativity Booster Method, the academic institution(s), competent for urban design, assumes the role of enabling leader (Lalović et al. 2019) through the application of several group communication techniques aiming to build trust and capacities of participants for co-creation in order to enable them to replicate the process in the future within the community.

6.2 Transformative Potential

This method stimulates the integration of a bottom-up decision-making approach in local urban design practice. The involvement of children and youth is a leverage of local community development and represents a "game-changing" potential for sustainable social transformation of local communities toward a more inclusive, innovative, and cooperative local governance model. To achieve effective results, the role of the academy is critical and yet not indispensable.

Also, the method's potential lies in establishing trust and partnerships between society's civil, public, and private sectors through collaboration, with demonstrable outcomes.

It disseminates best practices and motivates the wider community to participate in such processes.

Finally, the impact of Nature-Based Solutions on the environment is minimal, and the economic and social aspects are sustainable. It is about small-scale interventions that require minimal and affordable investment from the community and stimulate the community to cooperate.

6.3 Application

As creativity is limited in public procurement, this method represents an attempt to influence changes in practice in innovative ways. The method is developed through the cooperation of two research units of the University of Belgrade—Faculty of Architecture. These laboratories have rich experience with similar participatory processes.

The motive for applying this method was a need of one local community to find a way to articulate their idea about reaching the funds for the small-scale renovation of the public kindergarten courtyard. The "11. April" is a public preschool institution responsible for managing about 30 facilities, and does not have funding for such renovations. Then, through the local community association, the parents' council reached the Faculty of Architecture, which helped them to organize the workshop, which contained the following steps 1–4 and 6. However, step 5 is still an ongoing process.

1. The initial step is for the academic/scientific institution representatives to establish communication with the local community to strengthen their voices, articulate the needs of marginalized people, and initiate cooperation with critical decision-makers such as public authorities and public preschool/primary school utility companies. The main goal is to reach a mutual understanding of the initiative's purpose and a formal agreement and define the framework of the process.
2. The second step of gathering the youth's ideas about space design and changing the perspectives of other involved stakeholders is realized through the contextually tailored workshop facilitated by academic staff and students. The youth's creative group process is stimulated by applying different group communication, education, storytelling, and visioning techniques, to which the youth naturally responds positively. The main goal is to reach specific outcomes, such as narratives or drawings of an envisioned future when open public space fulfills all needs and is in harmony with nature. Conversations with youth and analysis of their drawings and stories reveal a completely different way of thinking and experiencing space. Parents and teachers are present and observe the process stimulated to change their perspective.
3. The next step, "shaping, materializing, visualizing ideas" refers to translating the collected ideas into conceptual open space designs enriched by students' creativity that easily emotionally and philologically relate to the youth involved. Under the guidance of teachers, associates, and representatives of other disciplines, students develop conceptual design proposals that are strongly related to the goatherd ideas but are realistically shaped for material realization. Conceptual designs need to have a preliminary calculation and specification of materials at a general level to prove the practicality of the proposal.
4. The broader local community should attend the public discussion with representatives of public preschool/primary school utility companies (especially teachers), private investors, and public administrators. The main goal of this step is to motivate the involvement of potential donors or contributors to support the realization of conceptualized design ideas. In this context, academic staff serve as facilitators of public events—such as exhibitions and discussions—where the main initiators are children and youth, now supported by their parents, teachers, and professors. The results of the public discussion are modalities for implementing and creating public–private partnerships.
5. When potential donors/contributors are recognized, the process initiators—youth's parents and guardians—are supported to form teams that will manage and take care of the realization of individual conceptual designs in cooperation with the donors and are supported by institutions. For every team, the first step is to find a construction firm that can adapt the designs to its capabilities. The result of this step is the completion of the open space design.
6. Public promotion of results and starting new initiatives represent the community's self-reflection event. Through it, the community shares experiences among themselves and with people who are interested in achieving the same or similar goals. The result of this step can build mutual trust and establish good practices and new initiatives (Fig. 6.1 and Table 6.1).

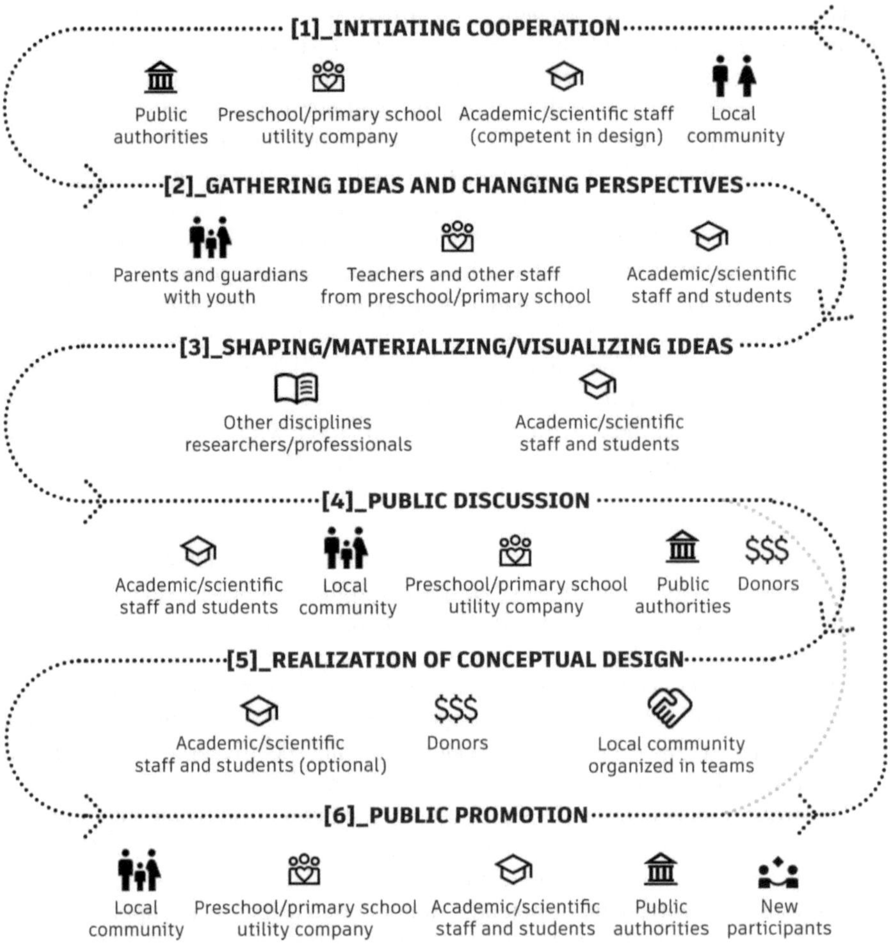

Fig. 6.1 The general process scheme

Table 6.1 Method overview

Main Purpose
Niche of innovation—innovation of social relations compared to usual practice. Orientation is toward the end user of a product of this process.
Gained competences
Learning good communication, establishing and fostering cooperation between different actors and institutions, and learning how to formalize those relationships.
Educational setting
This method uses the combined formal and informal educational format. The formal format (e.g., regular class) is taking place within the public institutions involved in the process. Informal formats (e.g., workshop and promotion of the conceptual project) are always taking place during the cooperation between different participants' categories.

(continued)

Table 6.1 (continued)

Space requirements/restrictions
If it is warm and dry weather, it is better to be outside within the space of future intervention. Otherwise, activities can take place within the public institution involved in the process.
Resources and necessary materials
For the workshops: paper and crayons. For the presentation: (1) a space to present the results, (2) a laptop with a projector or wider vertical surface for putting a poster on it.
Number of participants
It is group work with at least one representative of each participant category. It can be several groups.
Facilitator competences and skills
The process facilitator should be an educational and/or public scientific institution with recognized expertise in group communication. Also, there can be a team of facilitators consisting of experts from different fields.
Participants' skills/Age/Competences
Main categories of participants: Persons in need of care (e.g., children in kindergarten, primary school, persons with disabilities…) Legal representatives (of the persons within the first category) through the network of the public services of preschool and primary school education. Educational and/or scientific public institution, which is competent in the field of design (e.g., Faculty of Architecture). Other expert institutions (e.g., Institute of Social Sciences). Local community organization (formally or informally established).
Duration
The realization of steps 1–4 can last one to two months. On the other hand, the duration of project realization varies depending on its complexity.

References

Bauduceau N et al (2015) Towards an EU research and innovation policy agenda for nature-based solutions & re-naturing cities: final report of the Horizon 2020 expert group on 'Nature-based solutions and re-naturing cities'. Publications Office of the European Union, Bruxelles https://research.cbs.dk/en/publications/towards-an-eu-research-and-innovation-policy-agenda-for-nature-ba. Accessed 10 Sept 2024

Lalović K, Živković J, Radosavljević U, Đukanović Z (2019) An integral approach to the modeling of information support for local sustainable development—experiences of a Serbian enabling leadership experiment. Sustain For 11(9):2675. https://doi.org/10.3390/su11092675

Mitić-Radulović A, Lalović K (2021) Multi-level perspective on sustainability transition towards nature-based solutions and co-creation in urban planning of Belgrade. Serbia Sustainability 13(14):7576. https://doi.org/10.3390/su13147576

Živković J, Lalović K (2018) Place-based urban design education for adapting cities to climate change. In: Vaništa Lazarević E, Vukmirović M, Krstić-Furundžić A, Đukić A (eds) Keeping up with technonolgies to adapt cities for future challenges. Fifth International Academic Conference on Places and Technologies, Belgrade, April 2018. University of Belgrade—Faculty of Architecture, Belgrade, pp 641–651

Zivkovic J, Djukanovic Z, Radosavljevic U (2019) Urban Design education for placemaking: learning from experimental educational projects. In: Vaništa Lazarević E, Vukmirović M, Krstić-Furundžić A, Đukić A (eds) Keeping up with technologies to create cognitive city. Third International Academic Conference on Places and Technologies, Belgrade, April 2016. Cambridge Scholars Publishing, Newcastle Upon Tyne, pp 114–137

Chapter 7
Scale Shift & Mapping Drift: Transcalar Mapping for Participatory Planning

Eduardo Cassina

7.1 Introduction

We often look at notions of "regeneration" and "repair" at scales that are understandable and relatable to us, through a human lens, missing out on the more layered insights we would gain from approaching these ideas from different ranges, subjectivities, or entities. *Scale Shift & Mapping Drift* is a participatory workshop methodology designed to unravel and understand diverse perspectives within complex spatial systems such as a neighborhood, an institution, a building, a park, or an entire region.

Participants engage in an immersive experience that combines transcalar thinking (see Jaque and Office for Political Innovation 2020), Live Action Role-Play (LARP—a form of improvised embodied performance), card creation, and mapping. The workshop encourages participants to identify relevant (more-than-human) actors at various scales (such as a particular tree, an idea, a community, an ecosystem, or a policy), interpret them through a Life Action Role Playing (LARPing) exercise, design empathy-driven cards representing these entities based on their findings, and map their interrelationships.

This process provides an overview of the dynamics at play in a particular space, fostering a deeper understanding of the interconnected web of entities shaping the environment, unveiling areas that might require extra attention, and/or providing insights as to where regeneration can emerge from.

Here is an overview of the sequential steps:

E. Cassina (✉)
University of Agder, Kristiansand, Norway
e-mail: eduardo.cassina@uia.no

J. Bentz, J. Ristić Trajković (eds.), *Imagining, Designing and Teaching Regenerative Futures: Art-Science Approaches and Inspirations From Around the World*, Science for Sustainable Societies,
https://doi.org/10.1007/978-981-96-9029-9_7

1. *Identification of Entities*: Participants collaboratively identify various actors at different scales that are part of the *constellation* they want to work with (i.e., a given spatial context).
2. *Embodied Exercise*: Each participant selects one of the identified entities and engages in a Life Action Role Playing (LARPing) exercise, stepping into the shoes of the chosen stakeholder to understand its perspective, needs, and challenges.
3. *Card Creation*: Following the LARPing session, participants design cards individually reflecting on their gained insights. These cards encapsulate the characteristics, aspirations, and challenges of the entities they embodied in the previous step. Participants can swap the card method to include small objects, such as the creation of clay figurines, or other materials.
4. *Relational Mapping*: Using their created cards, participants collectively map the relationships and interactions between the identified entities, visually representing their corresponding interconnectedness across different scales.
5. *Reflection*: In the last step, participants share insights gained from the LARPing, card creation, and mapping exercises, identifying points of convergence, divergence, and potential areas for collaboration.

7.2 Transformative Potential

The methodology encourages a holistic understanding of spatial systems by considering entities at various scales. This approach goes beyond a singular perspective and unveils the complexities of interactions between diverse elements within a given environment. The improvised and embodied aspect of Life Action Role Play allows one to "borrow a position" (Kagovere et al. 2019) and develop empathy toward a diverse range of stakeholders. This visual aid helps participants grasp complex relationships, dependencies, and potential leverage points for intervention in a tangible and accessible way (Sachs Olsen 2022). By mapping the relationships between actors, the methodology highlights potential areas for collaboration and collective action. It is precisely, within this space of possibility and artistic collaboration, that the main contribution of this method resides: shifting scales, drifting connections, and encounters (De Kunst et al. 2020).

Scale Shift & Mapping Drift is useful in various spatial contexts where a comprehensive understanding of inter-actor dynamics is required. It is particularly beneficial when discussing larger issues such as how climate change is affecting a particular region, or the future of an inner city neighborhood, as this form of artistic collaboration allows for new imaginaries to emerge (see Eernstman et al. 2021).

7.3 Application

Imagine a community-driven urban resilience project aimed at addressing the large-scale transformation due to the gentrification, mass demolition, and eviction of several sectors of a historical inner-city neighborhood that is undergoing a profound turnaround. The project seeks to engage residents and local organizations in collaboratively designing regenerative solutions with the aim of preserving place identity and community networks amid the rapid changes that their neighborhood is experiencing.

7.3.1 *Identification of Entities (30–45 min)*

- *Preparation*: Assemble a diverse group of participants invested in the neighborhood, including long-time residents, business owners, and local organizations, as well as other human and social factors, gender, socioeconomic status, and neuromuscular abilities.
- *Workshop:* Participants collectively identify entities at various scales, such as historical landmarks, more-than-human entities (flora, fauna, geologic matter), intangible elements, local businesses, and municipal institutions. You may use group discussions, field observation, and storytelling to create a comprehensive list.

7.3.2 *Embodied Exercise - Life Action Role Play (1–2 hours)*

- *Workshop:* Each participant chooses one of the entities identified in the previous task, and announces it to the rest of the group. Once chosen, participants are invited to form a circle, close their eyes, and start grounding themselves, becoming aware of where they are. The facilitator invites participants to start thinking of their entities through a series of questions with the aim of inviting participants to enter into character (for example, *How are their entities part of that space? How do they feel right now?*). With each question, participants are invited to get more involved in their own entities. Eventually, participants are invited to open their eyes, and start interacting with one another and their surroundings.

- *How does a pigeon feel about the rise of tourist rentals? How is the smell of coffee from the local coffeehouses creating a sense of place? Where is the dust from the demolitions being accumulated? How does the dust affect the human and non-human population? And the smell of the coffee?*

- Participants are to act and interact with each other and their environment from the subjectivity they have chosen for the duration of this task.

- Participants are encouraged to move freely, and interact with one another, as it is through these interactions that new understandings between these different realities will emerge.

7.3.3 Card Creation (45 min)

- *Preparation*: Provide materials such as cardstock, markers, and other art supplies for card creation. Alternatively, instead of cards, you may create figurines, in which case obtain clay or modeling materials.
- *Workshop:* Participants design cards/figurines that encapsulate the characteristics, perspectives, and challenges of the entities they embodied, based on the insights they gained during the Life Action Role Play exercise. They design or build one card/or figurine per "insight." These insights can be based, for example, on findings resulting from particular interactions with other entities during the previous exercises.

7.3.4 Relational Mapping (45 min)

- *Workshop*: At this point, each participant has several cards or figurines representing emotions, feelings, tangible aspects, negotiations, etc., that they obtained during the previous exercise of Life Action Role Play. Participants place their cards and figurines on a surface and visually represent the relationships and interactions between the characteristics they have identified. These links can be either drawn or made with string, for example. The objective is to collectively create a rich relational map that provides a visualization of these interdependent relationships and a holistic understanding of the neighborhood.

7.3.5 Reflection (30 min)

- *Workshop*: Over the map, participants identify points of convergence, divergence, and potential areas for collaboration and action. They discuss how the workshop findings can inform regenerative solutions to navigate gentrification, community dissolution, and identity preservation. What new action plans have emerged from the session?

7.4 What Should Be Considered?—Some Tips!

- Participation should be as open and diverse as possible in order to have as many different perspectives.
- In the Life Action Role Play section, rather than embodying fellow humans, participants should be encouraged to explore other (non-human) entities. This avoids unnecessary problematic positionalities, but also enriches the conversation by adopting a radically different perspective (That's the whole point of the exercise!).
- Whenever possible, to further situate the workshop, it should take place at the location that is under study: this provides a great opportunity to engage with the materiality of the place itself (Table 7.1).

Table 7.1 Method overview

Main purpose
Empathy building/relational awareness.
Gained competences
Collaborative solutions, understanding interconnectedness in a given space, empathy-driven learning, and inclusive decision-making.
Educational setting
Workshop in participatory processes, educational environments, and informal settings.
Space requirements/restrictions
Can be done both indoors or outdoors (preferably outdoors, on the site that is under planning consideration!). A surface to draw on and to lay things on undisturbed and space to move around.
Resources and necessary materials
Card or clay (or modeling material), pens, string, scissors.
Number of participants
Individual and group work. Ideal group size under twelve people.
Facilitator competences and skills
The facilitator should be comfortable leading a group through a grounding exercise leading to the LARPing, as well as aiding discussion at different stages.
Participants' skills/Age/Competences
Community, students, policy makers, researchers, designers, practitioners, and residents. Of all ages and abilities (the more diverse the group, the better).
Duration
The entire workshop lasts around three hours.

References

De Kunst S, Babel J, de Andrés A, Paley V, Vervloessem G, Vuilleumier J (eds) (2020) This book is yours: recipes for artistic collaboration. Art&fiction publications, Vexer Verlag, Gallen/Berlin
Eernstman N, Pearson KR, Wals AEJ, Bjurström ÅE, De Vrieze A (2021) Designing collective artist residencies. Airea 3:17–34. https://doi.org/10.2218/airea.5314
Jaque A (2020) Office for Political Innovation. In: Graham J (ed) Superpowers of scale. Columbia University, Columbia
Kagovere O, Karvinen K, Vasko T (eds) (2019) Borrowing positions: role-playing design & architecture. Lugemik, Tallinn
Sachs Olsen C (2022) Co-creation beyond humans: the arts of multispecies placemaking. Urban Plan 7(3):315–325. https://doi.org/10.17645/up.v7i3.5288

Chapter 8
Enspirited Envisioning

Marilyn Mehlmann

8.1 Introduction

Enspirited Envisioning enables individuals and groups to create powerful visions of desired futures by accessing their tacit knowledge and intuition. The methodology includes subsequent steps to ground and actualize those visions to the point where they generate practical, feasible action plans.

Enspirited Envisioning was developed over several decades by Warren Ziegler, first in his role as leader of futures studies at the University of Syracuse, NY, and after retirement as a consultant and educator at Futures Invention Associates, Denver, CO. He himself used to say that Enspirited Envisioning was "only" a modern adaptation of ancient methods culled from Daoism (Biester & Mehlmann 2016).

This practice is useful for individuals, enabling new and powerful perspectives on the present and the past as well as the future. It is equally useful for groups, enabling consensus concerning desirable futures even in highly disparate groups, and offering tools to make concrete progress in a desired direction.

Having worked since the 1980s with scenario techniques I came to realize that, while powerful, they had a significant flaw: over-reliance on the facilitator not only for the process but also, at the critical point of synthesizing diverse scenarios, for the actual content.

Then I met Warren Ziegler, and that particular problem disappeared: the facilitation skills required for Enspirited Envisioning are concerned with the process and not the content, which the participants are enabled to develop themselves.

Working first for 10 years with Warren Ziegler and later alone or with other colleagues, I personally have facilitated Enspirited Envisioning in groups ranging from

M. Mehlmann (✉)
Legacy17 Cooperative Association, Stockholm County, Sweden
e-mail: marilyn@legacy17.org

J. Bentz, J. Ristić Trajković (eds.), *Imagining, Designing and Teaching Regenerative Futures: Art-Science Approaches and Inspirations From Around the World*, Science for Sustainable Societies,
https://doi.org/10.1007/978-981-96-9029-9_8

local government representatives (up to 90 participants, teams from different municipalities) to a local business club with 7 participants; as well as for groups of people participating on a purely personal basis.

8.2 Transformative, Regenerative Potential

One key to the efficacy of the program is that it builds on a key competence that is taught at all workshops: the practice of Deep Listening, both to others and—even more important—to oneself. This deceptively simple technique enables access to deep layers of intuition and genuine dialogue. And herein lies a huge transformative potential.

When leading a process for "Deep Listening to oneself," we invite participants to "visit" the different images of the future that they carry inside them. Everyone has at least three, and often many more. At a minimum, most people have a future image based on hopes, one based on fears or dissatisfactions, and one based on expectations.

Warren Ziegler used to say that *transformative change "is not something you plan. It's something that happens when there is a reasonable balance between dissatisfaction and hope."* No dissatisfaction: no need to do anything. No hope: no energy to do anything.

In today's society, there is a huge information imbalance: the very-real catastrophe scenarios dominate, often giving rise to despair or apathy—the "no-hope" end of the scale. It has become a cliché that while problems are increasingly global, the solutions must of necessity be local. The positive stories—the ones that give hope—are, in a globalized media world, less and less visible. And yet, in their eagerness to impress students with the urgency of the situation, educators often for instance take literally the words of Greta Thunberg: "I want you to panic!" They thus unwittingly contribute to exacerbating the situation.

It can be postulated that empowering people from all walks of life to envision viable, thriving futures—for themselves and for humanity—is one of the most powerful ways to inspire action toward sustainable and regenerative development. In this context, Enspirited Envisioning has an advantage in that it has not only been tested and refined over decades in a huge variety of circumstances but has also proved effective with people of widely varying cultures and educational levels: on all (inhabited) continents, from illiterate to academics, it has inspired untold numbers of personal and collective actions derived from huge numbers of futures images. Ziegler synthesized thousands of such images in his book, When Your Spirit Calls.

Important success factors are

- Deep Listening
- Engagement of art, music, body movement
- Simple decision-making tools for action planning

As to outcomes, organizations that follow through on the whole process have achieved notable results, including, for instance, revising municipal infrastructure planning to better reflect the wishes and needs of the residents. One municipality, for instance, created a strong movement to reinforce its character as in principle a network of villages, resisting the fashionable tendency to centralize.

The biggest challenge is probably securing a truly long-term process. A follow-up report 20 years later (2018) from that same municipality showed that the initial impetus, while very effective, had slowed down; the villages were intact, but enthusiasm had waned. Probably the work could have benefited from booster events, perhaps at intervals of 7–8 years—an eventuality that is not foreseen in the methodology.

8.3 Application

An Enspirited Envisioning program can take many forms, from a "study circle" format for local groups meeting once a week to a whole-week residential; or a set of sessions (ca 1.5 days) embedded in a longer event. The longest format, the whole-week model, was preferred by Warren Ziegler and is the most appropriate for groups whose members aspire to facilitate Enspirited Envisioning events in the future (New Economics Foundation 1999).

In all cases, there is a progression of activities, beginning with the individual/personal—and stopping there, if the workshop is for personal purposes. The general process is as follows:

8.3.1 Steps—Individual

- Listen inward to one's own images of the future: hopes, fears, and expectations that are present but not necessarily previously recognized or articulated.
- Listen to others' images, usually in threes: each person is allotted time to describe her or his images while the others Deep-Listen.
- Identify characteristics of a preferred future and make a personal poster "advertising" it.
- Visit the poster gallery and reflect on which futures one could happily live in.

8.3.2 Additional Step—Group

- Coalesce in groups around a common, combined desired future. This takes place via the poster gallery: participants who are drawn to each other's preferred futures form a group to work on and document an agreed common future, each of which is then added to the poster gallery.

The subsequent steps are versions of general action or project design: stakeholder analysis, identifying first actions and resource requirements, organization, risk analysis, planning, and not least: backcasting, or "future present."

Specific to this method is that the stakeholder analysis step includes identifying not only potential champions but also people or groups who may *not* appreciate your future, and incorporating these insights into the risk analysis.

Method overview

Main purpose
Access to intuition, decision-making, action design.
Gained competences
Deep Listening to self and others. Co-creation of desired futures. Reality check.
Educational setting
Any.
Space requirements/restrictions
Can be indoors, outdoors, hybrid, blended. Flexible spaces with room for a poster gallery.
Resources and necessary materials
Drawing materials, including for making posters.
Number of participants
6–90 participants (tested). Work: individual, group (often trios), plenary.
Facilitator competences and skills
Mastery of (teaching) Deep Listening to self and others.
Participants' skills/Age/Competences
Any from around 15 years and up.
Duration
Variations from 1.5 to 5 days, depending on ambition level.

References

Biester U, Mehlmann M (eds) (2016) A Transformative Edge. A publication in the Hosting Transformation library, https://legacy17.org/project/a-transformative-edge/. Accessed 2 June 2024

New Economics Foundation (1999) Participation Works! 21 techniques of community participation for the 21st century. https://neweconomics.org/uploads/files/e59722efbe227ca37e_4fm6b0lv9.pdf . Accessed 2 June 2024

Ziegler W (2002) When Your Spirit Calls. Xlibris Corp, FIA International, LLC

Chapter 9
Creative Partnerships for the Regeneration of Urban Rivers

Nagayamma Aragão, Luciana Bragança, and Carlos Smaniotto Costa

9.1 Introduction

In urban environments, water is an essential element for survival, embodying diverse meanings, paradigms, and usage patterns while also serving as a vital energy source. However, rivers are among the most vulnerable ecosystems within cities, particularly in the context of climate change (Aragão and Smaniotto Costa 2023). The intersection of social and environmental issues reflects contemporary urban society, where the nexus between cities, green spaces, and the natural heritage represented by urban rivers is pivotal for fostering sustainable urban environments. Through participatory and co-creative processes, *Creative Partnerships* leverage civic engagement in spatial development across varying scales and contexts. They develop tools and collaborative cartography to promote a socio-spatial agenda that advocates for social and environmental rights. By adopting a citizen science framework, Creative Partnerships aim to transform citizens into active political agents,

N. Aragão (✉)
Universidade Lusófona, Department of Architecture and Urban Planning & Research Centre for Architecture, Urban Planning and Design (University of Lisbon), https://urbanismo.ulusofona.pt/en/
e-mail: nagayamma.aragao@ulusofona.pt

L. Bragança
Universidade Federal de Minas Gerais, Architecture School, Departament of Projects, https://naturezapolitica.arq.ufmg.br/
e-mail: lubraganca4@ufmg.br

C. S. Costa
Universidade Lusófona, Department of Architecture and Urban Planning & Research Centre for Architecture, Urban Planning and Design, (University of Lisbon), https://urbanismo.ulusofona.pt/en/
e-mail: smaniotto.costa@ulusofona.pt

J. Bentz, J. Ristić Trajković (eds.), *Imagining, Designing and Teaching Regenerative Futures: Art-Science Approaches and Inspirations From Around the World*, Science for Sustainable Societies,
https://doi.org/10.1007/978-981-96-9029-9_9

engaging them in the assessment and preservation of local ecosystems. Central to this endeavor is the development of participatory mapping and design tools.

9.2 Transformative Potential

Creative Partnerships integrate citizen science, civic participation, and territorial education to cultivate sustainable urban spaces, working in collaboration with public policies and environmentally responsible urban projects. This initiative focuses on experiences in Lisbon, Portugal, and Belo Horizonte, Brazil (Bragança 2023; Sousa and Aragão 2025). Regenerating rivers under pressure from urbanization necessitates a multidisciplinary approach involving local governments, environmentalists, urban planners, water experts, and riverside communities.

The proposal seeks to redefine the urban river paradigm by improving water quality, enhancing biodiversity, enriching urban landscapes, and promoting the well-being of human and non-human communities. It serves as an invitation to reconnect with nature and demonstrates how teaching, extension, and research can generate positive impacts. Beyond environmental restoration, creative approaches empower participants and expand collective imagination. The development of tailored tools is vital to achieving these outcomes, as exemplified by the participatory mapping and design workshop described below.

9.3 Application

The participatory tools developed by *Creative Partnerships* are adaptable to various contexts. The Mapping and Participatory Design Workshop, first implemented in Belo Horizonte and refined in collaboration with Lisbon, illustrates this adaptability. The methodology involves presenting, collecting, and interpreting data and community demands, offering an understanding of territorial agents and their dynamics. Participants gain insights into the city and its riverbanks, share perceptions, and collaboratively produce spatialized proposals. These maps, reflecting the community's perspectives, serve as foundational documents for urban space projects and public policy initiatives.

Workshop participants create community maps by integrating perceptions, activities, agents, conflicts, and convergences into a cartographic base, thereby spatializing neighborhood relations, actors, and spaces of interaction. This process empowers participants to propose and visualize changes based on their unique experiences and interpretations.

- Workshop Materials
- Workshop materials include printed technical maps; large-format base maps of the area and surroundings; post-it notes, felt pens, colored pencils, crayons, and paper; and pre-prepared pictogram stickers representing activities and actors (both human and non-human). Aerial photography offers the ideal foundation. A neutral base, in particular, enhances opportunities for playful engagement, especially for children, though it is less effective for data collection.

- Workshop Methodology and Stages.
- Presentation of the Proposal: Participants are introduced to thematic mappings developed from quantitative surveys, official census data, georeferenced information, and qualitative research. The workshop facilitator outlines the objectives of the mapping exercise.
- Sensitization and Education (Optional): This phase, incorporated in recent workshops, enhances environmental awareness through walking tours, storytelling of local folklore, and group discussions. These activities deepen participants' understanding of urban morphology and environmental interactions, fostering imaginative engagement.
- Community Mapping: Participants independently create community maps, incorporating data into the cartographic base to reflect local relations, conflicts, and project sites. No external guidance is provided to ensure authentic community-driven outcomes.
- Practical Applications and Case Studies
- Initially developed in Belo Horizonte, the methodology was employed in the São Francisco River Park Project in Januária, Brazil, under a Technical Cooperation Programme between the Organization of American States (OAS) and Brazil's Ministry of the Environment. Two workshops were conducted with strategic stakeholders, involving 26 participants divided into two groups, and one workshop was held with 10-year-old public school children, comprising 10 participants (Bragança 2010).
- The first re-evaluation of the method in Belo Horizonte included non-human actors and socio-environmental conflicts. These were represented through pictograms and mappings. The workshop aimed to support public policy proposals for Recanto das Nascentes Park. It involved 12 members of the residents' association and 8 children.
- Subsequent refinements in Lisbon introduced interactive walks and folklore narratives to enhance participant engagement. Workshops conducted in Lisbon involved diverse groups, including researchers, government officials, preschool teachers, urban planning students, and 20 children (aged 3 to 6) (Sousa and Aragão 2025).
- Key Outcomes and Impact

- Creative Partnerships is supported by multiple entities demonstrating a robust network of interdisciplinary collaboration. Creative Partnerships foster the integration of technical and community-driven knowledge, leading to:

- The development and refinement of technical mappings from local perspectives.

- The creation of collective spatial information that informs urban river space planning.

By merging citizen science, civic participation, and territorial education, *Creative Partnerships* contribute to more sustainable urban environments. These activities address knowledge gaps while empowering communities, fostering inclusivity, and inspiring innovative solutions (Tables 9.1 and 9.2).

Table 9.1 Creative Partnerships: tools, objectives, and expected results

Creative Partnerships *Multi-actor approach to strengthen local decision-making structures toward regeneration of rivers*
River living lab
Instigate participants to reflect on the needs and potentials of rivers, also for more than human communities.
Awareness and sensitivity of complexity and challenges of urbanization, environment and rivers, and the human impact on them.
Collaborative cartography
Collectively map problems and opportunities.
Identified local realities and shared understanding of both material aspects present in the territory and the values attached to them.
Image bank
Support cartography and share views.
Co-constructed visual tools and mapped threats and opportunities.
Walking tours
Capture urban ambiance and generate knowledge about the relationship between space morphology and individual and group behaviors and practices.
Documented built-up, socio-cultural, and environmental components, and identified places of interest in the territory.
Game inventa(rio) cidade
Evaluate riverbanks quality, combining multispecies indicators.
Playful and visual (re)discovery of riverscape from local perspectives, leading to a reflection on the need to create sustainable urban development strategies.
Negotiation assembly
Increase capacity for empathy across multiple actors and species.
Links between rivers and communities, and identified needs as means to support behavior changes.

Table 9.2 Method overview

Main purpose
Creative Partnerships aims at transforming citizens into policy actors by involving them in the assessment and preservation of local ecosystems. It supports awareness building and behavior change in the context of rivers under urbanization pressure and addresses community engagement in river regeneration. It helps to strengthen the connection to rivers and participants' imaginary about the river's presence in their communities.
Gained competences
Individual and group motivation to actively participate in the changes; discover, understand, and value environmental assets that underpin behavior changes; imaginaries about rivers; ability to connect to urban water courses and to develop approaches to integrate them in the urban fabric in a more sustainable way.
Educational setting
The tools developed are to be applied with a riverine community, and in particular with non-traditional groups involved in the framework of urban planning (i.e., teenagers and children). The latest group should be involved through school, kindergarten, etc., to be able to contact them in a formal setting.

(continued)

Table 9.2 (continued)

Space requirements/restrictions
Creative partnerships foresee activities in both settings, indoor and outdoor, each setting with specific tools and with targeted objectives, the materials for both settings are slightly different.
Resources and necessary materials
The method does not demand expensive materials, but creativity and time efforts to tailor the tools to the audience. Suggested materials: Paper, colored pencils, markers; stickers. Stories, legends, and fairy tales about rivers.
Number of participants
There is no fixed number of participants. Just the tools should be enough to allow the full participation of all involved.
Facilitator competences and skills
The facilitator should: – Be able to clearly communicate the steps of the process to the participants. – Motivate all participants to actively share ideas. – Keep the group focused and on track. – Adjust the pace as needed.
Participants' skills/Age/Competences
Creative partnerships is primarily designed for community engagement, but it is also able to enlarge the audience and get students, policymakers, practitioners, etc., involved.
Duration
As a general rule, we recommend organizing three sessions, and each session is around 3–4 hr duration. Each successive session should wrap up the results.

References

Aragão N, Smaniotto Costa C (2023) A regeneração de rios urbanos, paralelos entre Portugal e África—Desafios e práticas. In: Smaniotto Costa C et al (eds) Rios Urbanos na Ibero-América: Casos, Contextos e Experiências. Cultura & Território 6. Lusófona University Press, Lisbon, pp 27–45. https://doi.org/10.60543/ecati/8tgb-ym88

Bragança LS (coord) (2010) São Francisco River Park in Januária – Community Biomapping Workshop. Januária, MG: OSLA Arquitetura. https://www.arqgrama.com.br/projetos/item/66-oficina-biomapas-comunitarios. Accessed 21 Nov 2024

Bragança L (2023) Jardins Possíveis e águas na cidade: pesquisa e ensino. Cadernos de Pós-Graduação em Arquitetura e Urbanismo 24(1):185–199. https://doi.org/10.5935/cadernospos.v24n1p169-184

Sousa M, Aragão N (2025) Plan for Flooding: Um jogo sobre rios e planeamento urbano para crianças dos 4 aos 8 anos. In: C Guimarães D M et al (eds) Marcos Conceituais e Metodológicos, USP, São Paulo, p. 245–250. https://doi.org/10.11606/9788566624410

Chapter 10
Foot(notes): A Walking Diary of Learning

Beatriz Moriano and Neus Lagunas

10.1 Introduction

In 336 BC, the Peripatetic (in Greek "walkway") school consisted of students at the Lyceum in Athens, who learned while walking outdoors with their teacher, Aristotle, believing that meaningful learning would occur by the method of observation, comparison, perception, and reflection of the world, even more so while strolling in nature.

The *Foot(notes)* approach is based on this concept. Walking is a primordial activity that has been studied by various fields of knowledge, which has, in turn, served and continues to serve as a source of inspiration and new understanding. From philosophy, sociology, anthropology, neuroscience, history, and art, this practice has been thoroughly analyzed, whose variants (marching, wandering, strolling, pilgrimage, exile) have political, religious, and spiritual connotations that can greatly vary depending on the context, the historical period, and the identities of the walking subjects. As an educational approach, moving on foot supports a reflective and interactive learning process that is useful in multiple interdisciplinary pedagogical contexts, due to the possibility it offers of articulating diverse content through the simultaneous practice of conversation in a mother tongue or, in our case, a foreign language, in particular, the teaching–learning of Spanish.

B. Moriano (✉)
Centre for English, Translation, and Anglo-Portuguese Studies (CETAPS)—Nova University of Lisbon, Lisbon, Portugal
e-mail: bmoriano@fcsh.unl.pt

N. Lagunas
Neus Lagunas, CHAM Centro de Humanidades - Nova University of Lisbon, Lisbon, Portugal
e-mail: nlagunas@fcsh.unl.pt

J. Bentz, J. Ristić Trajković (eds.), *Imagining, Designing and Teaching Regenerative Futures: Art-Science Approaches and Inspirations From Around the World*, Science for Sustainable Societies,
https://doi.org/10.1007/978-981-96-9029-9_10

Leaving the closed space of the classroom to move through public space, walking, instead of sitting during the learning session, is a practice with the potential to closely observe the environment and to question how we participate in the construction of the public space. In that sense, this approach can inspire regenerative futures, since putting the body in movement is to interconnect ourselves with a vital activity that offers creative experimentation and learning.

10.2 Transformative, Regenerative Potential

Walking is a free, ecological (nonpolluting), healthy (reduces tension and cholesterol), relaxing (de-stresses us), stimulating (awakens curiosity), and interactive way of learning, as it empirically relates us to others and to the world. Furthermore, it promotes relationships between people without power hierarchies. In many cases, each person can explain what they know and what they feel; the individual feels more freedom of movement and questioning. The barriers of communication in the classroom and the distance between the student, the teacher, and the rest of the group are diluted as we walk. The teacher is no longer the center of learning, as new pedagogical practices suggest, and the student becomes the driving force of their own learning. Learning is integrated into the environment yet is unpredictable like life itself, since we do not know what we may encounter and what we might notice. As we walk, thoughts occur and reality is presented to us with its inherent complexity.

This approach allows us to respond to students with different learning styles, such as those with special needs or those who need to move frequently, who respond better to more kinesthetic, visual, or auditory information, due to the reality of the public space, which can be explored while stimulating a variety of senses. Additionally, learning in a public space, with the analysis that this entails, allows self-knowledge, engagement, and participation on a small scale, which is a practice with considerable transformative potential that points toward regenerative futures.

The transformative potential of the different types of walking routes include (a) potential to develop awareness of languages that are present/absent in public spaces, that is, attention to configuration of the linguistic landscape, the current writing practices, and the communities that these practices (in)visibilize; (b) potential to bring attention to the noises and sounds produced by urban life, not only in the form of noise pollution but also conversations on the street as well as those among colleagues when speaking spontaneously in small groups without the discipline of the classroom, that is, attention to the soundscape and how it impacts our quality of life; (c) potential to study works of art in public spaces, their styles, and the claims of their authors, in other words, exploration of the artistic landscape and how it changes our lives; (d) potential to observe people who are on the street, what they do, what means of transportation they use, who they interact with, how they interact with the space (human ecosystem); and (e) potential to observe the interaction of humanity with oneself (reflection and self-knowledge). The potential of this teaching–learning practice lies in that it humbles us; it makes us vulnerable by depending on our body, by exposing us to the aggressions of the environment. Conversely, it also makes us resilient and integrates us into a place, inviting us to open ourselves to the unexpected, since, in public spaces, we can witness or intervene in multiple situations. Lastly, it forces us to look at the world beyond our screens, closed classrooms, and the interior of our cars.

10.3 Application

For its practical application, the profile of the group of students' interests and needs must be taken into account. Depending on this, the longer or shorter required duration of the walk or route is decided. In any case, the recommended duration is two hours, applied in small groups. As a learning task aimed at reflecting on the experience of walking and its transformative potential, each student individually writes a journal entry after each walking practice. This diary (written in the target language) can include guidelines the teacher/mediator considers useful. For example, the diary can be a multimodal text and, as such, can include photographs, drawings, sound recordings, videos, maps, notes found on the street, poems, personal written compositions, or links to other information related to the path traveled. Due to the combination of walking as a practice and writing a learning diary (in the target language) as an introspective tool that will be supervised by the teacher/mediator, with the student's permission, this approach is called *Foot(notes): a walking diary of learning*.

To illustrate this approach, the following activity serves as an example, which we carried out within the scope of the extensive reading program of the Department of Languages, Cultures, and Modern Literatures of the NOVA University of Lisbon—School of Social Sciences and Humanities. It involved an A1 level Spanish course focused on reading "Un caminar propio" (A Walk of Our Own), by Patricia Gosálvez (2019), published in the newspaper El País, in which the journalist walks for seven days through the surroundings of Madrid. After this reading, we proposed that the students take a walk following the previous approach, with an associated writing activity in the Moodle forum, a cooperative Diary of our *Foot(notes)*. We intended that the students, while reading in Spanish, would be more aware of a basic practice, such as walking through the city, simultaneously aware of the processes involved in walking: seeing things in another way, hearing new voices and noises, noticing elements that go unobserved, or simply being lost in thought. Subsequently, this experience was shared with positive results and was thus expanded so that oral interaction and conversation tasks could be carried out while walking: the conversations would be more fluid, more authentic, more natural, and more aware of the environment.

This method can be adapted and applied to other disciplines, for instance, Literature—the circuit of a literary novel; History—the history of mobility; Natural Sciences—trees and plants of the locality; Architectural—photographic diary of geometric forms in the city, etc. (Table 10.1). For further reading, we suggest considering the literature that examines the act of walking published by authors such as Iglesia (2019), Kage (2019), Le Breton (2012), Manuel Pedrosa and Estrela (2019), O'Mara (2020), Solnit (2001), and Thoreau (2018).

Table 10.1 Method overview

Main purpose
Awareness building.
Gained competences
Linguistic competences, authentic communication, possibility of self-knowledge, imagination, enjoyment of learning, local knowledge, spatial competences, resilience.
Educational setting
Formal (different levels)/informal (2 hours walking routes).
Space requirements/restrictions
Outside activities/Indoor activities.
Resources and necessary materials
Comfortable clothes, appropriate footwear, notebook, pen, colored pencils (optional), mobile to take photos or record sounds (optional).
Number of participants
Small groups of 4–12.
Facilitator competences and skills
General background Willingness to experience the place through walking Previous planning of the route (length and learning objectives)
Participants' skills/Age/Competences
Young and senior students.
Duration
4 h preparation (for the facilitator/s). 2 h walk (students and facilitator/s). 2 h diary writing (students and facilitator/s). 1 h discussion of diaries and balance of perspectives (students and facilitator/s).

References

Gosálvez P (2019) Un caminar propio (A Walk of Our Own). https://elpais.com/agr/un_caminar_propio/a. Accessed 28 Oct 2024

Iglesia AM (2019) La Revolución de las Flâneuses. WunderKammer, Girona

Kage E (2019) Walking. One step at a time. Viking, London

Le Breton D (2012) Marcher. Éloge des chemins et de la lenteur. Éditions Metáilié, Paris

Manuel Pedrosa M, Estrela J (2019) O que há neste lugar? Guia de exploração da paisagem. Museu da Paisagem, Lisboa

O'Mara S (2020) In praise of walking: a new scientific exploration. W. W. Norton & Company, New York

Solnit R (2001) Wanderlust. A history of walking. Granta, London

Thoreau HD (2018) Caminhada (M Afonso trans). Antígona Editores, Lisboa

Chapter 11
Futures-Inspired Embodied Transformation Theatre

Krisztina Jónás

11.1 Introduction

Futures-Inspired Embodied Transformation Theatre is a creative method that enhances imagination and enables a dynamic and embodied reflection about regenerative futures. Futures-Inspired Embodied Transformation Theatre combines existing methods of Social Presencing Theatre (Hayashi and Ricardo 2021) and the Three Horizons Framework (Sharpe et al. 2016). The Social Presencing Theatre is a method that belongs to a broader approach called Theory U (Scharmer 2009). The Theory U emphasizes that there is a greatly overlooked blind spot in initiatives that aim to facilitate any societal change: consciousness and awareness of the self and its relation to the qualities of open will, open heart, and open mind (Scharmer 2009). Social Presencing Theatre can make the invisible social structures and assumptions more tangible through a collective exercise of a facilitated theatre play, for instance, through the Stuck Exercise. The Stuck Exercise is a Social Presencing Theatre method (Hayashi and Ricardo 2021) where participants express themselves through different movements and by creating motionless sculptures of their bodies that represent their experience and feelings of anything they feel stuck with, which will be changed to a new sculpture that symbolizes the unstuck.

The Three Horizons Framework is a practice where various stakeholders individually and collectively reflect on how they think about the past, present, future and transition toward envisioned futures of a given phenomenon (Sharpe et al. 2016). Through fusing the Three Horizons Framework with the Stuck Exercise, the methods are enriched by reflecting on the past and present and how the self relates

K. Jónás (✉)
UiT The Arctic University of Norway, Norwegian College of Fishery Science,
Faculty of Biosciences, Fisheries and Economics, Tromsø, Norway
e-mail: krisztina.jonas@uit.no

J. Bentz, J. Ristić Trajković (eds.), *Imagining, Designing and Teaching Regenerative Futures: Art-Science Approaches and Inspirations From Around the World*, Science for Sustainable Societies,
https://doi.org/10.1007/978-981-96-9029-9_11

to the transition between tapping into their embodied wisdom. This way, the Three Horizons Framework is extended by a dynamic, embodied, and more personal layer.

In practice, in Futures-Inspired Embodied Transformation Theatre, participants first play the Stuck Exercise: participants go through three main phases and make three sculptures, one of the past or present, one of the future, and one of the transformations from past or present to future. The first sculpture symbolizes the current reality or past, where all participants shape an individual sculpture of their bodies. The next shows the envisioned future symbolized by the participants' body sculpture. In the third stage, the sculpture of current reality is transformed toward the future until it becomes the sculpture of the future. This is aligned with the Three Horizons Framework, where the three main steps are the past, future, and transformative pathways toward the future. The Three Horizons Framework exercise is done after an embodied sense-making process through the Stuck Exercise, enabling more relational, as well as personal insights into the past, present, future and transformations.

This combined futures and theatre method is powerful because it inspires participants to reimagine their relation to themselves, each other, and the future. Such an embodied sense-making process can help shift the focus on fixing the future with intellectual, cognitive means to an often overlooked, personal, emotional, and embodied, relational aspect of envisioned regenerative futures.

This practice is helpful for any context where regenerative futures are to be imagined. It supports visioning and sense-making of complex challenges while nurturing compassion through the embodiment of stakeholders and nature, offering insight into perspectives that might be otherwise difficult to relate to, with solely intellectual and mental approaches to the future.

11.2 Transformative Potential

This method can activate often overlooked and ignored situated, contextual, and personal wisdom that people carry. It might seem common sense that people and their experiences are embodied, which matters and should be considered. However, the current dominant understanding of what ways of knowing are valuable rarely recognizes it. In most dominant narratives, humans are depicted as separate from nature, mind from body, and humans from each other. The notion of embodiment highlights the embedded and complex relationalities between cognitive and emotional processes, human and nature. As such, humans and nature are interconnected and embedded (Weber et al. 2023). Acknowledging this interconnectedness can help envision regenerative futures for people and nature. Futures-Inspired Embodied Transformation Theatre might be used when complex challenges of futures are in focus, such as societal, environmental, cultural, and psychological challenges, to make them more tangible and relatable.

A significant potential of this method lies in its focus on envisioning futures, which can help to make past, present, and future experiences more tangible and relatable while connecting them to personal and collective perspectives and lived embodied experiences. This method emphasizes the contextual situatedness of the self and collective, drawing attention to the role of emotions, intentions, and cognitive processes.

The Three Horizons Framework has been widely used as a helpful future visioning practice with a mainly cognitive, intellectual approach to future imaginaries. This is why the Futures-Inspired Embodied Transformation Theatre can expand the Three Horizons Framework (Sharpe et al. 2016) with a more embodied perspective on past, present, future, and possible transformations between those, tapping into the lived experiences of the participants.

Embodied ways of knowing and knowledge co-creation, such as the Stuck Exercise, both consciously and subconsciously invite the present moment. It invites the wisdom of participants' embodied knowledge, including generational and intergenerational memories, as well as social, cultural, physiological, and psychological situatedness that cannot be generalized and should be valued and acknowledged in any transformation process. Such attention to these contextual nuances needs more attention in future visioning workshops, which is why the Stuck Exercise can enrich the Three Horizons Framework with embodied ways of knowing. At the same time, the Three Horizons Framework can contribute to the Stuck Exercise with a future vision of broader societal, regional, and global processes. Since humans have been changing the biosphere with consequences of unprecedented scales, they also have the agency to be stewards of the biosphere (Folke et al. 2011), contributing to regenerative futures. Such biosphere stewardship cannot come in isolation, and the interconnectedness between humans and nature, and between humans across social groups, is essential.

Through tapping into the embodied experience, participants are invited to explore their personal and intimate relation to their present, future, and nature while leveraging their involvement in the transformation process. This stands in contrast to the most common ways of engaging, mainly through the cognitive process of co-creational thinking about a shared future vision where relevance to the individual and collective embodied experiences might be overlooked.

11.3 Application

Facilitating this method might be easier with participants familiar with creative and embodied methods. However, this is not a prerequisite. Two hours are needed for its implementation: one hour of Stuck Exercise and another hour of Three Horizons Framework. Two facilitators are helpful for this method, and it is recommended that the facilitators know the method and ideally play a pilot workshop before the actual workshop to gather feedback and improve facilitation.

When it comes to materials needed for this exercise, flipcharts or big sheets of paper, colorful pens, and colorful sticky notes are helpful. The method can be applied individually, but it is recommended to be used in groups to facilitate collective learning. It is suitable for any group, from children to adults, professional to informal settings.

11.4 Step-by-Step Guide

1. Futures-Inspired Transformation Theatre Part 1 (inspired by Stuck Exercise):
    - Sharing in a circle: All participants sit in a circle and, one by one, reflect on the question: What is the past (or present) of the human–nature connection for you? Leaving the circle, removing chairs, and instructing for sculpture formation.
    - Sculpture 1 of Past or Present. What is the human–nature connection's past (or present) that feels "stuck" for you? Participants make a sculpture with their bodies that represents their feelings.
    - Collective reflection: Brief reflection on what everyone experienced in their sculpture.
    - Sculpture 2 of Regenerative Futures. What is the emerging regenerative future of the human–nature connection for you?
    - Sculpture 3 of Transformation. How do you move from the past (or present) to the envisioned regenerative future of your human–nature connection? This is a dynamic movement from Sculpture 1 (past or present) to Sculpture 2 (regenerative future) when participants take on the shape of Sculpture 1 and with a movement (that can consist of one movement or a series of movements) go to Sculpture 2.
    - Reflection: After all participants have shared their transition movement from Sculpture 1 to 2 (it can be done together or separately), they collectively briefly reflect, facilitated by guiding questions: How did experiencing the past, present, and future feel? What insights would you like to share?
2. Futures-Inspired Transformation Theatre Part 2 (inspired by the Three Horizons Framework):
    - The group collectively reflects on what regenerative future(s) mean(s) to them, tapping into the inspiration and insights they might have gained during the Stuck Exercise.
    - The participants individually take notes (e.g., on sticky notes) about how they envision regenerative futures, followed by sharing and placing the notes on paper (e.g., flipchart). On the paper, there can be a pre-drawn graph where three lines represent the three main stages of the exercise: past/present, future, and transformation, as used in the Three Horizons Framework (Sharpe et al. 2016). The sticky notes with ideas about regenerative futures can be placed on the scenario (lines) that symbolize the future. This may be guided by a particular focus, such as envisioning regenerative futures (taking notes individually and collectively on a shared board).

- After a regenerative future is envisioned, participants individually, then collectively, reflect on the past and that if it continues (as business as usual), it is not likely to enable regenerative futures (taking notes individually and collectively on a shared board).
- Individual and collective reflection on transformation pathways toward regenerative futures that can bring the currently dominant pathways toward the envisioned regenerative futures (taking notes individually and collectively on a shared board).

3. The participants' final collective reflection on the past/present, future, transformation between and how the embodied Stuck Exercise shaped their co-creational process, including reflection on their own individual and collective agency in the envisioned regenerative futures of human–nature connection (Table 11.1).

Table 11.1 Method overview

Main purpose
Awareness-building and sense-making.
Gained competences
A more profound sense of agency in envisioning and co-creating envisioned future(s); a sense of community and shared agency.
Educational setting
Informal (format—workshop).
Space requirements/restrictions
Indoors and outdoors are possible with spacious spaces where one can move freely. Initially, one chair is needed per person, so every participant can sit where the chairs are in a big circle.
Resources and necessary materials
Big sheets of paper, colorful pens, and Post-its.
Number of participants
2–30 (groups of 2, 4, or 6, depending on the whole group size).
Facilitator competences and skills
Familiarity and preparation with the methods are required, including a pilot workshop in advance. If desired, there are several official trainings offered for Social Presencing Theatre and the Three Horizons Framework, but these are not prerequisites for facilitating the described method
Participants' skills/Age/Competences
It is helpful if the participants have previous experience with embodied methods, but it is not a requirement.
Duration
Two hours.

References

Folke C, Jansson Å, Rockström J, Olsson P, Carpenter SR, Chapin FS et al (2011) Reconnecting to the biosphere. Ambio 40:719–738

Hayashi A, Ricardo DG (2021) A pattern language for social: field shifts. J Awareness-Based Syst Change 1(1):35–57

Scharmer CO (2009) Theory U: learning from the future as it emerges. Berrett-Koehler Publishers, Oakland

Sharpe B, Hodgson A, Leicester G, Lyon A, Fazey I (2016) Three horizons: a pathways practice for transformation. Ecol Soc 21(2):47

Weber EU, Constantino SM, Schlüter M (2023) Embedding cognition: judgment and choice in an interdependent and dynamic world. Curr Dir Psychol Sci 32(4):328–336

Part II
Artistic Expression and Experimentation for Environmental Change

Artistic expression and experimentation can be powerful catalysts for environmental change, transforming abstract ecological concerns into concrete, emotionally resonant experiences. Through diverse approaches, artists can challenge dominant narratives, provoke dialogue, reimagine humanity's relationship with nature, and open space for alternative ways of seeing and being. When embedded in participatory processes, art can become a tool for connection, resistance, and regeneration. The methods in this section invite communities to reflect, feel, and act together, building bridges between the individual and the collective, the local and the planetary. The creative methods, including poetry, mural-making, performance, eco-theatre, photography, and sensory installation, nurture the environmental aesthetic and affective aspects of sustainability learning. They engage learners through intuition, imagination, and embodied exploration. These approaches support transformative learning by awakening deeper ecological awareness and cultivating the creative capacities needed to envision and enact regenerative futures.

Part II
Antibody Expression and Experimentation for Fundamental Studies

Chapter 12
A-Lab Performed: Facilitating Aesthetic Exchange and Performance

Elsa Kosmack Vaara, Alina Östling, and Cheryl Akner Koler

12.1 Introduction

A-Lab Performed (A = aesthetic) is an exploratory method that builds upon the original A-Lab method developed by Akner Koler (Akner Koler and Ranjbar 2016). The original A-Lab method enables participants to share their aesthetic experiences of a phenomenon—such as time and temporality, nanotechnology, or vibrotactile technology—from an individual, emotional perspective within a playful setting. A-Lab Performed expands on this by guiding participants to compose and perform their experiences, adding a performative and expressive dimension to the process.

The A-Lab method consists of investigating and sketching with haptic technology while using aesthetics in artistic practice (for example, music) as a point of departure.

We have used A-Lab Performed in combination with different artistic practices such as music and culinary arts for many years, and we have invited researchers, practitioners, artists, musicians and industrial design students to use the method (Akner Koler and Ranjbar 2016; Kosmack Vaara and Akner-Koler 2021).

E. K. Vaara (✉) · A. Östling
RISE, Reseach Institutes of Sweden, Stockholm, Sweden
e-mail: elsa.vaara@ri.se; alina.ostling@ri.se

C. A. Koler
Konstfack University of Arts, Crafts and Design, Stockholm, Sweden
e-mail: cheryl.akner.koler@konstfack.se

J. Bentz, J. Ristić Trajković (eds.), *Imagining, Designing and Teaching Regenerative Futures: Art-Science Approaches and Inspirations From Around the World*, Science for Sustainable Societies,
https://doi.org/10.1007/978-981-96-9029-9_12

12.2 Transformative Potential

The A-Lab Performed method aids designers in exploring aesthetics and using artistic practices to evaluate and play with experiences with emerging technologies. In our aesthetic lab sessions, we worked with music and handheld vibration prototypes, offering perspectives that blend novel technical solutions with personal sensory engagement. This method helps designers apply artistic and aesthetic knowledge in sketching and designing with technology. The process is planned with the aim to let the outcome of the activities and participants' results guide the subsequent phase in the design process of products, services, and experiences. For example, in our case with vibration, the students continued to work with electric scooters and eventually designed suggestions on how vibration designs may have an impact on the scooter driving experience.

The method enhances our understanding of how technology can enrich lives, supporting aesthetic engagement and imagination as we shift from mono-disciplinary to inter- and cross-disciplinary methods for a sustainable society (Peeters and Trotto 2018; Höök 2020; Kosmack Vaara and Akner-Koler 2021). This transition fosters communication between engineering and cultural disciplines, blurring their boundaries.

Our example shows how the method helps bypass normative views of vibrations. Among other things, we have learnt that the students managed to refer their emotional experiences of music and vibration to real-life experiences. This indicates how the music can help dissolve the dualistic tension between "feeling" as an inner experience and "feeling" as a tactile act of experiencing material properties (Born 2011). We have seen that the method helps students and practitioners to engage in aesthetic bodily ways to understand what vibrations are and how vibrations may be used for sketching and how vibrations may be designed to evoke diverse experiences.

12.3 Application

In this section, we will describe how the *A-Lab Performed* method was used with students in industrial design at Konstfack—University of Arts Crafts and Design (For an overview of the method please see Table 12.1). The students explored vibrotactile technology with the final goal of conceptualizing aesthetic experiences with vibrations. The three main parts of the method A-Lab Performed are:

1. Artistic connoisseurship: Artistic connoisseurship means that an artist is brought into the sessions to share his/her knowledge and specific artistic skills, in, for example, dance, music, or culinary arts. This is similar to "somatic connoisseurs" (Shiphorst 2011). Through bringing the connoisseur into the design process, the participants learn about skills that are not necessarily part of a designer's training.

Table 12.1 Method overview

Main purpose
A-Lab Performed plays with haptic aesthetic phenomena such as vibration and lets the participants collectively explore and perform their learnings. The method emphasizes the importance of creating a connection to artistic practice and enables participants to share their experiences from their individual and emotional point of view.
Gained competences
Participants explore their haptic sensitivities and gain unique perspectives that blend haptic technology with aesthetic and artistic sensory engagement. This method enables participants to understand how haptic technology can be controlled and applied through, for example, vibration on diverse parts of the body. A-Lab Performed provides the possibility to support aesthetic engagements and imagination as we make the transition from traditional mono-disciplinary methods to today's science of collective, inter- and cross-disciplinary methods needed for transformations into a more sustainable society.
Educational setting
Design/art students, bachelor's, master's, PhD students, experienced professionals, artists, others interested in exploring various challenges and phenomena through aesthetic reasoning combined with artistic practice.
Space requirements/restrictions
The explorative and performative sessions are best led in a quiet and safe place with few possibilities of being interrupted. It can be performed lying down or sitting up, inside or outdoors.
Resources and necessary materials
1. One vibration controller for each group with a sliding scale for intensity from low to high and different vibrotactile patterns (can be other kinds of devices such as a high-quality sex toy or devices fit for purpose). 2. Templates for note-taking that portray the body. 3. Paper for graphically mapping the vibration composition with words, pictures, and time/body indications.
Number of participants
Around 6–18 participants are divided into groups of three people (the host, the blindfolded, and the note taker in part 2).
Facilitator competences and skills
Creativity, artistic skills, research skills.
Participants' skills/Age/Competences
A third of the participants should be students/practitioners with a design/art background. The others should see relevance in the haptic activities and phenomena.
Duration
Around 2–3 hours.

2. Collaborative sensitization of the chosen phenomenon, in our case vibration. (This part is described further down in the text.)
3. The participants use their experiences of the phenomenon to compose and perform. Reflection is done between each part.

The participants were introduced to vibrations by an artistic connoisseur in classical music. In our case, one of the co-researchers is a classical musician, so she could take on the role of artistic connoisseur.

The A-Lab Performed began with the musician showing how she manipulates her instrument (a viola) to create sounds and how she uses the bow to shape rhythms and volumes of sound with the instrument. She explained how she connects her own body to the vibrations of the resonating body of her viola. After that, the students were asked to lie down on the carpeted floor while the musician performed a piece of music for about 8 minutes. The students were asked while listening to the musician playing her instrument, to reflect upon how the vibrations of the music activated them (e.g., sensing something in their body, emotional reactions, a special place or memory).

As the *A-Lab Performed* progressed, the students were divided into groups of three to go through the second part of the method: collaborative sensitization. To explore vibrations, we used simple vibration prototypes that a colleague helped us construct (Windlin 2020). The prototypes consisted of small handheld vibration motors (ERM—eccentric rotating mass actuators) connected to a 9 V battery, and they were connected to a MIDI board programmed with 10 different vibration patterns.

The groups were handed a body template illustration with instructions, where they were asked to document their learnings. Each student in a group took turns doing one of three roles: 1, the host, 2, the blindfolded, and 3, the notetaker. The sensitizing practice involved a collective experience as the members were engaged with each person's unique aesthetic reaction when vibrations were tested on different body parts.

The practice with our handheld vibration prototype aimed to support the students to navigate their own embodied understanding of vibration, as well as sharing their aesthetic reactions to what it feels like, both in terms of negative and positive sensations and emotions. Following the collaborative sensitization activity, each group was asked to use their templates and descriptions of experiences of vibrations on different parts of their bodies, to compose and visualize (notate) a piece of vibration experience, and then perform it on their classmates. The activity aimed to inspire a shared vibration-form-giving process, allowing students to channel the creative energy they had gained through musical exploration and collaborative sensitizing practices. In sum, the A-Lab Performed can be used for an educational purpose with students and professional practitioners within design and/or artistic practice, PhD students, practitioners-researchers, and others interested in exploring new technologies through aesthetic reasoning combined with artistic practice.

In design, a vast quantity of new materials is continuously being invented and developed. The design process is flexible in particular in contexts in which there are no specific previously cultivated rules of composing, improvising, and performing. Designers can learn from artists in order to become stronger as expressive craftsmen with technology. Finally, we would like to point out that the method is not limited to vibration as a phenomenon but can be used to explore other phenomena with other kinds of tools. This chapter uses vibration as an example, and it is possible to perform the method without the vibration prototypes that were described here. In our earlier endeavors, the participants have used simple tools such as screwdrivers or wooden sticks, tuning forks, sextoys, and loudspeakers. Use the tools that you find suitable and purposeful!

References

Akner Koler C, Ranjbar P (2016) Integrating sensitizing labs in an educational design process for haptic interaction. FORM akademisk 8(2):1–25

Born G (2011) Music and the materialization of identities. J Mater Cult 16(4):376–388

Höök K (2020) Soma design-intertwining aesthetics, ethics and movement. In: Proceedings of the Fourteenth International Conference on Tangible, Embedded, and Embodied Interaction, pp 1–1

Kosmack Vaara E, Akner-Koler C (2021) Exploring and prototyping the aesthetics of felt time. J Artistic Res 22. https://doi.org/10.22501/jar.1039453

Peeters J, Trotto A (2018) Designing expressions of movement qualities. In: DIS '18: Proceedings of the 2018 Designing Interactive Systems Conference. DIS '18 Designing Interactive Systems, Hong Kong, June 2018, Association for Computing Machinery, New York, pp 679–690

Shiphorst T (2011) Self-evidence: applying somatic connoisseurship to experience design. In: Extended abstracts on human factors in computing systems, CHI '11: CHI conference on human factors in computing, Vancouver, May 2011. Association for Computing Machinery, New York, pp 145–160

Windlin C (2020) Designing with the body: addressing emotion regulation and expression. In: DIS' 20 companion: companion publication of the 2020 ACM designing interactive systems conference, the 2020 ACM designing interactive systems conference, July 2020. Association for Computing Machinery, New York, pp 557–562

Chapter 13
Circle of Life Workshop: Where Nature, Art, and Local Practices Meet

Andra Simanova, Dārta Unda Zālīte, Ieva Legzda, Līga Volkova, and Baiba Prūse

13.1 Introduction

The workshop *Circle of life as an arts-based environmental learning* links art, nature, and traditional and local culture and brings attention to the conservation of biological and cultural diversity. The workshop method is based on several years of experience and its central element is the application of plant specimens in creative artwork as a means of reconnecting with nature and local culture. The creation of plant artwork results in an aesthetically enjoyable outcome. During the series of workshop exercises, participants learn about plant species in their local context and collect and prepare these plants according to local practices. Fresh and dried plant artworks are created in a circular form, symbolizing the Sun's path in the sky, the solar calendar seasons, and traditional ethnographic patterns where the circle is one of the most universal motifs. Symmetrical patterns enclosed in a circle reflect cosmogonic ideas about the structure of the world (Laime 2006). The circular form of ethnographic patterns also corresponds to the shape of the mandala, used in art

A. Simanova (✉) · D. U. Zālīte
Society "Ecovillage", "Omrači", Aizkraukle County, Latvia
e-mail: andra.simanova@inbox.lv

I. Legzda
Society "The House of Fairy-tales "Undine"", "Omrači", Aizkraukle County, Latvia

L. Volkova
A. Upitis Skriveri Secondary School, Skriveri, Latvia

B. Prūse
Athena Institute Vrije Universiteit Amsterdam, Amsterdam, The Netherlands

University College Cork, Cork, Ireland

Ca'Foscari University of Venice, Venice, Italy

J. Bentz, J. Ristić Trajković (eds.), *Imagining, Designing and Teaching Regenerative Futures: Art-Science Approaches and Inspirations From Around the World*, Science for Sustainable Societies,
https://doi.org/10.1007/978-981-96-9029-9_13

therapy by Carl Gustav Jung to express the inner situation reflecting the transformation of the mind (Jung 2017).

13.2 Transformative and Regenerative Potential

Biologically diverse meadows and the cycle of solar calendar seasons are the sources of inspiration for the workshop "Circle of Life." A transformative experience combining a walk in the meadow with artistic activities and mythology concepts rooted in collective experience about the Sun allows the evolution of a person's attention to both realms: cultural and biological diversity. This is crucial in modern times when biologically diverse grasslands are almost disappearing. The experience gained through contact with nature and plants not only benefits physical, emotional, and intellectual human development (Klaar and Öhman 2014) but also increases interest and responsibility for preserving biodiversity (Witt and Kimple 2008).

Walking through a blooming meadow, we rarely imagine how rich it could be in terms of species diversity (for example, in the Baltic Sea region (Latvia), a third of the plant species found in the country—more than 400—could be found in grasslands (Rūsiņa 2017)). It's a space where life unfolds. Meadows are also a source of inspiration for folk songs, literature, and more. By working in the meadow or simply taking a stroll memories may arise. Human cultural identity and understanding of the surroundings are developed by inheriting traditions that have emerged in that environment. The biological diversity of childhood surroundings becomes a reference point for measuring environmental diversity in later life (Miller 2005). Losing contact with this diversity can lead to the societal extinction of cultural experience (Jarić et al. 2022). Arts-based learning invites participants to feel a direct connection with the subject of the activity (Van Boeckel 2013), and thus the method described provides the transformative and regenerative element of reconnecting with the natural surroundings.

13.3 Application

The workshop can be organized for small and medium-sized groups, suitable for both children and adults, including families with children. It can be implemented in various events, both in natural environments and urban areas, outdoors as well as indoors. The workshop involves two stages that can be done interchangeably. The first phase, namely *the meadow walk*, is implemented if there is an area for plant collection (e.g., grassland or garden) and at least one person in the group is knowledgeable about plant species and their harvesting methods as well as capable of sharing information about the local application of plants. If the first phase is not possible, only the second phase, namely *the art-based practice with the collected plant materials*, is implemented using pre-organized dried plant specimens (e.g., herbal teas).

13.3.1 The First Phase—Meadow Walk

The meadow walk involves physical activities, direct contact with plants, soil, and sunlight, and the opportunity to observe the surroundings and become aware of oneself in this environment. During the walk, all senses are actively engaged. The walk usually requires 2 hours and can serve as a preparatory phase for the second part of the workshop.

A prerequisite for the workshop is to choose a suitable area for plant harvesting. Together with the participants, collect the above-ground parts of plants—leaves and flowers, and gently cut them with scissors. For diversity purposes, select plants in different colors. Cut the collected plants and dry them by spreading them out on a paper away from direct sunlight. Another drying method is to bundle the plants with natural twine and hang them to dry. Store the prepared dried plants in paper bags or sealed glass jars, which can be reused. Label the packaging with the name of the plant.

Required materials: a basket or a bag made of natural materials, scissors, paper, and natural twine. It is recommended to harvest the plants under the guidance of someone who is familiar with the local flora. Plants should be collected in places where it is allowed, taking care to preserve their population. Pay attention not to accidentally harvest rare species.

13.3.2 The Second Phase "Circle of Life" Creative Session

The second phase of the workshop can be completed in 30 minutes to several hours. It is recommended for younger children to choose a smaller format or create the artwork together with a facilitator. As a workshop leader, do not restrict the participants and encourage creative freedom! It is encouraged to prepare aromatic herbal tea and enjoy it before starting the workshop session!

There are several methods that are applied as part of the creative session. One of them is arranging fresh plants in various patterns on a suitable surface—on the ground in a meadow, on a stone, on a table, etc. The used materials can later be dried or composted, observing how they change. Let each participant work at their own pace and rhythm; a carefully crafted work done with care and interest gains meaning in our lives. Allow everyone to feel herbal material through all the senses.

An additional technique is gluing dried herbs onto paper. Thus, the outcome of the workshop can be used as interior decor for an extended period. The artwork transforms just like the meadow changes across the seasons. You can observe how the shades of herbs transform, acquiring a lighter, silvery, or darker, brownish tone.

Four stages are involved when applying dried herbs:

1. Choose a suitable format for your work, and create a sketch of the circular ornament with a pencil, charcoal, or chalk. Prepare the herbs—cut the stems, and pulverize dried leaves and flowers crushing them between your palms.
2. Apply a small area of glue, put crushed dried herbs on it, and press with a cloth or fingers. It is important to apply a thick layer of glue and carefully cover it with plants, leaving no empty spaces. Repeat this process until the entire surface is covered.
3. Lift the work and shake off any excess herbs. Check if any additional touches are needed.
4. Place the artwork in a location where it is not directly exposed to sunlight.

Required materials: fresh or dried herbs, scissors, a working surface (ground, table, or cardboard), glue, and paint brush for gluing. If necessary, a pencil, charcoal, or chalk for sketching (Table 13.1).

Table 13.1 Method overview

Main purpose
Awareness building and arts-based environmental learning.
Gained competences
Creativity. Collaboration.
Educational setting
Informal learning.
Space requirements/restrictions
Outside activities. Indoor activities.
Resources and necessary materials
Required materials and technical equipment for the first phase of the workshop—meadow walk: a basket or bag made of natural materials, scissors, paper, and natural twine. Required materials and technical equipment for the second phase of the workshop—Circle of Life creative session: fresh or dried herbs, scissors, a working surface (ground, table, or cardboard), glue, and paint brush for gluing. If necessary, a pencil, charcoal, or chalk for sketching.
Number of participants
Small or medium-sized groups (5–15 participants).
Facilitator competences and skills
Knowledge about local plant species and their harvesting methods, the capability of sharing information about the traditional use of plants.
Participants' skills/Age/Competences
Children/adults.
Duration
Meadow walk—2 hours; "Circle of Life" creative session—30 minutes to several hours.

References

Jarić I, Roll U, Bonaiuto M, Brook BW, Courchamp F, Firth JA et al (2022) Societal extinction of species. Trends Ecol Evol 37(5):411–419

Jung CG (2017) Mandala symbolism : (From Vol. 9i collected works). Princeton University Press, Princeton

Klaar S, Öhman J (2014) Doing, knowing, caring and feeling: exploring relations between nature-oriented teaching and preschool children's learning. Int J Early Years Educ 22(1):37–58

Laime S (2006) Kosmoloģiskie priekšstati Latvijas klinšu rakstos un sprēslicu ornamentos. Cult Crossroads 3(2):53–65

Miller JR (2005) Biodiversity conservation and the extinction of experience. Trends Ecol Evol 20:430–434

Rūsiņa S (2017) Dabiskās pļavas un ganības. Dabas aizsardzības pārvalde, Sigulda

Van Boeckel J (2013) At the heart of art and earth. Res Arts Educ 3:32–36

Witt SD, Kimple KP (2008) 'How does your garden grow?' Teaching preschool children about the environment. Early Child Dev Care 178(1):41–48

Chapter 14
Herbarium with Poetry: How to Connect People and Plants

Baiba Prūse, Sandra Sarfo, Sarata Darboe, Julia Prakofjewa, Edy Fantinato, Andrea Troncoso, Cristina Flora, and Renata Sõukand

14.1 Introduction

Humanity might be in the midst of an epidemic of plant blindness (Parsley 2020; Jose et al. 2019; Krosnick et al. 2018). Plant blindness refers to the tendency of humans to perceive plants as one indistinguishable green background without differentiating any species or attributing any importance to plants. Yet, our food security depends on plants (Sõukand et al. 2021) and, even more, on our knowledge of how to use plants. To retain the knowledge accumulated through the generations, an urgent increase in interest in plants is needed. The method proposed here is based on our "Herbarium and Poetry workshop," which took place in Venice (Italy) during the ONA Short Film Festival (2021). This method merges biodiversity learning and creative writing: in the first stage, participants take part in collecting and displaying dried herbarium specimens; second, they custom their sample by writing a "Little Elfje," a short poem of 11 words that aims to describe their feeling for the natural

B. Prūse (✉)
Athena Institute Vrije Universiteit Amsterdam, Amsterdam, The Netherlands

University College Cork, Cork, Ireland

Ca'Foscari University of Venice, Venice, Italy

S. Sarfo · S. Darboe · J. Prakofjewa · E. Fantinato · R. Sõukand
Ca'Foscari University of Venice, Venice, Italy
e-mail: yuliya.prakofyeva@unive.it; edy.fantinato@unive.it; renata.soukand@unive.it

A. Troncoso
Freelancer Consultant in Creativity and Communications, A Mountain with Wings, Heidelberg, Germany

C. Flora
University of California Davis, Davis, CA, USA

J. Bentz, J. Ristić Trajković (eds.), *Imagining, Designing and Teaching Regenerative Futures: Art-Science Approaches and Inspirations From Around the World*, Science for Sustainable Societies,
https://doi.org/10.1007/978-981-96-9029-9_14

elements. This practice was applied during an outdoor nature and adventure film festival, but it can be easily adapted to diverse purposes, including research, awareness raising, and connection to the local environment.

A herbarium, as a collection of preserved plant specimens, holds a long history and has played a crucial role in transmitting knowledge in plant research and beyond. Poetry is one of the most ancient means of human artistic expression, and it can help us to bring plant stories into our lives. It can provide us with a new language that conveys the confluence of intellectual and emotional strands, creating a visceral connection to the environment. Throughout history, there have been several instances where the art of crafting poetry and the scientific practice of creating herbaria have been intertwined. One prominent example is Emily Dickinson, an American poet of the nineteenth century, who was known to have a deep interest in botany and created a herbarium with her own poems (Marshall 2015). Also, Charlotte Smith, an English novelist and poet, crafted a literary herbarium, intertwining the distinct characteristics of the plants depicted in her poetry. This collection highlights the ever-evolving material environments and connections between the nonhuman and human realms (Boettcher 2023). Particularly the Little Elfje (Elfchens in German, Little Elevens in English) provides the opportunity for anyone to express their story, which was for this particular exercise, about a plant. By combining both elements (the dried plant specimen and the poem), participants create a personal sharing and attachment to the chosen plant. This approach ensured that the workshop was not just an academic exercise but a holistic educational experience connecting the dots between botany, culture, and artistic expression.

14.2 Transformative Potential

Many things can happen when a person decides to embark on an internal journey that will lead them to a poem-writing experience in connection with a plant. We recommend that the participants pause, closely observe, and give themselves permission to express their creative self, in a language that might be dormant in them most of the time. We consider this act tremendously regenerative since it awakens a power that we often forget to have: to translate feelings, thoughts, and emotions into words, inspired by the connection with a plant. The plant works as a prompt for poetry, and the practice of sewing the dried specimen on paper creates the space for emotions to settle down and take the form of the Little Elfie. The local flora used in the exercise connects participants with the reality they likely experience on an everyday basis or gives them a hint for what to search for on later walks. As in ethnobotany, herbarium specimens serve as a link between emic and etic, between people's knowledge and scientific knowledge. The resulting artwork keeps the memories and the link vivid and may prompt them to explore more.

14.3 Application

The activity can be carried out both individually and in small or large groups (e.g., as an icebreaker for an event). This practice takes around 30 minutes, but it requires preparatory work, which adds an additional time requirement. Therefore, the authors have divided the application into two sections: the workshop session and the preparatory phase.

14.3.1 Workshop Session

The workshop was built upon a strong foundation of ethnobotanical fieldwork, meticulously prepared by the workshop facilitators, who collected plant specimens and dried them on newspaper sheets. For the workshop, the following tools were used: white cardstock paper, needles (for sewing the dried specimen on cardstock paper), white sewing thread, post-its, table/chairs, instructions for the poem, and pens or pencils to write the poem. Before the beginning of the workshop, the facilitators arranged the materials making them ready for participants to use.

Once the location is set up, participants are welcomed and briefed about the dynamic of the workshop. For the herbarium making, the dried specimens and required tools were handed to each participant. Afterward, a short briefing is provided on the Little Elfje poem, which involves five lines (or verses) and holds a distinct pattern (ECSA 2021):

A first line is made of 1 word (an adjective).
A second line is made of 2 words (a noun with an article).
A third line is made of 3 words (an action of the noun).
A fourth line is made of 4 words (a complement of the action).
A fifth line is made of 1 word (a noun that closes the story).

For the Festival workshop, we emphasized the attention to the plant as one of the elements for making the poem. After the poem-writing session, the participants were invited to use a post-it to note down the local name of the plant and the location where it was collected (provided by the facilitators), which created the "plant ID" for each specimen.

14.3.2 Preparatory Phase

In order for the dried specimens to be ready for the workshop, the facilitators collected the plants 2 weeks before the session. Smaller common plant specimens were also dug out with roots to create the matching element of the authentic herbarium collections. After the collection, the plants were dried following three steps. First, the specimens were cleaned from the soil; second, each plant specimen was placed between newspaper pages adding a post-it with the basic information ("plant ID"), including time and day of collection, place, and short location description, and Latin name of the plant when available (e.g., using available community-based platforms, e.g., the NatureSpots app running on SPOTTERON Citizen Science App platform, and for proper identification inviting a botanist is essential). Last, each newspaper with the specimen inside was covered with an empty newspaper page and placed under a heavy, flat object (e.g., a book). For the research process, the botanical press is used. The press usually consists of layers of cardboard, blotting paper, and newspaper to absorb moisture from the plant, with straps or other mechanisms to apply pressure and ensure the plant dries flat. Then, the plants undergo daily inspections throughout the drying process. It is crucial to store the collected plants in a dry environment.

14.3.3 Future Recommendations

When used together, herbariums and poetry can provide a well-rounded educational experience, combining scientific knowledge with emotional engagement and artistic expression. Ideally, involving the participants in the preparatory work, including the process of choosing, collecting, and drying the plant(s), would provide a more meaningful experience and connection with plants. The instructors of such activities can channel the choice of plants based on either their usability (wild food plants, medicinal plants, personal preferences of the students, etc.), their prevalence in the area (most widespread local species, all flora that grows around the school, etc.), or in a long-term perspective, to document the changes in the environment and relations between plants and people (Table 14.1).

Table 14.1 Method overview

Main purpose
Creative community engagement.
Gained competences
Creativity.
Educational setting
Workshop.
Space requirements/restrictions
Outside and indoor activity.
Resources and necessary materials
White cartoon paper, needles (for sewing the dried specimen on white cartoon paper), white sewing thread, post-its, table and/chairs, instructions for the poem, and pen/pencil for writing the poem, dried plant specimens.

(continued)

Table 14.1 (continued)

Number of participants
Group or individual.
Facilitator competences and skills
Experience with making herbarium and facilitating creative storytelling workshop.
Participants' skills/Age/Competences
Students, community, practitioners.
Duration
30 minutes to 1 hour (workshop). Up to 2 weeks (preparatory work for dried plant specimens).

References

Boettcher C (2023) "But poets have never been botanists": the literary herbarium in Charlotte smith's poetry. Eur Romant Rev 34(4):441–459

ECSA (2021) https://www.youtube.com/watch?v=AJz1O58YDF4. Accessed 2 June 2024

Jose SB, Wu CH, Kamoun S (2019) Overcoming plant blindness in science, education, and society. Plants People Planet 1(3):169–172

Krosnick SE, Baker JC, Moore KR (2018) The pet plant project: treating plant blindness by making plants personal. Am Biol Teach 80(5):339–345

Marshall Z (2015) What do collections tell us?--Teaching Dickinson's poems, the herbarium, and contemporary and historical material practices. Teach Am Lit 7

ONA Short Film Festival (2021) https://onafilmfestival.com/program/. Accessed 2 June 2024

Parsley KM (2020) Plant awareness disparity: a case for renaming plant blindness. Plants People Planet 2(6):598–601

Sõukand R, Kalle R, Fontefrancesco MF, Pieroni A (2021) Building a safety buffer for European food security: the role of small-scale food production and local ecological and gastronomic knowledge in light of COVID-19. Open Res Europe 1(10):10

Chapter 15
Ecotheatre as a Tool of Environmental Imaginary

Kitija Balcare

"Stories are ecological forces as potentially powerful as hurricanes." [May 2007]

15.1 Introduction

Ecological theatre or ecotheatre is an approach for the creation of an environmental imaginary through collaborative exploration and embodied storytelling while taking into consideration sustainability aspects of performance making. Ecotheatre is a collective practice that can help to reduce individual ecoanxiety, deepen personal ecological awareness, and foster engagement in environmental activism (Hudson 2020). Conventional anthropocentric or human-centered theatre is commonly searching for answers to the question "Who are we?" In contrast, ecotheatre seeks to find an answer to the question "Where are we?", reassessing both human attitudes to the environment and our place in nature. The strategy of ecotheatre is to ask questions, to encourage an active re-evaluation of the existing practices and systems, rather than to impose ready-made solutions, to develop a "what if" perspective as a precondition for environmentally conscious action (Balcare 2022).

With the escalation of the global climate crisis, especially since the beginning of the twenty-first century, new concepts continue to enter the theatre field, including ecodrama, ecodramaturgy (May 2007; Woynarski 2020), ecodirecting, and ecoscenography (Beer 2021)—the overarching concept of this theatrical movement is ecological theatre or ecotheatre (also eco-theatre). The concept of ecotheatre intertwines two meanings:

A theatrical production based on a narrative about the relationship between humans and nature, including the more-than-human world, in which the environment, whether imagined or real, plays an active role.

K. Balcare (✉)
Institute of Literature, Folklore and Art of the University of Latvia, Riga, Latvia
e-mail: kitija.balcare@lulfmi.lv

J. Bentz, J. Ristić Trajković (eds.), *Imagining, Designing and Teaching Regenerative Futures: Art-Science Approaches and Inspirations From Around the World*, Science for Sustainable Societies,
https://doi.org/10.1007/978-981-96-9029-9_15

A theatrical production that takes into account sustainability in its development, during its performance, and even afterward, being aware of its impact on the environment and trying to reduce it in a circular way.

15.2 Transformative Potential

Ecotheatre comes as a response to environmental issues through the performing arts, with the aim of using the tools of art to contribute to a sustainable society. While science typically warns of the climate risks and related issues, ecotheatre adds to science an emotional layer. Ecotheatre addresses environmental issues through storytelling, shifting from the collective experience to the individual experience. For instance, it tells the story of biodiversity loss through a close-up look at a single species and its role in the ecosystem.

Using ecotheatre as a tool in the field of socially engaged art is a technique that can deepen ecological awareness in an engaging and educational way.

- Ecotheatre is co-creative—it often requires the active participation of all those involved in the process, including the audience, not only mentally but also physically. It is often created in collaboration with people from different backgrounds, such as natural scientists, historians, journalists, families, and the local community, often involving them as performers.
- Ecotheatre allows for a substantive engagement with specific environmental challenges and can deconstruct particular issues, both scientifically and personally, allowing for deeper understanding and conscious action.
- The artistic space created by ecotheatre offers a variety of alternative solutions to the environmental issue, contributing to a change in attitude or even action by both theatre practitioners and audiences, thus transforming current practices into more sustainable ones.

Ecotheatre is therefore not only an art form but also a participatory tool for community work, looking at scientific facts with an emotional engagement and helping to alleviate ecological anxiety by channeling this energy into co-creative work. Through collaborative exploration of environmental issues, a particular group becomes open to change—at individual, community, and even wider levels.

An inspiring example is the ecotheatrical performance “All Birds Sing Beautifully” developed by the Latvian theatre director Krista Burāne in 2023. The touring performance with a participatory approach highlighted the rapid decline of bird populations, affected by intensive agriculture and forestry. The ecotheatrical performance drew on the latest ornithological research and alarming data on specific rapidly declining bird species in Latvia, with a poetic layer of local folklore folk songs about the affected species. Song was the connecting element between people and birds. The performance took place in urban green areas in several cities across Latvia. The audience was involved by singing in a choir led by bird conductors. Birds were dressed in impressive folk costumes made of recycled fabrics, ribbons, and yarns. Birds told stories about their species, while engaging the audience in physical activities that illustrated bird-friendly farming practices, such as sustainable mowing practices. Through active participation, the audience was able to incorporate this knowledge into their physical memory, becoming more aware of their connection to nature while experiencing a closer interaction with the more-than-human world.

15.3 Application

Theatre practitioners implement various forms of ecotheatre that can be used by others, for example, in community activities, schools, NGOs, and team-building events. When used in pedagogical work, the ecotheatre approach can promote the acquisition of interdisciplinary skills by combining theoretical knowledge with practical experience. The method is inclusive, as it can be used for individual tasks as well as for groups of different sizes and ages. It can be a one-hour task to find an idea and share it with others through storytelling in a comforting, reassuring way, or it can be a longer-term project culminating in a site-specific performance involving the local community (Table 15.1).

What do we care about? If you want to create your own performance, you can start by identifying a specific environmental problem that you want to address with the tools of ecotheatre. Pollution in the local river? The effects of climate change in

Table 15.1 Method overview

Main purpose
Awareness building/activism enabling.
Gained competences
Storytelling/collaboration/environmental literacy.
Educational setting
Individual or group workshop.
Space requirements/restrictions
Outside activities/access to the natural environment.
Resources and necessary materials
Access to information (library, archive, Internet, etc.). Recording device. Writing tools (digital or physical).
Number of participants
From one to unlimited.
Facilitator competences and skills
Media literacy/environmental literacy/sensory literacy/mentoring skills.
Participants' skills/Age/Competences
Community/students/NGOs/volunteers.
Duration
Not fixed.

the town? The impact of forestry on wildlife habitats? Lack of climate justice in the community? To analyze the problem, it is advisable to research scientific findings, articles in the local media, local archives, and folklore. It is advisable to approach the local community by asking questions about personal experiences related to the issue. On the basis of the information gathered, ecodramaturgical material should be produced, fusing all these voices into one story or choosing another strategy—to build a personal story based on research.

Where are we? Once the content or ecodramaturgical material has been developed, the next step is to find a location that will play an important role in conveying the message of the performance across. If a very local theme is chosen, the search for an idea can also start from the other side—by choosing a location that contains a sensitive story. It is possible to choose between an imaginary setting and a real, specific place in nature. It can be a place in nature, a museum, an archive, a zoo, the shadow of a tree next to the school, a hiking trail, or any other place that resonates with the story or which plays an active role in it. It is also important to include elements of audience participation during the performance. For example, the opportunity to be in physical contact with nature helps to deepen ecological awareness and improve people's physical and mental health.

How will we tell the story? When creating a performance, consider sustainability aspects by creating a green action plan that will contribute to an environmentally friendly end result and reduce the ecological footprint of the performance (Theatre Green Book 2024; Sustainable Production Guide 2024). In ecotheatre, costumes should also be made from natural materials, recycled materials, or upcycled objects. The performance can take different forms, such as a performance-lecture, a performance-walk, a performance-forum, a performance story evening, etc. One task could be to find a place in nature that is important to the group or local community and create a story about it, inviting others for a theatrical walk. You could also choose a place in nature that is important to you, and give it a voice by writing a monologue using the voices of nature, or by recording an audio file and marking the route for others to follow as an audio-walk. A task for the students could be to observe a particular animal in the neighborhood, e.g., a beetle, a bird, or a horse, to understand its structure and to mimic and follow its movements without trying to caricature the animal. If folklore is part of the school curriculum, local stories or songs that reflect humans' relationship with nature can be used in themed sketches and performed in a suitable setting outside the school.

What is the next step for us? The emphasis of ecotheatre is to move toward action—ideally in a way that brings people and nature closer together and improves mental and physical health. By exploring the possibilities of tomorrow, ecotheatre practitioners and participants become activists for an uncomfortable yet irresistible change.

References

Balcare K (2022) Ecotheatre: changing perspective from who we are towards where we are. Culture Crossroads, 21:56–65

Beer T (2021) Ecoscenography. An introduction to ecological design for performance. Palgrave Macmillan, Singapore

Hudson J (2020) The environment on stage. Scenery or Shapeshifter? Routledge, London/New York

May TJ (2007) Beyond a Bambi: toward a dangerous ecocriticism in theatre studies. Theatre Topics, 17(2):95–110

Sustainable Production Guide. https://juliesbicycle.com/resource/sustainable-production-guide/. Accessed 15 June 2024

Theatre green book: sustainable productions. https://theatregreenbook.com/book-one-sustainable-productions/. Accessed 15 June 2024

Woynarski L (2020) Ecodramaturgies. Theatre, performance and climate change. Palgrave Macmillan, New York

Chapter 16
Ecostories for Children: An Art–Science Approach

Lena Pozdnyakova

16.1 Introduction

In an era where sustainability is crucial, international agreements like the Paris Agreement or the United Nations Framework Convention on Climate Change often fail to emotionally engage the public. Scientific research is typically inaccessible to those outside academia, creating a communication gap between cutting-edge research and society. One solution is the Ecostories/Eco-comic method. This approach creatively explains connections between natural phenomena and culture, encouraging children to take daily actions that can lead to societal transformation and climate resilience. Through collaboration with artists, professionals can write kids' ecostories or create comic books that convey these ideas in compelling, inspiring, and accessible ways.

16.2 Transformative Potential

Viewing the world through the lens of children, the surroundings are usually full of wonder, playful potential for transformation, and joy. Looking at the world through the lens of art, scientists, and the environment around us often presents a network of connections full of information and meaning. To connect the worlds of children to those of scientists and artists requires building bridges. In this context, art and its impact on human perception arguably becomes one of the most effective forms of

L. Pozdnyakova (✉)
Institute of Art History, Free University Berlin, Berlin, Germany
e-mail: pozdnyakoy85@zedat.fu-berlin.de

J. Bentz, J. Ristić Trajković (eds.), *Imagining, Designing and Teaching Regenerative Futures: Art-Science Approaches and Inspirations From Around the World*, Science for Sustainable Societies,
https://doi.org/10.1007/978-981-96-9029-9_16

communication and a potential way to build bridges among different groups, including children.

Communicating environmental issues through children's stories allows us, as scientists, researchers, and global citizens, to precisely determine what we want to convey to future generations inheriting the planet from us. The approach of simplifying complex phenomena is challenging but beholds transformative potential as the medium of illustrated books requires the creator to provide concise and clear messages while taking constructive and hopeful positions.

It is safe to say that, throughout our lifespan, books deeply touched many of us, changed our worldviews, taught us empathy, and guided our behavior throughout life. Often, the seeds of our potential are planted in childhood. With this, we can imagine that illustrated books for children featuring heroes and role models who, in today's context, have the potential to take action and address environmental and social challenges have tremendous significance for our culture. We can create new stories that inspire kids and young adults and, most importantly, allow them to believe they can make a difference and be part of positive change in the world. Many children's books now emphasize diversity and inclusion, showcasing characters from different backgrounds and abilities, helping to create a sense of belonging and respect for all members of ecosystems while promoting a more equitable society and kin-centric worldview.

Addressing challenging issues in a playful and accessible way could be inspiring and lead to practical changes in behavior within the whole family. Why? Because reading stories, picking books from the store, and changing habits at home start with parents. In other words, easy, hopeful literature integrated into households can serve as a catalyst for change.

16.3 Application

An exciting aspect of this method is its accessibility and potential to integrate seamlessly into collaborative processes within interdisciplinary and transdisciplinary teams. Imagine a world where children, working alongside researchers and artists, can create such books, feeling empowered to contribute to positive change. The good news is that some projects are already embracing and supporting these initiatives.

The experience of creating such a book emerged from the Global Research Academy, a collaboration between King's College London, Free University Berlin, and the University of São Paulo, centered on sustainability. The project took place in three cities, bringing together early-stage researchers to conduct site-specific, interdisciplinary, and transdisciplinary research. The aim was to gather knowledge on-site, engage with local communities, and then analyze and share the findings in a form that is both accessible and engaging. As an artist, my involvement was driven by the intention to foster alternative methods of knowledge dissemination—specifically, through a children's book that would convey the story our research group focused on: the narrative of water justice in São Paulo, Berlin, and London.

The process of creating a children's book based on scientific research was divided into four stages:

16.3.1 Interdisciplinary Research On-Site

The first core element of the project was to bring together individuals from diverse disciplines—hydrologists, ethnographers, sociologists, anthropologists, biologists, activists, artists, and other early-stage researchers. Organized into smaller groups, they spent a week in each of the three cities. Their primary objectives were to gather data, observe the environment, establish connections, and, most importantly, engage with local communities. Each group included at least one local expert who could offer in-depth insights into the area.

During these sessions, the researchers explored the cities on foot, took photographs, interacted with various community groups, and recorded extensive field notes. The success of this stage lies in bringing together curious and motivated individuals, each approaching the subject from their own angle, and giving them the freedom to explore the environment while maintaining an ongoing dialogue with one another.

16.3.2 Choosing a Shared Point of Interest and Synthesizing Findings

Given that science-driven information is often dense with terminology and context-specific details, the next step for the group was to identify shared points of interest (a common theme) and deepen their understanding through ongoing communication. At this stage, the involvement of an artist becomes invaluable. Techniques like free-hand drawing and visual mapping exercises help illustrate connections between findings, highlighting similarities and differences in a clear and accessible way. Simple sketches can accelerate the process and provide initial concepts for the book.

This phase can last from several weeks to several months, depending on the depth of exploration. The key is to draft intersections of findings from each expert's perspective, with the artist continuously creating connections, maps, and sketches to guide the narrative and shape the emerging themes.

16.3.3 Interdisciplinary Storytelling Phase

In this phase, the group identifies the most critical aspects of their findings and weaves them into a cohesive thematic narrative. Storytelling ideas are tested, protagonists are chosen, and the structure and flow of the story take shape. The process is deeply collaborative: the artist creates initial sketches for the book, which are then shared with the group, leading to a dynamic exchange. Scientists help refine complex language, while the artist brings attention to overlooked details. This iterative dialogue ensures the narrative is both engaging and accurate. Depending on the group's size and the project's scope, this phase can last from several weeks to several months.

16.3.4 Transdisciplinary Storytelling Phase

The transdisciplinary storytelling phase is an inclusive process where different communities are invited to contribute to and refine the draft version of the project. This involves workshopping the initial draft through community feedback sessions with the target audience—children, parents, and local communities. Each session begins with a brief introduction of the section's theme, followed by discussions and drawing activities. Participants, in this case, kids and parents, pause after each section to sketch their ideas and provide feedback. Protagonists, environments, and situations in the book are adapted based on these contributions, shaping both the content and stylistic elements of the book. Simplified forms and more diverse environments often emerge from the children's input.

This phase is highly productive, significantly influencing the content, production, and final outcome. It demonstrates that involving the community in creating a story about sustainability benefits everyone involved, making the process both entertaining and educational. For the GRA book, this workshopping took place at a local community project in Berlin, where a Saturday daycare in Moabit, hosting kids aged 5 to 14, provided the perfect setting for this interactive and inclusive phase.

16.3.5 Finalizing the Book

The final stage involved refining drawings, adding text, and ensuring all elements were aligned. With the narrative and style established, editing became an enjoyable task. The book was then prepared for printing or online publication, culminating in a public presentation celebrated with stakeholders either on-site or online.

16.4 Conclusion

As the climate changes, children are among the first to recognize the urgency of adapting. Books can guide them naturally toward change rooted in knowledge and action. Children's books on sustainability serve as a child's introduction to complex issues like daily habits, human impact, and civic responsibility. The GRA project showed that engaging content, accessible language, and collaborative processes benefit everyone involved and inspire further sustainable action (Table 16.1).

Table 16.1 Method overview

Main purpose
Communication/awareness building/empowerment.
Gained competences
Environmental awareness, empathy and emotional engagement, and communication skills.
Educational setting
Formal and informal formats—workshops, seminars, community sessions, etc.
Space requirements/restrictions
Indoor and outdoor environments can be practiced on the go (during field trips, expeditions, etc.).
Resources and necessary materials
Art supplies, space to work, printing/zine-making resources.
Number of participants
Groups of people, from 3 members to 12 members per group ideally.
Facilitator competences and skills
Facilitator skills: writing and storytelling/drawing, painting and sketching/environmental science knowledge/critical thinking/pedagogy/communication skills/editing and proofreading.
Participants' skills/Age/Competences
Communities, students, kindergarten kids and school kids, scientists and researchers, policymakers, designers, activists, and others. Any age and ability.
Duration
Depending on the scale and ambition of the project, it can range from a simple zine (created in a single 3–4 hour session) to a full book, which may require at least a month with 1–2 sessions per week.

Chapter 17
The Inner Beauty of Art-Science Installations

Martina R. Fröschl

17.1 Introduction

The Science Visualization Lab at the University of Applied Arts in Vienna specializes in blending scientific visualization with artistic expression to create accessible and emotionally resonant messages about environmental sustainability. By combining methods such as documentary filmmaking, art–science installations, collaborative research, and efficient computing, the lab addresses complex ecological challenges in creative and impactful ways. This contribution focuses on art–science installations as a method for educational and advocacy contexts. It provides a practical guide to using this method, detailing the steps, tools, and considerations necessary for readers to apply it effectively in their own settings. Please keep in mind that this is only one method of many, applied in the Science Visualization Lab of the University of Applied Arts Vienna, since limited space does not allow to describe all the methods, also the methods and approaches are constantly adapted to contemporary circumstances.

17.2 Transformative Potential

Art–science installations are powerful tools for fostering environmental awareness and inspiring action. Their transformative potential lies in awareness raising, accessibility, emotional engagement, community involvement, and versatility. By integrating scientific data with artistic expression, installations make abstract concepts

---

M. R. Fröschl (✉)
University of Applied Arts Vienna, Vienna, Austria
e-mail: martina.froeschl@uni-ak.ac.at

J. Bentz, J. Ristić Trajković (eds.), *Imagining, Designing and Teaching Regenerative Futures: Art-Science Approaches and Inspirations From Around the World*, Science for Sustainable Societies,
https://doi.org/10.1007/978-981-96-9029-9_17

tangible and relatable for diverse audiences. Immersive experiences, interactions, and visual storytelling resonate with audiences on a personal level, sparking curiosity and a deeper connection to the subject matter. Information is conveyed with a low threshold and accessible in a playful manner. Installations often invite active participation, fostering dialogue and encouraging collective action. Exhibition side events are encouraged to motivate more people to see the respective art–science installation. It is an adaptable approach that can be applied in various settings, from classrooms, galleries, and museums to public spaces.

17.3 Application

This guide provides a step-by-step approach to creating an art–science installation. It focuses on the practicalities and considerations needed to implement this method successfully in various contexts.

17.3.1 Step 1: Identify the Topic

Choose a theme that aligns with your educational or advocacy goals, such as biodiversity loss, climate change, or water conservation. Ensure the theme is relevant to your audience and location. Use scientific data to ground your project in credible research.

17.3.2 Step 2: Collaborate Across Disciplines

Form a team that includes scientists, experts, designers, scholars, and artists to combine technical expertise with creative perspectives. Create opportunities for open dialogue and idea exchange. For instance, organize a meetup in which participants are led to brainstorm. Define roles and responsibilities to streamline collaboration.

17.3.3 Step 3: Design the Installation

Develop a concept that integrates scientific information with engaging artistic elements. Use visuals, soundscapes, and interactive components to create a multisensory experience. Simplify complex data without compromising accuracy. Additional information can be provided in publications but should not compromise the main artistic or entertaining experience.

17.3.4 Step 4: Develop the Components

Create the visual and physical elements of the installation, using tools and materials suited to your concept. Tools: Digital equipment (projectors, screens, VR setups), art supplies, and scientific visualization software. Materials: Recyclable or sustainable materials whenever possible. Testing: Iterate on prototypes to ensure technical functionality and audience engagement.

17.3.5 Step 5: Install and Engage

Set up the installation in a venue that suits your audience, such as a school, museum, or community space. Accessibility: Ensure the space is inclusive and easy to navigate. Engagement: Include opportunities for audience interaction, such as Q&A sessions or hands-on activities.

17.3.6 Step 6: Evaluate and Reflect

Gather feedback from participants to assess the installation's impact. Use surveys or discussions to collect insights. Reflect on what worked and identify areas for improvement.

17.4 Example: Butterfly}Pieris{Effect Installation

The example Butterfly}Pieris{Effect is an art–science installation that invites participants to see the world through the eyes of caterpillars, chrysalis, and butterflies, highlighting the ecological importance of insects, emphasizing that even pest species have their important role in ecosystems.

17.4.1 Design Elements

- A soundscape featuring expert insights on insect conservation and the ecology of butterflies and moths.
- Virtual Reality goggles that let you view the world through the three metamorphosis stages (caterpillar, chrysalis, butterfly) of Pieris *brassicae*.

- An inflatable egg with electron microscopical scanned texture on it with projection mapping of magnetic resonance imagery blended in computer-animated stages of the development of a caterpillar in a butterfly egg.
- A small publication about the main ecological messages of the art–science installation.
- A huge ultra-high resolution print of a Pieris *brassicae*, which shows fascinating details of butterflies that you can hardly see with the naked eye.

Art–science installations offer a compelling method to communicate environmental challenges and inspire sustainable actions. By integrating scientific rigor with artistic creativity, this approach not only informs but also motivates, leaving lasting emotional impressions on participants and opening the way to tacit knowledge. With this guide and the right people coming together, readers are equipped to design and implement their own installations, fostering a deeper understanding and appreciation for the natural world among their audiences (Table 17.1).

Table 17.1 Method overview

Main purpose
Awareness building/behavior change.
Gained competences
Scientific and ecological knowledge and team work competences.
Educational setting
Informal (workshops, learning by doing).
Space requirements/restrictions
Outside activities and indoor activities.
Resources and necessary materials
Necessary materials depending on the project are: Art supplies, digital equipment, scientific data, and exhibition materials.
Number of participants
Best outcome would be in small groups.
Facilitator competences and skills
Ecologist, scientific imaging expert, and (media) artists.
Participants' skills/Age/Competences
Students, researchers, and practitioners of different age groups and skill levels.
Duration
Two to three years.

References

Brown T (2013) Breaking the fourth wall. Edinburgh University Press, Edinburgh
Butterfly}Pieris{Effect. https://www.scivizlab.com/flybutterfly. Accessed 1 Oct 2023
Fröschl M (2019) Computer-animated scientific visualizations of tomographic scanned microscopic organic entities [doctoral thesis]. University of Applied Arts Vienna, Vienna
Fröschl M (2020) Computer-animated scientific visualizations in the immersive art installation NOISE AQUARIUM. In: O'Conner L (ed) 2020 IEEE conference on virtual reality and 3D user interfaces abstracts and workshops (VRW), The IEEE virtual reality conference, Atlanta, GA, March 2020. Institute of Electrical and Electronics Engineers (IEEE), New Jersey, pp 183–187
Haraway D (2016) Staying with the trouble: making kin in the Chthulucene. Duke University Press, Durham
Noise Aquarium. http://noiseaquarium.com. Accessed 27 Sept 2023

Chapter 18
Photography with Intervention

Miriam Mary Brgles

18.1 Introduction

The Photography with Intervention method is a way to engage, affirm, and give an active voice to participants via creative methods. "The visual image is used to furnish evidence and to promote an effective, participatory means of sharing expertise and knowledge" (Wang and Burris 1997). It is a participatory research method that involves refining a photo with new creative content. The basis of this method is the photovoice method, but with new additions that bring more flexibility, creativity, and reflexivity. Photovoice is also known as a process by which people can identify, represent, and enhance their community through a specific photographic technique (Wang and Burris 1997). Through Photography with Intervention, participants give an additional voice, reflection, or meaning to a photograph while capturing reality, their environment, specific events, or artifacts. Photography with Intervention can expand the initial number of participants, as the dissemination provides a new opportunity for intervention. The photograph can be infinitely refined and recycled by different participants once it is published, shared, or exhibited in a virtual or public space. This expands the documentary and artistic content of photography. Photography with Intervention has a great potential for self-transformation.

As cameras are built into mobile phones, the method is available to anyone who owns a mobile phone and knows how to use a camera. Interventions can also be edited on a mobile phone, and various (mostly free) photo editing software programs are available. In a broader sense, the photovoice method has a variety of applications, and in the last thirty years, it has been most commonly used in health, education, and social work research (Brenny and McMorrow 2021). It is well suited

M. M. Brgles (✉)
Catholic University of Croatia, Zagreb, Croatia
e-mail: miriam.brgles@unicath.hr

J. Bentz, J. Ristić Trajković (eds.), *Imagining, Designing and Teaching Regenerative Futures: Art-Science Approaches and Inspirations From Around the World*, Science for Sustainable Societies,
https://doi.org/10.1007/978-981-96-9029-9_18

to research with young people and youth communities, as well as vulnerable social groups. Photovoice is not only a community-based participatory research method but also a social justice method that places community participants, organizations, and policymakers at the center of this research method (Brenny and McMorrow 2021).

Most often, photovoice uses words added by participants in addition to the photo, but in Photography with Intervention, participants can add some other visual content to the photo, such as emoticons, diagrams, AI-generated stickers, interactive images or maps, timelines, etc., to express their opinion, question, or interpretation. It is a kind of photomontage with meaning. The aim is to transfer certain knowledge to the audience and to empower the participants of the group that creates the intervention, as well as the audience that can participate in the intervention.

18.2 Transformative Potential

Using Photography with Intervention to engage participants, present a project, or disseminate research results can be transformative in itself because the data are works of art, created as inspirational material in collaboration between individuals or groups. Participants can contribute to a fairer future by respecting different socio-demographic and socio-economic statuses such as gender, age, education level, level of social exclusion, etc. For example, by working with vulnerable groups, we contribute to their inclusion in society, highlighting their problems, issues, but also their strengths and potential. By working with local communities, we also have the opportunity to highlight the sustainability needs and ecological potential of a place or space where these groups live or work. For Bourdieu (1990), photography had an integrative function. Photography with Intervention focuses less on answering questions and more on raising questions and giving a voice to participants. This means that the presentation can have a profound effect on individuals or groups, but (perhaps) also on decision-makers. Photography with Intervention can be conducted in an online environment or as a face-to-face event in nature/city/neighborhood.

All participants should be aware that they are not claiming their work as original, but sharing it as living material that will be transformed and given new meanings by new interventions.

18.3 Application

Preparation and main steps of the application:

1. *Determine the topic* such as youth engagement in promoting sustainable habits in the community, the role of local people in making the neighborhood more inclusive for vulnerable social groups, the environment and health priorities of

the community (Hergenrather et al. 2009), the inclusion of students with autism (Hergenrather et al. 2009), etc.

2. *Select the participants and organize a short discussion about the topic* (give the instructions about the method and ethical issues.)
   The recruitment of participants will depend on the topic, but the facilitator should be aware of his/her position as an insider, outsider, or in-betweener. "The themes are often developed in partnership with the participants. At minimum, themes are revised and validated by participants" (Hergenrather et al. 2009, p. 688).
3. *Organize the photographic work*
   The photographic work is defined as an engagement of participants at a location that is significant for the topic under discussion. This may take the form of a public square, a schoolyard, a local medical institution, and so forth. Participants are encouraged to engage actively with the subject matter and document their observations through photography. Additionally, participants may take written or verbal notes to supplement their observations.
4. *Organize group discussions*
   Provide a space where the group will gather and work on the interventions, but also talk about the experience of working in the field.

   The facilitator poses questions pertaining to the participants' experiences on the selected topic, with particular reference to the photos and notes. The main goal of the group discussion is to facilitate listening and empowerment through the exchange of diverse perspectives and reflections. Following the discussion, participants engage in the creation of interventions utilizing photo editing software (such as Canva, GIMP, BeFunky, PhotoDirector Essential, and others) or printed photographs. Additionally, participants may engage in the creation of a collage or poster in a group setting. This constitutes a pivotal stage of the process, and it is essential to allow sufficient time for the implementation of interventions. The utilization of Photography with Intervention facilitates the convergence of community engagement, creativity, and reflection. Participants identify and represent the community's problems or concerns from their own positions as active individuals engaged in the process of raising questions and bringing changes to the community.
5. *Provide a space for dissemination and further interventions*
   Suitable spaces can be libraries, schools, municipality buildings, public squares, parks etc.

   Collaboration with local institutions could facilitate the provision of a space for an exhibition or a public event. It is essential that the event is inclusive, whereby the role of the audience is transformed into that of participants, with their interventions representing the final step of the method. It is essential to prepare the requisite materials, such as paper, markers, stickers, and labels, or alternatively, a QR code for a virtual exhibition. The primary objective of the dissemination and subsequent creation of interventions is to facilitate dialogue and drive social change or promote social justice within the community.

The use of this method can be illustrated by the following example:

In our research on a neighborhood in Zagreb (Croatia), we gave students the task of using cameras to record changes in the neighborhood, focusing on sustainable living according to the concept of the 15-minute city (Moreno et al. 2021). We divided them into groups of four or five. Each group was to go to a different part of the neighborhood. Their time was limited, but in the group discussion, they expressed a desire to take photos beyond the suggested hours because they wanted to capture the dynamics of the city at different times of the day. This turned out to be very interesting and enriching. This is why time flexibility is important.

In the second phase, they used (free) photo editing software. Again, flexibility proved important, as some people printed the material, then made a collage, and then took new photos of the collage. In this way, the method can have an iterative or spiraling character, and the intervention can contain an almost infinite number of interventions. The photographs with the intervention were exhibited in the local library, where people from vulnerable groups (such as migrants or users of the Caritas soup kitchen, young people, people with disabilities, etc.) were invited as audience.

Bringing the public closer is also part of the method. Papers and markers were placed next to the QR codes for the virtual exhibition. Visitors could become participants. After a brief presentation by the facilitator(s) on the project theme and the possibilities for additional interventions, visitors could add their interventions on the spot (with markers and papers) or virtually by photo editing software and sending them to a virtual exhibition space. This created new visual descriptions of the neighborhood, representing different voices. The photographs had undergone multiple interventions and had become signs of a shared understanding of the sustainability of the neighborhood and the city.

Given the spontaneity and discovery that occurred during the process, the facilitator did not prepare all the steps in advance, but let the method happen or create it on the spot through the creative activities of the participants.

For some topics, public squares, bus stations, or shopping centers are ideal places to display results, while for others, a local gallery, library, or school hall may work well. This approach has transformative potential because it involves individuals or groups who are often far removed from academic work, but whose interests and needs may be closely related to the problem. We can work with our participants to create new ones, combine existing ones, or derive them from the creative process itself. It is necessary to be creative, to use different resources, and to have the courage to think outside the box (Table 18.1).

Table 18.1 Method overview

Main purpose
Participation of different social groups. The aim is to integrate vulnerable social groups and to bring them together with the researchers. Researchers can be academics, but also children, youth, students, etc.
Gained competences
With the method, people learn that people's needs and perspectives are different and how their own perspectives and needs can be transformed.

(continued)

Table 18.1 (continued)

Educational setting
All environments where they can formally participate in the creative process of the research. Formal: academic/school projects. Informal: small community projects, NGOs.
Space requirements/restrictions
Flexible, depends on the topic. Could be planned as: Outside activities: fieldwork as a first phase, for example, exploring the setting of the topic. Indoor activities: interventions as a second phase.
Resources and necessary materials
Basic: Camera and photo editing software. Other: recycled paper, printed photos, etc.
Number of participants
Group work: from 3 to unlimited.
Facilitator competences and skills
People who like photography and have a sensitivity for details and social issues
Participants' skills/Age/Competences
Vulnerable people/community/students/policymakers/researchers/designers/practitioners, etc.; all people of all ages.
Duration
From 2 hours to unlimited.

References

Bourdieu P (1990) Photography. A middle-brow art. Polity Press, Cambridge

Brenny JM, McMorrow SL (2021) Photovoice for social justice. Visual representation in action. SAGE, Thousand Oaks

Hergenrather KC, Rhodes SD, Cowan CA, Bardhoshi G, Pula S (2009) Photovoice as community-based participatory research: a qualitative review. Am J Health Behav 33(6):686–698. https://doi.org/10.5993/ajhb.33.6.6

Moreno C, Allam Z, Chabaud D, Gall C, Pratlong F (2021) Introducing the "15-Minute City": sustainability, resilience and place identity in future post-pandemic cities. Smart Cities 4(1):93–111. https://doi.org/10.3390/smartcities4010006

Wang C, Burris MA (1997) Photovoice: concept, methodology, and use for participatory needs assessment. Health Educ Behav 24(3):369–387. https://doi.org/10.1177/109019819702400309

Chapter 19
The Living Canvas: Community Mural Creation

Serhat Burmaoglu

19.1 Introduction

The Living Canvas is an engaging, hands-on method that involves community members in the collaborative creation of a large-scale mural with a focus on regenerative themes. This approach is grounded in the belief that art not only reflects society but also shapes it, creating a shared vision for a thriving, sustainable future.

The method works by gathering a diverse group of participants to work together on a mural that depicts their collective vision of a regenerative future. This process involves multiple stages, including workshops, discussions, and collaborative design sessions, where participants brainstorm ideas and refine them into a cohesive mural design. The mural is then painted, ideally on a public building or a community space where it can be widely seen.

This practice is powerful because it transforms a blank wall into a conversation starter, an educational tool, and a daily reminder of the community's commitment to regeneration. It fosters a sense of ownership and pride among participants, as they see their contributions become a permanent part of the community's landscape. This method is particularly useful in urban environments where concrete and steel often dominate, as it can reintroduce a sense of the living environment into the heart of the city. By engaging community members in this creative process, the Living Canvas promotes environmental awareness, artistic expression, and social cohesion.

S. Burmaoglu (✉)
Faculty of Economics and Administrative Sciences, Department of Health Management, Cigli, Izmir, Türkiye

Izmir Katip Celebi University, Faculty of Economics and Administrative Sciences, Department of Health Management, Cigli, Izmir, Türkiye
e-mail: serhat.burmaoglu@ikcu.edu.tr

J. Bentz, J. Ristić Trajković (eds.), *Imagining, Designing and Teaching Regenerative Futures: Art-Science Approaches and Inspirations From Around the World*, Science for Sustainable Societies,
https://doi.org/10.1007/978-981-96-9029-9_19

19.2 Transformative Potential

The Living Canvas method is transformative because it creates a tangible, visual representation of the community's aspirations for the future. By engaging a diverse group of community members in the creation of a mural, this method encourages ownership and stewardship of both the art and the regenerative concepts it represents. Community murals have historically been a tool for social and political expression, transforming urban spaces into canvases of community identity and activism (Greaney 2002). The process of mural creation involves rigorous organization and collaboration, teaching participants not only artistic skills but also fostering social capital and community bonds. This collaborative effort helps reclaim public spaces for public discourse, reflecting local biodiversity, cultural narratives, and shared values around regeneration.

The mural serves as a daily visual cue for those who pass by, reminding them of their connection to each other and the living world. Beyond traditional education, the mural makes regenerative concepts accessible and engaging to all, regardless of age, background, or education level. It is transformative because it combines the regenerative potential of art with social engagement, creating a lasting impact on the community's cultural landscape. Public art projects like these have been shown to enhance urban vibrancy, build a sense of belonging, and foster social interactions among residents (Furtado and Payne 2022).

19.3 Key Transformative Elements

- Community Ownership and Stewardship: When community members contribute to the mural, they feel a sense of ownership over the artwork and the regenerative ideals it portrays. This ownership fosters a deeper commitment to maintaining the mural and promoting the values it embodies.
- Public Discourse and Awareness: The mural serves as a public space for discourse, sparking conversations about regeneration and sustainability among viewers. It becomes a daily visual cue for passersby, reminding them of their connection to each other and the living world.
- Educational Tool: Beyond traditional education, the mural makes regenerative concepts accessible and engaging to all, regardless of age, background, or education level. It visually communicates complex ideas in a way that is easy to understand and relate to.
- Social Engagement and Inclusion: The process of creating the mural brings together people from different backgrounds, fostering social engagement and inclusivity. It combines the regenerative potential of art with social interaction, creating a lasting impact on the community's cultural landscape.
- By fostering a sense of community ownership, providing a platform for public discourse, serving as an educational tool, and enhancing social engagement and

inclusion, the Living Canvas method creates a powerful, transformative experience for all participants and viewers.

19.4 Application

To apply the Living Canvas, a facilitator needs to meticulously plan and execute several key steps:

- *Community Engagement*:
  - Introduction: Organize initial meetings with community members to introduce the concept of the Living Canvas. Explain the goals and benefits of creating a regenerative-themed mural.
  - Inclusive Input: Gather input on the mural's design from all parts of the community. Use reflection questions such as "What does a regenerative future look like for our community?" and prompts like "Depict an element of nature that inspires you."
  - Outreach: Ensure broad community involvement by reaching out to schools, local organizations, and diverse groups within the community.
- *Design Workshops*:
  - Idea Refinement: Conduct multiple workshops to refine the collected ideas into a coherent design. Facilitate brainstorming sessions where participants can share their visions and collaborate on the mural's theme and composition.
  - Guidance: Lead the workshops with an experienced artist or designer who can help translate ideas into a unified visual narrative. Provide visual aids and examples of other community murals for inspiration.
  - Feedback Loops: Implement iterative feedback rounds where participants can review and adjust the design, ensuring it reflects the collective vision.
- *Materials and Space*:
  - Wall Selection: Secure a visible and accessible wall for the mural. Obtain necessary permissions from property owners or local authorities.
  - Materials Procurement: Gather all necessary materials, including high-quality paints, brushes, and protective gear. Consider environmentally friendly and durable materials to ensure the mural's longevity.
  - Safety Considerations: Arrange for scaffolding or ladders for larger murals and ensure all participants are briefed on safety protocols.
- *Painting the Mural*:
  - Organized Sessions: Schedule community painting days where participants can help bring the design to life. Organize these sessions to accommodate different time slots, ensuring maximum participation.

  - Facilitation: Have the facilitator and assisting artists guide the painting process, ensuring the design is followed accurately. Assign specific sections of the mural to different groups to manage the workflow effectively.
  - Documentation: Document the painting process with photos and videos to capture the community's engagement and the mural's evolution.

- *Education and Celebration*:

  - Completion Event: Once the mural is complete, host a celebratory event to unveil the mural. This event can include speeches, educational workshops on regenerative practices, and activities for attendees.
  - Educational Programs: Use the mural as an educational tool by organizing tours and discussions around its themes. Develop educational materials that explain the regenerative concepts depicted in the artwork.

- *Considerations for Effective Use*:

  - Time Commitment: The entire process, from planning to completion, can take several weeks to months. Plan accordingly and set realistic timelines.
  - Materials Needed: Ensure all necessary art supplies, including paints, brushes, and scaffolding, are procured. Opt for sustainable and durable materials to ensure the mural's longevity.
  - Group Dynamics: The method is best applied in groups to foster a sense of community and collective ownership. Encourage collaboration and open communication throughout the process.
  - Preparatory Work: Engage the community early, secure design approvals, gather materials, and schedule sessions well in advance to ensure smooth execution.
  - Adaptability: Tailor the project to reflect local environmental and social issues. Ensure the mural resonates with the community's unique identity and aspirations.

By following these detailed steps and considerations, the Living Canvas can be successfully implemented in any community, transforming public spaces and promoting a collective commitment to regeneration and sustainability (Table 19.1).

Table 19.1 Method overview

Main purpose
Awareness building /behavior change; "the Living Canvas focuses on engaging community members to actively participate in creating a mural that reflects and promotes regenerative themes, thereby building awareness and potentially changing behaviors towards sustainability."
Gained competences
Artistic expression, collaborative skills, environmental awareness: Participants gain skills in artistic design and execution, learn collaborative practices during the mural creation, and enhance their understanding of regenerative and sustainable concepts
Educational setting
Informal (format—Workshops, seminars, community meetings): The method uses informal settings such as community workshops, brainstorming sessions, and design workshops to engage participants and facilitate the creation of the mural

(continued)

Table 19.1 (continued)

Space requirements/restrictions
Outdoor activities (mural painting on a public or community building): Requires a large, visible, and accessible wall, with permissions from property owners or local authorities. Workshops and planning sessions may be held indoors
Resources and necessary materials
High-quality paints, brushes, protective gear, scaffolding or ladders for larger murals, and visual aids for design workshops. Environmentally friendly and durable materials are preferred to ensure the mural's longevity
Number of participants
Group work: Size and number of groups can vary, tailored to the community size and the mural's complexity. The project encourages broad community participation
Facilitator competences and skills
Experienced artist or designer, skilled in community engagement and project facilitation. The facilitator should have expertise in art and design, as well as experience in leading community projects
Participants' skills/age/competences
Community members/students/local organizations: Participants of all ages and backgrounds are encouraged, fostering inclusivity. No specific artistic skills are required, as the project aims to be accessible to everyone
Duration
Several weeks to months: The process includes initial community engagement, workshops for design refinement, the mural painting itself, and final educational and celebratory events. Timelines should be planned realistically to accommodate all stages

References

Furtado LS, Payne JM (2022) Inclusive creative placemaking through participatory mural design in Springfield (MA). J Am Plan Assoc 89(3):310–323. https://doi.org/10.1080/01944363.2022.2083008

Greaney ME (2002) The power of the urban canvas: paint, politics, and mural art policy. N Engl J Public Policy 18(1):6. https://scholarworks.umb.edu/nejpp/vol18/iss1/6

Chapter 20
The Echographic Poiesis Approach

Stephanie Leite

20.1 Introduction

Echographic Poiesis is an arts-based method intended to increase individuals' capacity to process and respond to life in the Anthropocene. Echographic Poiesis draws from several different practices, namely Jack Mezirow's (1997) transformative learning theory, Roger Simon's (2004) historiographic poiesis, and Joanna Macy's (2007) "Work That Reconnects."

Transformative learning theory is increasingly applied to sustainability education for its aim to shift the way learners think and act in the world through disruption and dialogue (Mezirow 1997). In a process of critical reflection, learners may transform mindsets—for example, dualistic, anthropocentric worldviews—toward more sustainable, regenerative worldviews.

Echographic Poiesis couples transformative learning with Roger Simon's (2004) practice of historiographic poetics, which involves deep engagement with the testimony of those who have been subjected to violence and marginalization. This engagement allows practitioners to relate to "counselors" of the past and present by taking in their testimony and then responding through action. Historiographic poetics can be described as a pedagogical practice of remembrance for learning how to live relationally and justly with the living and the dead. In this way, it may be used as a way of convening with the "ghosts of the Anthropocene" identified by Gan et al. (2017): if we see our actions in the present as an act of reconciling with those beings who have suffered in the past, we move toward a future that is more relational and just.

S. Leite (✉)
McGill University, Montréal, Canada
e-mail: stephanie.leite@mail.mcgill.ca

J. Bentz, J. Ristić Trajković (eds.), *Imagining, Designing and Teaching Regenerative Futures: Art-Science Approaches and Inspirations From Around the World*, Science for Sustainable Societies,
https://doi.org/10.1007/978-981-96-9029-9_20

In describing historiographic poetics, Simon stresses that to receive the gift of testimony is effectively to be given a task,. Echographic Poiesis provides one means for carrying out such a task, by—as Greenway (2019) describes it—shifting the locus from *poetics* to *poiesis*, or from *doing* to *making*. Furthermore, Echographic Poiesis builds upon historiographic poetics by moving from an anthropocentric practice to one that also includes testimony from more-than-human counselors. Therefore, "echographic poiesis" engages with the echoes of ghostly presences of nature with the aim of disrupting the Western cultural assumptions of individualism, progress, and anthropomorphism identified by Donald (2019) in order to cultivate a greater sense of relationality in learners.

The Echographic Poiesis process roughly emulates the four steps of Joanna Macy's Work That Reconnects. The Work That Reconnects process includes: (1) expressing gratitude, (2) honoring our pain for the world, (3) seeing with new eyes, and (4) going forth. To engage with testimony is an emotional investment, potentially leading to catharsis and reconnection but also evoking feelings of grief, guilt, and despair. Experiencing such emotions may be necessary for healing and, in turn, building capacity for compassionate action. Instead of suppressing or denying the emotions associated with receiving testimony, Echographic Poiesis invites these feelings as guides on a journey toward empathy. Our emotional responses to the suffering caused by climate change and environmental destruction are normal (Macy 2007). Echographic Poiesis may create space for acknowledging, mourning, and empathizing with ecological harm; responding through *artistic making* helps process the testimony and imagine compassionate responses that foster alternatives to our destructive pasts.

20.2 Transformative Potential

The difficult act of radical reimagining is at the core of Echographic Poiesis. This is challenging work because, as speculative fiction writer Neal Stephenson asserts, it is easier to destroy that which has already been constructed than to imagine alternative worlds (Marchese 2022). For those of us socialized within modern society, our ability to imagine alternative ways of being in the world has been restricted (Andreotti 2021). Such constrained thinking keeps us from considering the possibility of humans living in beneficial relationships with the environment. Within this scarce imaginary, climate change acts as a wall beyond which we cannot conceptualize.

One objective of Echographic Poiesis is to break through these conceptual barriers. Through the process of grieving and observing in new ways, the intention is to restore connections and cultivate a sense of responsibility to past and future ancestors. In this way, Echographic Poiesis practitioners invite counsel from multiple sources to imagine possible futures.

20.3 Application

Echographic Poiesis has been facilitated for students in the 2022 graduate-level course *Education in and for Social and Environmental Transformation* and the 2024 undergraduate-level course *Twenty-First Century Learning* at McGill University, and, with the collaboration of Daphne Chalmers and Sage Comstock, for participants in a workshop at the *2023 Artful Inquiry Research Group Symposium*. The sequence of steps requires a minimum of 45 min but could be expanded over several hours.

The sample practice focuses on visual arts and entails receiving testimony in the form of landscape observation, photo elicitation, or another form of encounter with a nonhuman relation. Choose a "counselor" to receive testimony from (this could be a visual image, an outdoor landscape, etc.) and set up a space where you may observe and draw. Assemble art materials for the activity, such as paper, colored pencils, and clipboards.

20.3.1 Expressing Gratitude

Begin with an expression of gratitude. This may include immersing oneself in nature, practicing mindful breathing, or another activity that grounds the learner in an appreciation for the world and the relationships that support life. If you are working as a group, you may wish to turn to someone next to you and share one or two things you are grateful for, or just spend several minutes taking in your surroundings and quietly meditating on gratitude.

20.3.2 Honoring Our Pain for the World

Once a connection has been established with the place and those with whom you are experiencing the activity, then the work of grieving may begin. This may happen through observing an image or a place. Such images and interactions inundate our senses regularly, but in our hurry to scroll through information or move from one place to another, we rarely afford time to process the images and spaces that document the suffering of our world. This practice of *interacting* without *feeling* can lead to repression—we know the destruction is happening but are not equipped to acknowledge it. As part of honoring our pain for the world, spend several minutes engaging with your selected "counselor"—perhaps the photograph of a scorched redwood tree, or the skyline of a cityscape—and sketch what you see or feel.

20.3.3 Seeing with New Eyes

From here, in an attempt to "see with new eyes," we adapt the common artistic practice of drawing negative space. Artists use this activity to trick the brain into seeing objects differently by outlining the space that surrounds the object. For example, the brain might think it knows how to draw a building, but when attention is focused instead on drawing the shapes around the building—the negative space—the building emerges in a new way. This exercise disrupts the brain's logic of what a building should look like, therefore requiring more detailed concentration and rethinking of the assumed architectural features. One may apply this concept to landscape drawing, not only drawing around objects, but also imagining what is missing, i.e., attempting to see, hear, smell, taste, and touch what is not there—listening to the echoes in order to imagine the potentialities of the land. In short, in this stage, attempt to encounter the landscape *ghosts* and, from their counsel, re-envision the landscape. What do the ghosts tell you about this image or landscape? What can they tell you about history, and what future might they desire? Add these counselings to your sketch.

20.3.4 Going Forth

If working in a group, share your sketch and your experience. Discuss how this activity may inspire real-world action. For Echographic Poiesis, this means honoring the ghosts with whom we have convened by taking concrete action in the world. Some possibilities for action include lobbying for a green space or garden, or manifesting your sketch in the form of a public art project, for example, by posting images of the reimagined landscape around the area as a provocation for community engagement and feedback. Regardless of the path taken, it is essential that the artistic process is utilized as a point of departure for making social change (Table 20.1).

Table 20.1 Method overview

Main purpose
Awareness building
Gained competences
Futures-thinking, systems-thinking
Educational setting
Formal or informal; secondary or postsecondary
Space requirements/restrictions
Outdoor activity or a view of the outside
Resources and necessary materials
Outdoor access, paper, drawing materials, solid surface to draw on
Number of participants
Group setting with 4–20 people.

(continued)

Table 20.1 (continued)

Facilitator competences and skills
Ability to build group trust. No artistic skills needed, but the ability to view and debrief artistic works in a group setting is an asset
Participants' skills/age/competences
Community/students/policy makers/researchers/designers/practitioners, etc.
Duration
20–60 min.

References

de Andreotti VO (2021) Depth education and the possibility of GCE otherwise. Glob Soc Educ 19(4):496–509

Donald D (2019) Homo economicus and forgetful curriculum: remembering other ways to be a human being. In: Tomlins-Jahnke H, Styres SD, Lilley S, Zinga D (eds) Indigenous education: new directions in theory and practice. University of Alberta Press, Edmonton, pp 103–125

Gan E, Tsing A, Swanson H, Bubandt N (2017) Haunted landscapes of the anthropocene. In: Tsing A, Swanson H, Gan E, Bubandt N (eds) Arts of living on a damaged planet. University of Minnesota Press, Minneapolis, pp G1–G16

Greenway K (2019) Historiographic poiesis and adoption ephemera: journeys in arts-based research. In: Andrews BW (ed) Perspectives on arts education research in Canada, volume 1: surveying the landscape. BRILL, Boston, pp 47–65. https://doi.org/10.1163/9789004405202

Macy J (2007) World as lover, world as self: a guide to living fully in turbulent times. Parallax Press, Berkeley

Marchese D (2022) A Sci-Fi Visionary Thinks Greed Might Be the Thing That Saves Us, New York Times, 10 January. Available at: https://www.nytimes.com/interactive/2022/01/10/magazine/neal-stephenson-interview.html

Mezirow J (1997) Transformative learning: theory to practice. In: New directions for adult and continuing education, vol 74, pp 5–12

Simon RI (2004) The pedagogical insistence of public memory. In: Seixas P (ed) Theorizing historical consciousness. University of Toronto Press, Toronto, pp 183–201

Chapter 21
Overcoming Otherness: An Artistic Approach

Marina Kalashyan

21.1 Introduction

This approach involves the active use of the arts, particularly theatrical art, allowing learners to express themselves and their knowledge through performances. Its main goal is to cultivate learners' emotional sensitivity and empathy toward each other and the world in general. As a result, the approach helps to overcome feelings of superiority and arrogance toward entities that are somehow different (Adams 2009). It contributes to the development of a sense of solidarity and a desire to achieve equality and justice.

The method consists of incorporating playful practices into the educational process (Huizinga 2016) by encouraging learners to organize and participate in theatrical performances based on the course material. For this purpose, the educational content must be transformed into a script with appropriate roles for all participants to perform during class. It does not matter whether the subject is on biology, history, physics, or literature — almost any topic can be enacted. In fact, the less usual and familiar the assigned roles are, the better. For example, someone may take on the role of a molecule, a stamen, Plato, a thorn, a peasant, young Werther, a mammal, or carbon.

Such practices contribute to the development of learners' imagination (Dewey 1990) and enable them to embody their particular roles, presenting issues not from the perspective of a third party or observer (scientist, writer, etc.), but from the viewpoint of the character they have assumed. As a result, they learn to overcome narrow individualistic interests, cultivating empathy for entities different from themselves. Learners begin to perceive other human and non-human beings as possessing intrinsic value, their own interests, and perspectives, which makes them worthy of respect

M. Kalashyan (✉)
Department of Philosophy, Russian-Armenian University, Yerevan, Republic of Armenia

J. Bentz, J. Ristić Trajković (eds.), *Imagining, Designing and Teaching Regenerative Futures: Art-Science Approaches and Inspirations From Around the World*, Science for Sustainable Societies,
https://doi.org/10.1007/978-981-96-9029-9_21

and attentive care. Such imaginative play helps to experiment "with the idea of otherness in ways that are less threatening than the direct encounter with another may often be… The presence of the other, which can be very threatening, becomes, in play, a delightful source of curiosity, and this curiosity contributes toward the development of healthy attitudes in friendship, love, and, later, political life" (Nussbaum 2016, pp. 99–100).

Furthermore, the proposed approach stimulates learners' critical thinking. This becomes especially significant in the contexts of undemocratic or authoritarian regimes, whose educational systems do not facilitate freedom of thought and speech. In such circumstances, these practices can greatly assist in nurturing independent individuals while bypassing certain political risks.

21.2 Transformative Potential

The transformative potential of this method lies mainly in cultivating positive changes in contemporary norms and priorities, which is crucial for creating regenerative futures, achieving peace and sustainability. While today's ethics tend to be more individualistic and profit-oriented, the artistic approach to education can become an efficient means of developing an understanding of the deep interconnectedness of all beings and create new, non-profit forms of communication with the world. This can be achieved through the following effects of the proposed method.

First, by adopting a particular character, learners develop their *imagination*. They learn to envision the situation of the Other from that entity's perspective, even if it is very different from their own. Eventually, the Other begins to be perceived as possessing intrinsic value. In other words, each entity becomes an end in itself, not just a means. The needs and rights of others, as well as the opportunities to achieve their goals, may eventually become an important part of a learner's worldview, fostering appropriate motivations.

Second, learners expand their horizons and develop *empathy* — and then *sympathy* — toward entities that differ from themselves. Unfortunately, contemporary education is mainly focused on teaching science and technology, mostly valuing rationality while disregarding the nurturing of feelings and emotions. The proposed method creates a balance between the rational and emotional aspects of personality (O'Connell 2002).

Third, learners begin to exercise *critical thinking* more extensively. They come to understand that various life situations are not as unambiguous as they may have seemed before. Much depends on the chosen point of view, the goals, and the interests of a particular entity — and therefore needs to be critically examined.

Finally, by developing learners' critical thinking, this approach also enhances their *independence* and sense of *responsibility*. Children begin to express opinions they might not otherwise always dare to speak out. By being involved in an active learning process, they participate in discussions on important issues. This compels participants to act responsibly, as they need to convincingly portray their roles and justify their opinions during class activities.

21.3 Application

The approach is suitable for both primary/secondary and college/university classes and is most effective with groups of up to 20 students. The application of the method requires:

1. *Educator's preparatory work*: Creating a script and dividing the educational material into appropriate roles. It is up to the educator to choose which part of the text is to be performed. It is preferable that assigned roles are less obvious and contain an element of surprise. Anything can become a role—animals, inanimate objects, scientific abstractions, etc. Such work can be done with any educational text, including scientific ones.
2. *Assigning roles in advance*: Providing these roles to learners 1–2 weeks before the class so they have time to thoroughly consider and familiarize themselves with them. For example, they may receive a fragment from *Don Quixote* to introduce the beliefs of the main character or a passage from a biology textbook describing infusoria to illustrate its life stages. In some cases, it is possible to allow students to creatively modify the text.
3. *Performing the script*: Executing the performance in the classroom. The main task of the participants is to fully immerse themselves in their roles and strive to be convincing. Participants can use the classroom space or other areas agreed upon in advance with the teacher. Depending on the topic, appropriate inventory (decorations, costumes, etc.) can also be utilized.
4. *Post-performance discussion*: Engaging in the discussion at the end of the class. This stage is very important because it facilitates a free exchange of opinions, helping everyone to better understand different forms of life and worldviews (Todd 2020).

As a philosophy teacher, I have repeatedly implemented these theatrical mini-performances. For example, when studying ancient philosophy, one group of students is tasked with representing the views of the Stoics, while another explores the ideas of the Epicureans. In discussions on ethics, one student might act as Immanuel Kant, defending the ideas of deontology, while another portrays Jeremy Bentham, justifying utilitarianism. By applying this methodology, even students who are not initially interested in philosophy begin to emotionally defend the theses of a particular philosophical school, emphasizing its strengths. The approach stimulates student engagement in the educational process, develops tolerance, and enhances the ability to listen. Students learn to see the Other not as an adversary to oppose but as someone to understand and appreciate (Winnicott 2005) (Table 21.1).

Table 21.1 Method overview

Main purpose
Reshaping identity
Gained competences
Cultivating imagination, empathy, critical thinking, independence, and responsibility
Educational setting
Formal (primary/secondary school and college/university education)
Space requirements/restrictions
No space restrictions
Resources and necessary materials
Appropriate educational materials (textbooks, primary sources, etc.) and an elaborated scenario
Number of participants
Group of approximately 10–20 pupils/students
Facilitator competences and skills
Ability to work creatively with texts, ability to work with a team, and basic knowledge in the fields of ethics and psychology
Participants' skills/age/competences
Pupils, students
Duration
2–7 days (on elaboration), an academic hour (on presentation)

References

Adams AW (2009) Human subjects and animal objects: animals as other in the law. J Anim Law Ethics 3:29–51

Dewey J (1990) Froebel's educational principles. In: Dewey J The school and society and the child and curriculum. University of Chicago Press, Chicago, pp 116–131

Huizinga J (2016) Homo Ludens. Angelico Press, New York

Nussbaum M (2016) Not for profit. Why democracy needs the humanities. Princeton University Press, Princeton

O'Connell MK (2002) Rabindranath Tagore: the poet as educator. Visva-Bharati, Kolkata

Todd S (2020) Creating aesthetic encounters of the world, or teaching in the presence of climate sorrow. J Philos Educ 54(4):1110–1125

Winnicott D (2005) Playing and reality. Routledge, London/New York

Part III
Envisioning Tomorrow: Scenarios and Possibilities

The journey toward regenerative futures begins with our ability to imagine and design beyond the current state. The methods presented in this part invite educators and learners to look beyond the present moment and envision tomorrow through scenario planning, speculative storytelling, visioning workshops, and creative exercises. These approaches cultivate the ability to anticipate, explore, and critically shape emerging possibilities and scenarios. In the face of uncertainty and disruptions, the capacity to imagine alternative paths becomes an essential skill for navigating complexity. These methodologies offer structured yet imaginative spaces to rethink responses, question dominant assumptions, and imagine and design just, regenerative, and life-enhancing futures. They encourage learners to dream collectively, not as an escape from reality, but as an act of empowerment and transformation. Firmly grounded in the principles of sustainability science, these approaches foster anticipatory thinking and participatory visioning, as competencies that are crucial for engaging with long-term challenges and for designing solutions that restore ecological balance and social equity. They can foster agency in times of crisis, offering ways to move toward regenerative futures.

Chapter 22
Future-Oriented Creative Workshops

Beata Poteralska, Joanna Łabędzka, and Katarzyna Brożek

22.1 Introduction

The proposed tool is a future-oriented creative workshop that applies foresight to support creative and long-term planning and decision-making processes. Foresight is a recognized concept used to model, explore, and test variant visions of the future, those that are desirable and unexpected, possible and probable. The proposed future-oriented creative workshops are built around such visions and use the scenario method. A scenario is a "story" that illustrates visions of the possible future or aspects of the possible future (Vesnic-Alujevic and Störmer 2024).

Scenarios are the focal point of future-oriented creative workshops. These workshops are based on the assumption that although the future is unknown and uncertain, we may try to "forecast" it and imagine it. By doing so, we may be better prepared for it. Our imagination of the future can contribute to more efficient decision-making in the present.

Future-oriented creative workshops aim to demonstrate the possibilities of using the future in planning. Running workshops that apply the foresight approach and the scenario method supports the ability to assess possible consequences, anticipate problems before they occur, and reflect on the present implications of possible future events.

The workshops prepare participants for further work on various future-related issues characterized by uncertainty and unpredictability. Identification and development of future-oriented competences needed to cope with climate change or imagination of the functioning of a future regenerative city may serve as examples here.

B. Poteralska (✉) · J. Łabędzka · K. Brożek
Jan Kochanowski University of Kielce, Kielce, Poland
e-mail: Beata.Poteralska@ujk.edu.pl; Joanna.Labedzka@ujk.edu.pl;
Katarzyna.Brozek@ujk.edu.pl

J. Bentz, J. Ristić Trajković (eds.), *Imagining, Designing and Teaching Regenerative Futures: Art-Science Approaches and Inspirations From Around the World*, Science for Sustainable Societies,
https://doi.org/10.1007/978-981-96-9029-9_22

22.2 Transformative Potential

Thinking about a climate-resilient future is, in a natural way, related to putting the emphasis on looking ahead. Futures tools, including future-oriented workshops, are considered extremely useful in regenerative practice. The usefulness of applying the scenario method during the future-oriented creative workshop stems from the fact that scenarios enable one to create holistic, integrated images of what the future might look like. These visions become a context for planning, testing ideas, or a stimulus for development. Scenarios are a way of forcing ourselves to find our way in a rapidly changing world, both in terms of the present and the future (Poteralska et al. 2022). Scenarios present possible alternatives to emerging events. They provide a framework for exploring alternative futures and prepare us to answer the question "What should we do if something happens?" Using the scenario method is of great value when thinking about creating regenerative futures.

22.3 Application

Future-oriented creative workshops allow the future to be used in present planning. In the proposed framework, they are a first step in preparing for more advanced or detailed work with the future.

Two types of future-oriented workshops using the scenario method are proposed: a short introductory online workshop and a full-scale face-to-face workshop. The short online workshop, using foresight and Futures Literacy methods, aims to get participants to think about the future and to get used to the uncertainty involved. Participants are asked to create individual visions of the future within two given contrasting foresight scenarios, e.g., for the year 2030, which provide a backdrop for discussions and show how certain assumptions, often unconsciously, influence our visions of the future. The workshop lasts about 2 h, is conducted online, and is supported by an IT tool. There should be at least 2 groups of participants, 6–8 people in each. The groups are supported by facilitators (who guide the workshop) and observers (who act as rapporteurs). The workshop is a starting point for further discussions, e.g., on detailed aspects covered by the scenarios, policy implications in the future, competences needed in the future, etc. Such a workshop was organized in the framework of the FUTURES project: "Future laboratories for professional and personal development" (ERASMUS+), aimed at developing tools that use foresight and Futures Literacy methods for professional and personal development needs in an unpredictable labor market situation (Kononiuk and Rollnik-Sadowska 2022) (Table 22.1).

The full-scale face-to-face workshop, also using the scenario-building method, aims at empowering and harnessing the collective intelligence of citizens/researchers/students/etc., to, e.g., explore the role of research and innovation and to identify future-oriented competences through future scenarios and visions of desirable

Table 22.1 Method overview

Main purpose
Supporting creative and long-term planning and decision-making thanks to creating, sharing, and analyzing visions of the future
Gained competences
Ability to "look into the future" and better cope with uncertainty, thinking outside the box
Educational setting
Formal (primary, secondary, and higher education) and informal (face-to-face and online workshops).
Space requirements/restrictions
Indoor activities
Resources and necessary materials
Face-to-face workshop, round table discussions: a room with a few tables (1 table for each group), a set of inspirational pictures, templates for developing ministories and developing joint visions of the future, post-its, flipcharts (1 for each table) Online workshop: developed 2 contradictory scenarios of the future, IT tools facilitating group work, e.g., Google Jamboard
Number of participants
Group work, each group composed of 6–10 members, the number of groups: 2–6
Facilitator competences and skills
Experience in workshop facilitation (face-to-face and online), experience in applying foresight and Futures Literacy approaches
Participants' skills/age/competences
Preferably a diverse group of participants (with respect to age, sex, educational background, place of living, etc.). Possible for application also in homogeneous groups, e.g., within the formal education system
Duration
Online: ca 2 h. Face-to-face: ca 8 h

sustainable futures. The workshop lasts about 8 h and aims to create visions of the future on the basis of visions prepared by individual citizens. Visions are understood as imaginations of a desirable future with a time horizon of 30–40 years. These individual visions form the basis for the creation of common visions of the future. The workshop is conducted face to face. There should be at least 2 groups of participants, 6–8 people in each. The groups are supported by facilitators (who guide the workshop), and the participation of observers (acting as rapporteurs) is optional. These visions can be further developed. Such a workshop was organized in the framework of the CIMULACT project "Citizen and Multi-Actor Consultation on Horizon 2020" (Dagorne and Gudowsky 2018). In the case of this project, initial visions were created and directly used to co-create scenarios. And then, from the validated and prioritized scenarios, a set of policy options, possible topics of research calls, and recommendations for European funding were proposed. The structure of the workshop is shown in Figs. 22.1 and 22.2.

In both presented examples, workshops enabled participants to create, share, and analyze visions of the future, which became an inspiration and a starting point for further discussions about the future.

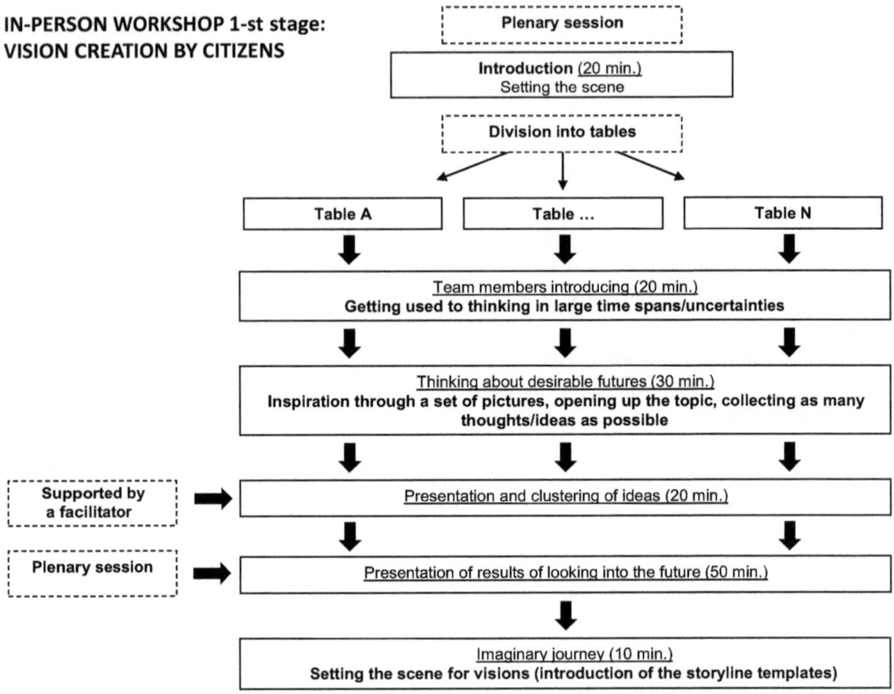

Fig. 22.1 Structure of the face-to-face future-oriented creative workshop, first stage

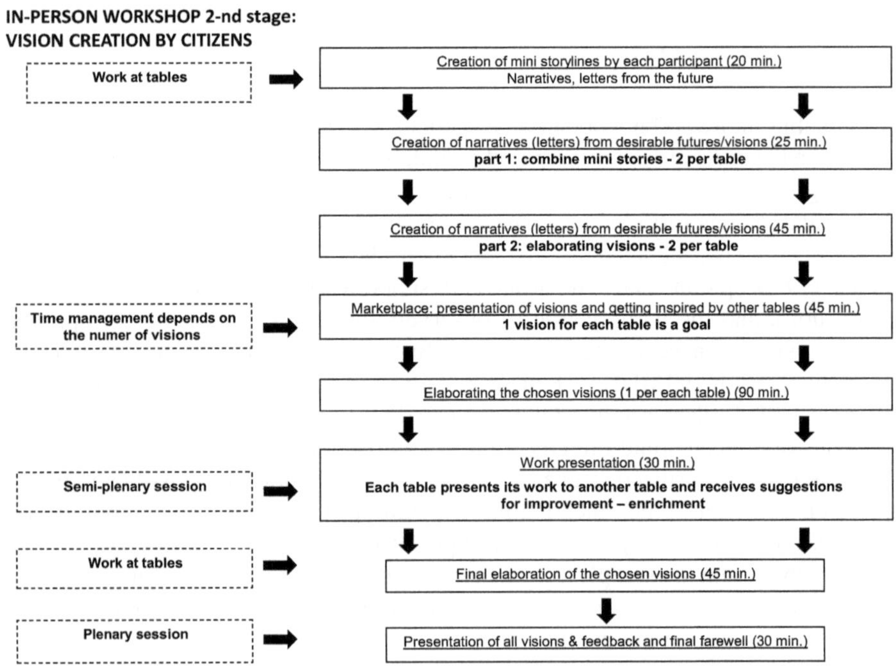

Fig. 22.2 Structure of the face-to-face future-oriented creative workshop, second stage

References

Dagorne E, Gudowsky N (2018) Deliverable 5.1 Inspiration Catalogue for consulting different groups. CIMULACT project: Citizen and Multi-Actor Consultation on Horizon 2020 http://www.cimulact.eu/wp-content/uploads/2018/04/D5.1_Inspiration-Catalogue-for-consulting-different-groups-compressed.pdf. Accessed 12 June 2024

Kononiuk A, Rolnik-Sadowska E (2022) Replay your futures – labs for exploring undiscovered pathways. Łukasiewicz Research Network – Institute for Sustainable Technologies, Radom

Poteralska B, Łabędzka J, Brożek K (2022) Development of competences for the needs of tomorrow's labour market. Zeszyty Naukowe Politechniki Częstochowskiej. Zarządzanie 48, pp. 144–155. https://doi.org/10.17512/znpcz.2022.4.12

Vesnic-Alujevic L, Störmer E (2024) Stress-testing of policy options using foresight scenarios – A pilot case. Publications Office of the European Union, https://data.europa.eu/doi/10.2760/292666

Chapter 23
Narrating New Future Visions: Collective Storytelling About the Climate Crisis

Cara Sophie Meffert and Censu Caruana

23.1 Introduction

This method aims to enhance agency in the climate crisis, to motivate (collective) action, and to foster critical thinking on dominant climate narratives. It is designed for anyone seeking hopeful stories and visions of the future. It is based on the observation that current climate narratives are predominantly scary and apocalyptic (Toivonen 2022). While it is important to be informed about the dramatic consequences of climate change, these stories often lead to feelings of guilt, grief, apathy, and frustration, which hinders climate action (Meffert 2023; Wamsler et al. 2022).

The method of Narrating New Future Visions is trying to bridge this knowledge-action gap by using tools of storytelling, transformative learning, education for sustainable development, and active hope. It centers around narrating hopeful future visions where we try to shift the focus—instead of telling stories of what we need to avoid, we tell stories of what we want to achieve and how we want to live. The method is designed as a two-day workshop for 3–12 people with a focus on questioning current stories about climate change, reflecting personal dreams and utopias, and then weaving these utopias into a collective vision that they can work toward. A regenerative future requires collective stories where everyone finds their agency to change both their own and society's ways of being (Veland et al. 2018). Therefore, this method centers around group processes and storytelling as a collective act.

C. S. Meffert (✉)
EMJMD Adult Education for Social Change, University of Glasgow, Glasgow, United Kingdom
e-mail: cara.meffert@okeanos-stiftung.org

C. Caruana
Centre for Environmental Education and Research, University of Malta, Msida, Malta
e-mail: censu.caruana@um.edu.mt

J. Bentz, J. Ristić Trajković (eds.), *Imagining, Designing and Teaching Regenerative Futures: Art-Science Approaches and Inspirations From Around the World*, Science for Sustainable Societies,
https://doi.org/10.1007/978-981-96-9029-9_23

23.2 Transformative Potential

When looking at climate change through stories, it becomes "a social, cultural and ideological dilemma" (Hendersson and Wamsler 2020, p. 346) rather than a scientific or knowledge issue. Stories can provide pathways from unsustainable ways of living toward more sustainable ones. This opens up new perspectives and ways to address climate change. Instead of learning about the science of carbon footprints, the participants question the ways we live as a society. By uncovering societal narratives, the participants become aware that social norms and rules and the structure of institutions and organizations are grounded on stories we tell ourselves about how life works. That means that we can question those stories and tell new ones. This insight gives the participants back their agency.

Collective imaginations can either reinforce or transform social systems, as "they have the potential to (re-) direct collective behaviors, institutions and structures" (Milkoreit 2017, p. 6). That makes them a powerful tool for social transitions. The collective story that the participants create gives them a pathway to act on and immediate ideas on how to make the first steps, thereby bridging the knowledge–action gap. Being in a space with other people who care can make the climate crisis less overwhelming. The experience of collectiveness can build a future vision that is inclusive of multiple perspectives and creates a shared responsibility toward this future. This experience not only motivates collective action but also mitigates climate anxiety. Furthermore, enabling people to become part of creating and shaping future visions democratizes the socio-ecological transformation process toward a climate-just future.

Stories can further create entangled empathy between humans who are not yet negatively affected by climate change and already affected beings. That makes them an important tool for motivating action as well as for enhancing communication with and about more-than-human beings (Celermajer et al. 2021). By addressing the emotions connected to climate change and the societal narratives, the participants get room to express and listen to their feelings. Additionally, they learn that these emotions are shared by other people concerned about climate change. This insight helps participants to address and also mitigate their climate anxiety.

Envisioning future scenarios aligns with the Education for Sustainable Development competency of future thinking, also known as the anticipatory competency. This skill enables participants to anticipate and shape preferable futures, fostering proactive, rather than reactive, mindsets. Additionally, the workshop has the potential to build empathy, not only toward current generations already affected by climate change but also toward future generations. By considering the impacts on future livelihoods, well-being, and opportunities, participants cultivate a sense of intergenerational responsibility, solidarity, and justice. Furthermore, the workshop can be structured to include feedback loops for participants to receive input on their visions. These loops enable iterative improvements and deeper engagement, ensuring that the collective story evolves and strengthens over time. This continuous reflection and adaptation process enhances the resilience and relevance of the envisioned futures, making them more empowering and transformative.

23.3 Application

This workshop outline was designed and evaluated in the context of the first author's Master's thesis (2023). It can be adapted to a wide range of target groups. The more heterogeneous the group is, the more credibility/validity the stories will have. Educators can and should experiment with possibilities to engage diverse voices, including more-than-human-beings, in the stories.

23.3.1 Block 1: Recognizing Current Narratives

Goal: Framing the workshop, understanding the context, and becoming aware of the current narratives.

Participants write their version of the current climate change story up until today, using the following prompt: "Write the story of climate change as you see it unfolding until today. Begin with 'once upon a time'." They note the main events in their stories on paper slips. In small groups of 2–3 people, they then weave their individual stories together into a comprehensive narrative by creating a shared timeline.

The facilitator then provides context on the power of stories and narratives and their role in society. Participants, within their groups, explore the underlying narratives of their combined stories, such as the narrative of eternal growth.

23.3.2 Block 2: Understanding the Functions of Narratives

Goal: Understanding the role and influence of narratives on our own lives, understanding the emotions connected to climate change and narratives.

Participants are invited to write about a day in their lives, focusing on the narratives discussed in the previous session: "Pick one of the current narratives we identified earlier. Write about a day in your life and consider how this narrative influences you." In a group reflection, they each share one influence that surprised them.

Next, participants take differently colored paper slips and add their emotions to the timelines from the first session and the paper slip collections in the room. The prompt is: "Emotions are deeply connected to how we think and act but are often neglected in climate change communication. By reflecting on and sharing these emotions, we bring them back into the discussion."

This session concludes with a group reflection on the activities.

23.3.3 Block 3: Writing New Narratives

Goal: Imagine and create a desired future together.

Following an approach by Luise Tremel, the participants first individually create a story of a desired future by free-writing about what they would see, what would be in their refrigerator, and what celebrations they would have in their personal utopia. Keywords from these stories are then noted on paper slips. The group subsequently gathers to narrate the future story, arranging the paper slips in a timeline using the backcasting technique. This method begins with the desired outcome and works backward toward the present to identify the steps necessary to connect the future with the present. This approach makes the group's envisioned utopia both tangible and achievable. The group then reflects on the underlying narratives within the new story. Participants are asked to write a short story about themselves or the group as characters in the future story they have created. The objective is to connect their own lives to the story and to take ownership of it. The exercise concludes with a group reflection.

Storytelling in this context provides a means to express, explore, and reshape dominant climate narratives. We encourage readers to experiment with this approach and adapt the method to other arts-based pedagogies, such as visual arts or theatre, to inspire and mobilize individuals and communities to co-create regenerative futures. We would love to hear from you and listen to your stories (Table 23.1).

Table 23.1 Method overview

Main purpose
Enhance agency in the climate crisis, motivate collective action, critical thinking on societal narratives
Gained competences
Critical thinking, anticipatory competency, understanding the role and use of narratives in societies, emotional awareness, empathy, collective storytelling, reclaiming agency in the climate crisis
Educational setting
Informal
Space requirements/restrictions
Indoor space with enough room for timelines on the floor or walls
Resources and necessary materials
Paper in different sizes, pens
Number of participants
3–12 people
Facilitator competences and skills
Knowledge on the function and role of stories and narratives in our society as background. Basic ecological and climate literacy
Participants' skills/age/competences
Everyone interested in social change. Age 16+. Structure can be adapted to younger kids as well
Duration
The workshop consists of three blocks, each lasting 2–2.5 h.

References

Celermajer D, Schlosberg D, Rickards L, Stewart-Harawira M, Thaler M, Tschakert P, Verlie B, Winter C (2021) Multispecies justice: theories, challenges, and a research agenda for environmental politics. Environm Polit 30(1–2):119–140. https://doi.org/10.1080/09644016.2020.1827608

Hendersson H, Wamsler C (2020) New stories for a more conscious, sustainable society: claiming authorship of the climate story. Clim Chang 158(3–4):345–359. https://doi.org/10.1007/s10584-019-02599-z

Meffert C (2023) Narrating new future visions—the role of collective storytelling and authorship in tackling the climate crisis [International Joint Erasmus Mundus Master Thesis]. University of Glasgow, Glasgow

Milkoreit M (2017) Imaginary politics: climate change and making the future. Elem Sci Anth 5:62. https://doi.org/10.1525/elementa.249

Toivonen H (2022) Themes of climate change agency: a qualitative study on how people construct agency in relation to climate change. Hum Soc Sci Commun 9(1):102. https://doi.org/10.1057/s41599-022-01111-w

Veland S, Scoville-Simonds M, Gram-Hanssen I, Schorre AK, El Khoury A, Nordbø MJ, Lynch AH, Hochachka G, Bjørkan M (2018) Narrative matters for sustainability: the transformative role of storytelling in realizing 1.5°C futures. Curr Opin Environ Sustain 31:41–47. https://doi.org/10.1016/j.cosust.2017.12.005

Wamsler C, Osberg G, Panagiotou A, Smith B, Stanbridge P, Osika W, Mundaca L (2022) Meaning-making in a context of climate change: supporting agency and political engagement. Clim Pol 23(7):829–844. https://doi.org/10.1080/14693062.2022.2121254

References

Chapter 24
Regenerative Futures with Children and Arts: The Radio Approach

Catarina Cadima

24.1 Introduction

The "*Regenerative futures with children and arts: the radio approach*" is an activity with three main objectives: (i) to promote transformative learning by using arts and critical thinking to stimulate a radical reassessment of worldviews among students; (ii) to transform feelings of isolation and powerlessness into love, empowerment, communication, and community engagement; and (iii) to share the resulting meaningful messages or narratives of love for the planet on social media and radio, to inspire and empower others. Regenerative thinking argues that the world functions as a whole and that we cannot solve the ecological problem in isolation. Recognizing that the environmental, economic, and social spheres are inseparable. We can consume less, learn to care and appreciate, and strive for sustainable development. We need to find a new paradigm in which the problems arising from globalization, such as human rights and civil rights, gender equality, or the fight against racism and the right to work, are a priority. It is essential to transform the mindset of "*we can do nothing*" into a sense of a regenerative future that we can create through small gestures of love, sharing, and belonging. In Lefebvre's words: "*Humanity only raises problems that it can solve itself*" (Lefebvre 2008: 141). The potential for meaningful learning comes from bringing individuals and communities together to share, discuss, and reflect on understandings of local and global social-ecological issues.

C. Cadima (✉)
Research Centre for Territory, Transports and Environment (CITTA), University of Porto, Porto, Portugal
e-mail: ccadima@fe.up.pt

J. Bentz, J. Ristić Trajković (eds.), *Imagining, Designing and Teaching Regenerative Futures: Art-Science Approaches and Inspirations From Around the World*, Science for Sustainable Societies,
https://doi.org/10.1007/978-981-96-9029-9_24

24.2 Transformative Potential

The transformative potential of this approach lies in its ability to transform the fear and anger associated with the climate crisis into feelings of love and empowerment. Broadcasting messages that support regenerative futures related to climate change aims to give children and young people a public voice and show that their opinions matter. It helps participants reflect on their role and inspires others to rethink. The radio approach is an interconnected process where the health of all life forms is linked to the well-being of the environment, emphasizing the synergy of all living beings and the importance of small daily gestures. The activities are structured to work with children and young people in the first and second cycles of education. Between 7 and 12 children can participate, depending on their ages. The skills acquired at the end will enable participants to think about climate change in a positive way, with empathy and a sense of empowerment. They will also identify and learn different ways of working together and how to write and produce a podcast. The approach is simple and can be used in various settings by different facilitators with small groups of participants. The potential for transformation is in all of us and in our small everyday actions. Developing this knowledge at an early age is crucial.

24.3 Application

The facilitator develops this approach in three sequential activities: REFLECT, TRANSFORM & BUILD, and SHARE (please consult Table 24.1). In the first phase, “Reflect,” the question (a) *What do you want?* is posed without any further information being given. Then, using art and nature as catalysts to generate discussion and build new thoughts, we ask *(b) how do we feel about the climate crisis and political inertia?* The second activity, “*Transform and Build*,” centers around a card game called “*I love the planet, so...*,” in which students think about simple daily tasks that we can all do in our daily lives to love the planet and share kindness (www.msseriousgames.com). It creates new learning opportunities and ideas. The trainer can also ask the participants (c) how to involve the local community and create a new game with a similar objective. The third and final activity is to share and connect with the local community and produce a podcast for dissemination on social media and radio. This can be a starting point for future events. The facilitator should be capable of listening, moderating speaking time fairly, and letting the group flow. Each facilitator should prepare this humanist and liberating approach according to their convictions and the working group.

The educational setting should be conducive to recording sounds, writing notes on Post-it notes, playing a game, and observing and appreciating the selected materials (works of art and objects from nature). Children and the facilitator should be distributed in a circle so everyone can see and hear each other. This circular distribution is essential for the creation of a sense of belonging. It is recommended that each

Table 24.1 Summary and application of the "Regenerative futures with children and arts: the radio approach"

Objective	To transform feelings of fear, powerlessness, and anger typical of eco-anxiety into feelings of empowerment, community, love, and ecological sharing	
Participants	Children and young people, 7–12 participants	
Materials	Images, photographs, installations, performances, music, and objects from nature. Post-its and pens or pencils for writing and coloring. Stopwatch to monitor timings. Sound recorders	
Context	In the classroom or outside, in nature. It should be possible to write on Post-it notes and play the game. The space should also be quiet enough to record sounds	
Time	The activity is divided into three blocks: depending on the group, these could be 3 h. There should be short breaks between blocks	
Methodology	The moderator gives the directions and explains to the group that there is no right or wrong, no one is judging anyone, it will be a group effort, and the more trustworthy we are, the easier it will be to carry out the tasks. A game called "*I love the planet, so…*" is played, in which participants transform and think of simple daily tasks that we can all do in our day-to-day lives to love the planet and share kindness. Five questions are discussed: two before and three after the game. Before the game, *(a) What do we want? (b) How do we feel about the climate crisis and political inertia?* Then, after the game: *(c) How can we engage a local community? (d) How can we share our messages? (e) Would you like to invite an artist to work with our materials?*	
Activities	**(i). Reflect**	WARM UP—Walk, stretch, dance, twirl around the circle as a group, and find peace in meditation with movement and music DESIRE—What do I desire, without limits, without judgment? Are they things/objects? Are they situations? Write 3 wishes on a Post-it note and keep it for yourself. Listen to a sound (e.g., melting ice), look at an image (representative of the climate crisis), or watch an installation or performance (e.g., work of Jenny Holzer, project—Truisms, from survival—PROTECT ME FROM WHAT I WANT) CHANGE—Write 3 thoughts and feelings on a Post-it note and keep it yourself. Close your eyes, think about what we must change, and 3 alternatives to our current actions. Take an object from nature and share your thoughts with the group, what they symbolize for you, and how they contrast. In the end, put your feelings on the wish poster, your worries, and what you want to throw away on the anger poster PAUSE (a few minutes)

(continued)

	(ii). Transform & Build	PLAY—In a circle, walk around the cards; when the music stops, take one of the cards from the game "I love the planet, so..." (repeat 3 times). Together, discuss your choices LOVE—Sitting down, closing your eyes for a few seconds, and thinking to yourself 3 examples from the game: 'I love the planet, so...' They can be repeated or make new examples; use imagination and creativity. If they want, participants can create new phrases. Share the examples with the person to your right and write the two messages that make the most sense to you on two new Post-it notes DISCUSS—Share the messages with your classmates. Discuss each message separately CHOOSE—Stick the various messages on a poster: "I love the planet, so...." each participant should choose a message that makes them feel good PAUSE (a few minutes)
Activities	**(iii). Share (radio approach)**	DRAW—DRAW a new card with the message, which could be used by other groups in the future. You can illustrate it. You can do this together or on your own RECORD—Record the message of the card you drew in groups of two or three COMPOSE—Put the recording's voices together with the message "I love the planet, so...." TOGETHER—Create a story, or start a dialogue among group members and do interviews; invite an artist to comment on their feelings when listening to your voices; or discuss how you could involve the local community. Compose again, recycle, mix, and create a new idea IDENTIFY—A radio station, an association/NGO, and a community organization partner. CONSTRUCT—A podcast. You will need a title and an objective. To help the group decide, ask the questions: Why do you want to share? Why can listening to small messages help the planet? Should we share a new message every day or make a big event? Children's voices have more impact than adults: why? PUBLICIZE—In more than one place SHARE—Don't forget to share the experience process. You can challenge the group to choose an artist and challenge them to use your materials in new creative ways Enjoy your work!

Table 24.2 Method overview

Main purpose
To transform feelings of fear, powerlessness, and anger typical of eco-anxiety into feelings of empowerment, love, and ecological sharing
Gained competences
At the end, participants should be able to reflect on climate change in a positive way, with empathy and a sense of empowerment. Identify different ways of working collaboratively. Write and produce a podcast
Educational setting
Informal (format—Workshops, seminars, etc.) collaborative work
Space requirements/restrictions
Outside activities or indoor activities. In the classroom, outside, in a field, or in nature. The children and the facilitator should be distributed in a circle so everyone can see and hear each other. It should be possible to write on post-it notes, play the game, and see/feel/hear the materials. The space should also be calm and quiet enough to record
Resources and necessary materials
Images, photographs, installations, performances, music, and objects from nature Post-its and pens or pencils for writing and coloring. Stopwatch or pendulum to help fulfill timings—sound recorders
Number of participants
Children and young people, a small group of 7–12 participants
Facilitator competences and skills
The moderator should be capable of listening, distributing the word fairly, and letting the group flow. As Paulo Freire understands it, each facilitator should prepare this humanist and liberating approach according to their convictions and the working group (Freire, 2014). Each moderator should select their works of art or objects from nature, bearing in mind that we are all different, and if some are more sensitive to some things, others will be to others. The works of art and objects from nature serve as catalyzing tools for the discussion; the important thing is that they can generate a reaction and discussion on the climate crisis and political inertia. They are essential to the approach process, and time is needed to reflect on the artists and works of art chosen
Participants' skills/age/competences
Children and young people can be of different ages, but the group should be balanced so that older participants can help write down the ideas of the younger ones
Duration
One morning or afternoon. Three times for each activity, depending on the group, it could be 3 h or more. There should be short breaks between the three activities

facilitator select their works of art and objects from nature. The works of art and objects from nature serve as catalysts for discussion, and it is important that they can generate a reaction and discussion on the climate crisis and political inertia. They are essential to the approach process, and time should be allowed for reflection (Table 24.2).

This work has been a starting point. With each activity, new voices will come, different perspectives will appear, and additional reflections will allow us to explore and reflect on some of the concerns these themes raise. Furthermore, our small part will become more synergistic in the end. Transform isolation and fear into a feeling of love that we can change daily in small gestures. The end idea of the activity is to share the constructed and meaningful love for the planet on social media and radio, inspiring and empowering smiles in the young participants.

References

Jenny Holzer. *5 ways Jenny Holzer brought art to the streets.* https://www.tate.org.uk/art/artists/jenny-holzer-1307/5-ways-jenny-holzer-brought-art-streets. Accessed 13 July 2025

Freire, Paulo ([1967] 2014). *Pedagogia do oprimido.* Paz e Terra, São Paulo. ISBN 9788577532285

Games that make the difference. https://www.msseriousgames.com. Accessed 13 July 2025

Lefebvre, Henri ([1968] 2008) *O direito à cidade.* Centauro, São Paulo. ISBN 9788588208971

Chapter 25
Innovation Know-How Workshop

Danica Stojiljković and Ivan Spasojević

25.1 Introduction

Our team at the University of Belgrade has developed a microalgae-based alternative concept of urban greening named urban photo-bioreactor LIQUID3. This concept represents a nature-based solution that targets urban areas and pockets, which cannot accommodate traditional greening due to the lack of space, high value of land, excessive air pollution, or harsh climate, such as squares and pedestrian areas, streets with intensive traffic, public indoor spaces, and cities in the regions with arid climate. The solution relies on the high efficiency of photosynthetic microalgae in CO_2 fixation and O_2 production and their capacity to sequester heavy metals from particulate matter of highly polluted air (Spasojević and Stojiljković 2021). Microalgae are grown in a combination of flat-panel and airlift bioreactors, which are integrated with a bench, solar USB charger, and a night light and may also serve as a translucent urban barrier or info panel.

We started with a concept in 2018 and produced a prototype (technology readiness level (TRL) 7) in 2021. The concept attracted much positive attention from the public and media, including TV stations, newspapers, and social networks. LIQUID3 is an excellent example of a viable innovation and could be used as a model for learning the key aspects of the innovation development process. The dissection of the innovation process produced the innovation know-how that could be transferred to other innovators through interactive workshops. "Innovation know-how workshop" participants can learn about the process of innovation development and identify the aspects essential for a viable innovation, i.e., for bringing the concept to the public and the market.

D. Stojiljković (✉) · I. Spasojević
University of Belgrade – Institute for Mutlidisciplinary Research, Belgrade, Serbia
e-mail: danica@imsi.rs; ivan@imsi.bg.ac.rs

J. Bentz, J. Ristić Trajković (eds.), *Imagining, Designing and Teaching Regenerative Futures: Art-Science Approaches and Inspirations From Around the World*, Science for Sustainable Societies,
https://doi.org/10.1007/978-981-96-9029-9_25

25.2 Transformative Potential

Innovations have been and continue to be motivated to a large extent by the forces of economic development and market competition. Nevertheless, anthropogenic environmental degradation processes pose a great challenge and call for innovations that can mitigate climate change and promote sustainability. Innovations are integral to building adaptive capacity and bringing new capabilities to decision-makers and communities. Every economic sector is affected to some degree by climate-related risks, and innovations are needed to create regenerative opportunities (World Economic Forum 2024). Promoting and supporting innovations and innovators focused on regeneration, integration, synergy, and sustainable living is essential for a regenerative future.

The method "Innovation know-how workshop" has the potential to facilitate and improve the viability and the commercialization success rate of innovations. The transfer of knowledge and the emphasis on specific aspects are essential to bringing the concept into the real world. Workshops on innovations are expected to attract people of common interest and different experiences and expertise. By sharing experiences and analyzing challenges together, participants enhance their innovative capacities. The participants are stimulated to critically rethink concepts or innovation development, consider different scenarios, envision the future, and simulate the process of innovation. Such problem-solving activities have a transformative potential on the participants' understanding of key phases of bringing innovations from the concept level to the market entity. Furthermore, learning about specific aspects of innovation development can reshape the broader innovation landscape. When participants understand that examples only become meaningful through analysis and discussion, they can better adapt these lessons to new challenges. Identifying and sharing key stages of innovation development through interactive workshops represents a powerful and transformative educational approach.

25.3 Application

The "Innovation know-how workshop" method is applied by sharing and discussing the key aspects of an innovation process. The participants will acquire innovation development skills through a two-day "Innovation know-how workshop" with 5–15 participants (students, researchers, and designers with innovation with at least TRL3). The participants can attend individually or as small teams of innovators. During the workshop, examples, such as the LIQUID3 photo-bioreactor, are used to illustrate the key aspects of an innovation process. Other examples can be used, and other aspects of innovation can be underlined. This depends on the lecturer's experience in the field of an innovation application, and the profile of the participants.

The "Innovation know-how workshop" method is applied by the following steps:

1. Use an example of a viable innovation to illustrate the key aspects of developing such innovation. Show your path from concept to prototype to product and the market.

   We use LIQUID3 as an example to show the importance of aspects that we have identified first-hand:

- Multidisciplinarity
  The team that developed LIQUID3 involves bio-scientists, an architect, and a commercial adviser. Such multidisciplinarity was essential for the development of the concept. The involvement of social science researchers in innovation development in technical sciences is considered a prerequisite in solving the grand challenges of energy, food, water, climate, and health (Ledford, 2015).

- Outsourcing and overview
  LIQUID3 was developed and optimized through a collaboration with two small enterprises in Serbia. Outsourcing increases flexibility and R&D capacities. Understanding the capabilities of your industrial partner will benefit the future start-up. The partners may bring to the table expertise, manufacturing know-how, resources, market knowledge, understanding of export and other legal procedures, and contacts.

- Sustainability and low maintenance cost
  The innovation production aimed at tackling environmental issues has to be sustainable. "Green"/recyclable materials that, on the other hand, do not compromise long lifespan have been used in the development of LIQUID3.

- Compatibility, multifunctionality, and design
  LIQUID3 has been developed to integrate with highly urbanized environments. It is recommended to have multiple functions in order to appeal to more markets and induce additional interest. The market analysis showed that LIQUID3 fits three markets—air purifiers, street furniture, and nature-based solutions, which leads to additional opportunities and improves the product's viability.

- Public acceptance
  Public acceptance is the key to the viability of the innovation. It generates spontaneous project promotion and commercial opportunities. LIQUID3 represents an installation that promotes environmental awareness, sustainability, high-quality urban life, and the daily fight against climate change. Its beneficial function, innovative profile, and thoughtful design all contribute to its broad public acceptance.

2. Invite participants to openly discuss the implementation of the outlined aspects to their own problem, i.e., to development of their innovation and the work of their team.

   The lecturer and other participants discuss their views through cross-disciplinary dialogue, critical thinking, scenario planning, and role-playing. Stimulate participants with different areas of expertise to take part and openly present their opinions. Write down key parts of the discussion and disseminate them to the participants.

3. Invite participants to prepare brief presentations on their innovations and how they can utilize the outlined key aspects of innovation development.

   On the second day of the workshop, the participants shall present their concepts or innovations, emphasizing newly obtained innovation know-how skills. They will use scenario planning, envisioning for sustainable futures, simulation, design thinking, and transformative learning. Each presentation will be discussed within a group through moderation by the lecturer.
4. Use the workshop as a think tank for developing new ideas and for transformative learning.

   Nominate subjects/problems depending on the group's profile and invite participants to propose their solutions. Open discussion on the most viable ideas through critical thinking and scenario planning. The lecturer will use the best transformative learning examples to prepare takeaway messages.

   Each workshop has the potential to add to the innovation know-how, deepen the understanding, and broaden the spectrum of relevant aspects of the innovation development process. Overall, the "Innovation know-how workshop" provides a vehicle for students, researchers, and designers to consider the improvements of their approaches and new channels of thinking that could materialize and truly set the foundation of a regenerative future (Table 25.1).

Table 25.1 Method overview

Main purpose
Promotion of innovations; learning key aspects of the innovation process.
Gained competences
Innovation development skills.
Educational setting
Workshop; seminar.
Space requirements/restrictions
Any.
Resources and necessary materials
PC with a projector.
Number of participants
Small groups 5–15; work: individual or small team.
Facilitator competences and skills
At least one developed innovation from concept to technology readiness level (TRL) 8/9. Understanding of aspects that are key to the success of the innovation process.
Participants' skills/age/competences
Students, researchers, and designers with an innovation with at least TRL 3.
Duration
Two-day workshop.

References

Ledford H (2015) How to solve the world's biggest problems. Nature 525:308–311

LIQUID3. www.liquid3.rs. Accessed 20 June 2024

Spasojević I, Stojiljković D (2021) Urban photo-bioreactor Liquid3 – synergy of design and biotechnology. In: Matić Bujagić I et al (eds) Šesti naučno-stručni skup Politehnika – Zbornik Radova [The sixth scientific and professional conference of the Polytechnic – Proceedings]. Šesti naučno-stručni skup Politehnika [The sixth scientific and professional conference of the Polytechnic], Belgrade, December 2021, Akademija tehničkih strukovnih studija, Belgrade, pp 904–909

World Economic Forum (2024) Innovation and adaptation in the climate crisis: technology for the new normal insight report. World Economic Forum, Cologny/Geneva. www3.weforum.org. Accessed 20 June 2024

Chapter 26
The BEEHIVE: Method to Explore Alternative Futures

Ger Pepels, Celiane Camargo-Borges, and Annalena Schmidt-Faber

26.1 Introduction

The BEEHIVE is a method designed at Performatory (Breda University of Applied Sciences, Netherlands) to foster transformative change in real-life contexts. It builds on Transformative Social Innovation and aims to explore possible futures "beyond reform" by addressing the root causes of the challenges at hand and exploring alternative transformative future possibilities together.

The framework builds on a process-based approach to change, taking the time and inviting participants to understand the current situation, experiment and experience vulnerabilities, uncertainties, and complexity, and be open to potential losses of existing interests and related pain (Andreotti 2021). Transformative change might occur through ongoing experiments in new relations and be supported by reflexive monitoring (Aviles JR 2023). The Beehive framework has four phases.

1. Break the Sky: Decide on the challenge that speaks to you from the world. Identify and invite the "system" into the room, including those "whom it concerns."

The original version of the chapter has been revised. A correction to this chapter can be found at https://doi.org/10.1007/978-981-96-9029-9_67

G. Pepels (✉) · C. Camargo-Borges
Breda University of Applied Sciences, Breda, The Netherlands
e-mail: ger.pepels@home.nl; camargo-borges.c@buas.nl

A. Schmidt-Faber
Eberswalde University for Sustainable Development, Eberswalde, Germany
e-mail: annalenasf@posteo.de

J. Bentz, J. Ristić Trajković (eds.), *Imagining, Designing and Teaching Regenerative Futures: Art-Science Approaches and Inspirations From Around the World*, Science for Sustainable Societies,
https://doi.org/10.1007/978-981-96-9029-9_26

2. Open Time: Creating time–space for participants to step out of routine and develop a shared understanding of their challenges and desire for changes.
3. Find What Gives Life: Encouraging participants to identify and appreciate elements from the previous step that contain energy and offer hope and possibility.
4. Build New Relationships: Co-designing new relationships among all involved to develop innovative ideas and methods for alternative futures.

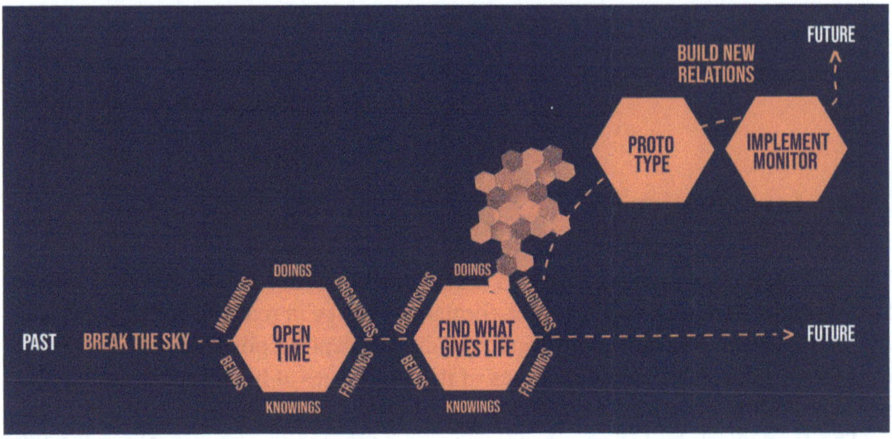

26.2 Transformative Potential

The latter three phases contain six spaces. The six spaces of the participatory process foster a comprehensive understanding of a topic and encourage all involved to engage in creative and transformative problem orientation. The last two spaces especially challenge the participants to explore (more-than-human) relational connections, creating a deep understanding of the values and beliefs foundational to our positions (Framings) and using beyond-rational analysis (Beings).

1. Imaginings space: Participants start by individually envisioning their initial ideas on solutions to the challenge, which can be defined by the researchers or the community/organization. These initial ideas are rooted in tacit knowledge based on personal experience, intuition, empathy, and compassion and offer the basis for later reflection on the transformative potential of alternatives.
2. Doings space: Participants critically assess all the stakeholders directly and indirectly involved in the challenge and describe their concrete activities and relations. By going back and forth, the participants will discover more actors.
3. Organizings space: Consider the organizational and institutional settings that shape the challenge, as well as the various scales at which these conditions are established and reestablished within the societal context.

4. Knowings space: Participants explore the knowledge and skills available to the stakeholders. Critical evaluation helps identify the knowledge and skills that may have been overlooked.
5. Framings space: Understanding the various worldviews, values, beliefs, and mental models of those involved. Exploring the narratives driving all these perspectives might offer insights into alternative discourses that can open new directions.
6. Beings space: In the final space, participants delve into ways of knowing that go beyond the rational. They explore embodied, spiritual, and emotional experiences that raise awareness of what's happening and invite consideration of different directions. This phase transcends pure rationality and adds a deeper dimension to the overall understanding of the challenge.

After Break the Sky, the three phases contain these six spaces. Participants work in these spaces, preferably in pairs or small groups, in diverse orders, and they are invited to revisit spaces multiple times. The comprehensive exploration of these spaces empowers a shared and holistic understanding, paving the way for the next phase. It supports reframing the challenge and offering a starting point to "Build new relationships" around concrete experiments exploring potentially transformative directions.

The BEEHIVE methodology fosters transformative journeys by uncovering previously invisible interconnections, leading to a deeper understanding of challenges. It encourages novel collaboration, openness to all voices, and receptivity to new ideas, catalyzing innovative solutions and transformative change.

26.3 Application of the BEEHIVE

The BEEHIVE method is versatile and applicable to brief 2-h sessions up to multi-day collaborations to explore alternative futures. Its four phases support identifying future directions and defining, implementing, and monitoring concrete projects. If time is limited, progressing through the first three phases (Break the Sky, Open Time, and Find What Gives Life) can help identify future directions and guide the next steps. Below is an example of a 3-h session.

26.3.1 Break the Sky (Preparation)

Begin by exploring the value of a complexity-based approach to the challenge. For instance, use the Cynefin framework (see references) to distinguish between simple, complicated, *complex*, and chaotic problems. If applicable, invite stakeholders from across the system, including those directly affected by the challenge, in addition to experts, policymakers, and academics. Ideally, 10–20 participants collaborate to have diverse perspectives available.

26.3.2 Open Time

Prepare the space by mapping the visual on the floor. Ensure it is large enough for participants to move around, and provide large post-its and pens for adding ideas.

- Start by welcoming participants and explaining the BEEHIVE approach. Elaborate on the challenge (from Break the Sky) and invite everyone to develop a deep understanding and explore future directions collaboratively in this session (15 min).
- Next, participants enter the Imaginings space to write down their initial ideas for solutions to the challenge silently. Allow time for initial questions. Once the post-its are shared, everyone reflects on the diversity of inputs (15 min).
- Participants then visit the other spaces in pairs or small groups and leave post-its on the floor. There is no set order for visiting the spaces. By sharing, the participants affect and are affected by others. Multiple rounds encourage interaction and reflection (30 min).
- When energy drops, share insights as a group. Each participant is invited to give meaning to the inputs. The participants will already have experienced the change in their perspectives (as shared at the start in the Imaginings space). This will inspire new ideas and the need to add more contributions. Allow participants to go to the diverse spaces again (20 min).
- Finally, the participants collaboratively select in an open dialogue the most relevant inputs that could contribute to exploring regenerative futures. And importantly, following these efforts, the participants collaboratively reframe the challenge generatively (20 min).

26.3.3 Find What Gives Life

The participants move to the generative phase itself.

- The selected inputs are mapped in the respective spaces of the BEEHIVE: Find What Gives Life. Starting from the reframed challenge, participants explore the findings in pairs again and add more thoughts to the spaces (except for the Imaginings space) (20 min).
- As the final step here, participants move together to the Imaginings space to generate multiple ways of addressing the reframed challenge in dialogue (20 min).
- This generative phase ends with collaboratively selecting the most inspiring transformative future idea (10 min).

26.3.4 *Build New Relationships*

In the final phase of the approach, the participants use the diverse spaces again to design a prototype for approaching the challenge holistically. They also share ideas on how to prepare for implementation. Finally, they consider how to monitor the outcomes reflexively in the future (30 min) (Table 26.1).

Table 26.1 Method overview

Main purpose
To collaboratively find pathways to alternative futures
Gained competences
Develop competencies concerning system intelligence, deep listening, using the power of language, and mobilizing cocreation
Educational setting
Applicable in both formal and informal settings; preferably bringing "the whole system" related to a real-life social practice into the room, especially giving voice to "whom it concerns." To gain an understanding of the tool, learners can use it in a formal learning setting
Space requirements/restrictions
The application of the method requires the physical space to map the BEEHIVEs (at least: Open time and find what gives life) on the floor. The space needed depends on the number of participants; the participants need to have the space to walk around in the spaces of the BEEHIVE, leaving post-its or other signs in the spaces. It can be executed indoors and outdoors
Resources and necessary materials
Map the BEEHIVE with adhesive tape to the floor. Each axis of the BEEHIVE is seen as a BEESPACE To support the participants, put prints of the core concepts (doings, organizings, ...) in the respective SPACES of the BEEHIVE Participants require a pen and post-its to write down their thoughts to leave these in the spaces
Number of participants
Ideally, 10–25 diverse participants join a session as the methods aim to make participants work collaboratively, exploring alternative futures
Facilitator competences and skills
The facilitator needs to have experience to manage the dynamics of the group, inviting participants to interact in open ways with one another and with the contributions of participants. The facilitator also needs to have strong system intelligence
Participants' skills/age/competences
If the diversity of the group is large enough, then there are no special requirements for the skills/compentences
Duration
A minimum of 2 h is needed. The duration of the application in real-life practices can be at multiple lengths, from a minimum of 2 h to spanning multiple months

References

Aviles JR et al (2023) Transformative research toolkit. Othering & Belonging Institute, University of mCalifornia, Berkeley. https://www.otheringandbelonging.org/transformative-research-toolkit. Accessed 16 Mar 2024

Cynefin framework. https://thecynefin.co/about-us/about-cynefin-framework/. Accessed 18 Nov 2024

de Andreotti VMO (2021) Hospicing modernity: facing humanity's wrongs and the implications for social activism. North Atlantic Books, Berkeley

TRANSIT Resources. http://www.transitsocialinnovation.eu/downloads. Accessed 18 Nov 2024

Chapter 27
Tales of Transformation: Working with Narratives for Climate-Just Futures

Justus Wachs

27.1 Introduction

The climate crisis is, in part, a storytelling crisis. This idea has motivated my work as a climate justice activist in Germany and the UK for the past decade. Here, I introduce how I have combined creative practices (e.g., Theatre of the Oppressed, mindfulness, storytelling) into a method—"Tales of Transformation"—that works with stories and narratives for climate justice. This approach helps deconstruct existing, often limiting, climate narratives and their internalized effects and facilitates an imaginative and reflective exploration into alternative, empowering, and regenerative narratives, allowing activists to craft imaginaries where environmental degradation has been overcome. The method includes embodied, emotional, intuitive, or spiritual knowledge in these processes of deconstruction and world-making.

The method unfolds in four main stages (see Tables 27.1 and 27.2 for summaries). First, participants explore a range of creative, reflective methods, including movement, theatre, creative writing, drawing, and storytelling. Second, they inquire into dominant climate narratives surrounding them—or persistent ways of thinking about the climate crisis. They identify these narratives and explore how they make them feel emotionally and physically. Third, participants envision a world where climate justice has been achieved, including felt senses, values, and responsibilities in such a world. They do this individually at first and then come together in small groups to identify similarities and differences in visions. Finally, all groups come together in a celebratory forum of world-making, presenting their shared vision through a performance, story, poem, or meditation.

J. Wachs (✉)
McGill University, Montreal, QC, Canada
e-mail: johann-justus.wachs@mail.mcgill.ca

J. Bentz, J. Ristić Trajković (eds.), *Imagining, Designing and Teaching Regenerative Futures: Art-Science Approaches and Inspirations From Around the World*, Science for Sustainable Societies,
https://doi.org/10.1007/978-981-96-9029-9_27

Table 27.1 Method overview

Main purpose
Deconstruct existing and co-create new climate narratives.
Gained competences
Critical thinking, narrative analysis, creative expression, collaborative problem-solving, emotional intelligence, and agency.
Educational setting
Informal: Workshops, seminars, grassroots spaces, community settings, art galleries, and festivals.
Space requirements/restrictions
Indoor activities; a spacious room for movement and group work.
Resources and necessary materials
Art supplies (paper, markers, flipchart), ceremonial items such as a bowl of water, and open space for movement and performance.
Number of participants
Group work, ideally 12–30 participants.
Facilitator competences and skills
Required: Experience in climate justice, creating safer and inclusive spaces, ability to hold emotional content. Useful: background in creative arts (theatre, storytelling).
Participants' skills/age/competences
Suitable for activists, community members, students, policymakers, researchers, professionals, no age limit or specific prior skills required.
Duration
Minimum 2.5 h, preferably 4 h for a full session.

Table 27.2 Stages of the 'Tales of Transformation' method

Stage	Description	Suggested activities	Goals
Prep: Setting the space	Create a supportive environment that separates the workshop from outside distractions and facilitates emotional safety.	Creating a circle of chairs, mindfulness moment, co-elaborating a safe space agreement, and removing shoes.	Participants feel grounded, safe(r), and ready to engage with vulnerable and creative work.
Stage 1: Creative exploration	Participants explore various creative, reflective methods to unlock imagination and emotional connection.	Be led by the methods you know best; embodied games, movement, sculptures, creative writing, drawing, storytelling, and more.	Participants relax, have fun, become playful, access wider ways of knowing, and begin emotional and intuitive engagement.
Stage 2: Inquiry into dominant narratives	Participants identify dominant climate narratives and reflect on their emotional and physical impacts.	Build on methods introduced in stage 1; cycle through different ways of knowing (e.g., if narratives are explored with the body first, encourage participants to reflect on the experience by free-writing or drawing after).	Increased awareness of how dominant narratives affect emotions and bodies; this awareness often engenders a sense of agency.

(continued)

Table 27.2 (continued)

Stage	Description	Suggested activities	Goals
Stage 3: Envisioning alternative futures	Participants individually and collectively envision a world where regenerative futures/ environmental justice have been achieved.	Craft prompts that invoke participants to time-travel to alternative futures with all their senses; begin with individual visioning; then pairs, groups of four, eight, and so on; explore vision commonalities and differences.	Common ground is found in visions of the future, fostering creativity and collective ownership of new narratives.
Stage 4: Forum of world-making	Groups present their shared visions through artistic and performative expressions, celebrating collective imagination.	Performances, storytelling, guided meditations, poems, or other forms of artistic expression.	Celebration and inspiration through the sharing of multiple possible futures, leaving participants energized and inspired.
Closing the space	Participants reflect on the experience, share final thoughts, and formally close the session to return to everyday reality.	Group questions, reflections and impressions, final mindfulness or grounding exercise.	Participants leave the session with a sense of closing together and harvested insights to apply in their activism or lives.

27.2 Transformative Potential

The "Tales of Transformation" method addresses two critical observations from my experience in the climate justice movement in the UK and Germany. First, specific narratives gain hold in the collective consciousness and guide how the climate crisis is portrayed. In my context, I see the "technocratic narrative" (viewing the climate crisis as a technical/managerial challenge) and the "urgency narrative" (viewing the climate crisis as a struggle that demands immediate action and is already lost). These narratives legitimize some solutions and ways of knowing, feeling, and acting over others, obscuring the plural narratives through which the crisis is experienced.

The second observation is that many activists, myself included, were so used to fighting that we started to find it challenging to remember what we were fighting for. Solnit (2023) suggests, "We have to find stories of a livable future, stories of popular power, stories that motivate people to do what it takes to make the world we need."

The "Tales of Transformation" method tackles these observations. Participants engage their bodies, feelings, intuition, and thinking to recognize dominant climate narratives and explore how they make them feel and what actions they inspire and inhibit. Viewing dominant climate narratives as one of many ways to respond to the crisis can be liberating. Using embodied, emotional ways of knowing helps surface elements of these narratives that participants might not have been aware of. For example, the body might hold unrecognized anger or stress about the crisis. Bringing these feelings into conscious awareness can inspire a fresh sense of agency to respond more mindfully.

After playing and exploring, participants are invited to dream up a world in which the climate crisis is overcome and explore it with all their senses. They initially dream individually and then form groups to look at commonalities and differences in their visions, building common ground while not erasing plurality.

Finally, all groups come together in a celebratory forum of world-making, presenting their shared vision through a performance, story, poem, or meditation. The act of sharing multiple futures can leave participants empowered and energized with their imagination unleashed.

27.3 Application

To illustrate the method, I will describe facilitating a *Tales of Transformation* session at the People's Summit 2021 in Glasgow. Hosted by the Climate Justice Alliance, the summit occurred alongside the COP26, providing workshops, discussions, and cultural events for frontline communities to build solidarity. Activists from all over the world engaged in protests alongside the People's Summit.

The workshop took place in an art gallery in central Glasgow, accommodated 30 participants, and lasted 4 h. These conditions were ideal. I recommend between 12 and 30 participants and no less than 2.5 h. Participants ranged from the early 20s to 70s and included public sector workers, NGO professionals, COP delegates, Indigenous activists, private sector professionals, and artists.

27.3.1 Setting the Space

To create a container separate from the intensity of COP and the People's Summit, we placed a bowl of water with flower petals in the middle of a large chair circle. Participants removed their shoes to mark the transition into the imaginative bubble of our session. We opened with a moment of silence, followed by introductions and a check-in.

Working with climate narratives can bring up challenging emotional content. To co-create a holding container, we co-elaborated and committed to a safer space agreement. We invited participants to go as far as they felt comfortable and offered availability for emotional support if needed.

We encouraged an atmosphere of playfulness, inviting participants to co-create and rebel against our structure. Warm-up activities included walking around the room, greeting others by looking them in the eyes, moving at different speeds, and playing games to stimulate the imagination.

27.3.2 Deconstructing Existing Climate Narratives

We used Boal's technique of *statues* to explore existing climate narratives and their effects. Participants represent objects and ideas as statues in their bodies, starting with simple ones like *tree* or *politician*. Participants search for the representation of the notion in their bodies, adding a repetitive movement and sound to make the object come to life and deepen its exploration. There is no right or wrong—the focus is on searching for a form that feels true in the moment.

We moved on to more abstract ideas, such as *change*, before embodying the climate change narratives. The emotions and reflections this exercise stirred up were tangible, almost as if physically present in the space.

Participants discussed their experiences in pairs, then engaged in free-writing and drawing central words or ideas. Finally, we gathered as a whole group and reflected on the process. Recognizing the impacts of dominant narratives on emotions and bodies gave participants a sense of agency to respond differently.

27.3.3 Co-constructing Alternative Climate Narratives

After a break, participants used the methods learned (statues, movement, writing, drawing) to investigate the following prompt: *Imagine a world where environmental degradation has been overcome. What would it feel like to live in such a world? What would we value? What are personal and shared responsibilities? How can people contribute to the emergence and maintenance of such a world?*

Participants explored these questions individually, then in pairs, and eventually in groups of four and eight, identifying commonalities and differences in their visions.

27.3.4 Celebratory Forum of World-Making

Each group artistically represented their common visions, choosing forms such as a movement, a short scene, a story, a poem, or a guided meditation. We built an improvised stage, and each group performed their piece. One group conducted a guided meditation; another performed a scene where disconnected and chaotically moving actors converged and joined hands; another told a story where some characters had a significant realization and changed their ways. It was uplifting and touching to hear these fruits of people's explorations, testified by the resounding applause that followed the performances. We shared overall reflections before closing the space together.

27.3.5 Conclusion

The Tales of Transformation workshop format creates a sense of ownership over how climate narratives affect activists, fostering agency to envision alternative narratives and tap into the power of collective imagination.

While my experiences stem from climate justice activism in Western Europe, this method could suit civil society or public and private sector organizations. The People's Summit 2021 highlighted its strength in allowing participants from different contexts to deconstruct and co-create climate narratives. Creating a safe space and ensuring consent is crucial, especially for participants who have experienced profound traumas and injustices.

Reference

Solnit R (2023) 'If you win the popular imagination, you change the game': why we need new stories on climate. The Guardian, 12 January. https://www.theguardian.com/news/2023/jan/12/rebecca-solnit-climate-crisis-popular-imagination-why-we-need-new-stories. Accessed 5 Dec 2024

Chapter 28
The Future You: Visioning and Empathizing for Futures Literacy and Long-Term Thinking

Justyna Doherty

28.1 Introduction

Timothy Morton (2013) describes our current environmental crisis as challenging both our control and understanding of the world. The "Future You" activity aims to address this crisis by transforming abstract concepts into tangible, empathetic connections with future generations.

By envisioning future contemporaries as vividly as our loved ones today, participants can develop a deeper sense of empathy, expanding their capacity to care across time. This shift in perspective fosters a proactive responsibility for the well-being of those who will follow, embedding the principles of sustainability into daily life. The activity also encourages a broader, non-anthropocentric worldview, acknowledging the interconnectedness and agency of the natural world. Through personal, cognitive connection to future beings and ethical reflection that follows, participants deepen their commitment to planetary stewardship, understanding their role in sustaining a thriving ecosystem for generations to come.

Inspired by Indigenous principles of intergenerational responsibility, including those of First Nations in Canada, this activity emphasizes that today's decisions must ensure a balanced and harmonious future (Clarkson et al. 1992). It challenges participants to extend their ethical duties beyond the present moment, urging them to consider how their actions today will shape tomorrow's world.

J. Doherty (✉)
Institute of Art, Design + Technology, Dublin, Ireland
e-mail: Justyna.Doherty@iadt.ie

J. Bentz, J. Ristić Trajković (eds.), *Imagining, Designing and Teaching Regenerative Futures: Art-Science Approaches and Inspirations From Around the World*, Science for Sustainable Societies,
https://doi.org/10.1007/978-981-96-9029-9_28

28.2 Transformative, Regenerative Potential

The activity aims to catalyze personal and societal transformation and shift in thinking about the future. By cultivating a personal connection to future generations, participants are encouraged to see their own actions as part of a broader narrative of planetary care. Magnason (2021) suggests that this understanding, or mindset, "brings greater purpose and joy to our own individual efforts and transforms the task of planetary renewal into a motivating and positive endeavor".

This activity also enhances futures literacy and long term thinking, enabling participants to understand the broader impacts of their decisions. By adopting systems thinking, they can recognize the complex interplay between human actions and natural systems, fostering ecological stewardship.

Moreover, it empowers to initiate changes in behavior, as the practice of envisioning future generations and the natural world as part of our ethical community encourages a shift toward more sustainable practices in everyday life. To summarise, engaging in this activity and deep reflections that follows, underscore the urgency of adopting a futures-thinking mindset in our collective journey toward a sustainable and flourishing future, and highlights the importance of imagination, creativity and art in it.

28.3 Application

Time 3 h/half day

Materials Newspapers, magazine cut-outs, natural elements (such as leaves, dry flowers, found materials), scissors, glue or masking tape, blank postcard templates (optional).

Participants are required to bring scanned photos (not the originals) and magazines suitable for use in collages.

Group Size Suggested 20 people max.

28.3.1 Part A: Individual Timelines

28.3.1.1 Introduction (10 min)

In this section, participants will create personal timelines spanning several generations, moving fluidly between reflections on ancestors and imaginative representations of future individuals. The collage-making serves as a conversation starter, inviting participants to share memories, experiences, facts, struggles, and reflections, while visualization activities foster richer insights and storytelling. This activity aims to deepen understanding of intergenerational interconnectedness and the importance of responsible decision-making for the well-being of both past and future generations.

To provide context, it's worth introducing the Seventh Generation Principle, rooted in Indigenous wisdom from the Haudenosaunee (Iroquois) Confederacy (1992). This principle urges us to make decisions today that will sustain a viable world for seven generations ahead. By embracing this perspective, participants can cultivate a sense of responsibility and connection across time.

28.3.2 Part A: Reflection on Generational Connections and Gratitude (30 min)

Participants start from reflecting on their memories and feelings about past generations, exploring the connections they share with them. While connecting with people from the past may be easier, forming a similar bond with future generations requires deliberate effort.

28.3.2.1 Creating a Timeline of Gratitude

Start by drawing a simple timeline from today's date. Consider significant individuals from the past: parents, grandparents, siblings, and their pets or other significant non-human beings you remember. Use a collage and old photos to visually represent these individuals and non-human beings, placing them chronologically on the timeline.

Note: This exercise has been meaningful for groups, including war refugees who may have lost family albums. Using photos, visuals from others, or magazine cutouts allows for recollection and shared memories.

28.3.3 Part B: Looking Toward the Future (30 min)

The second stage involves looking toward the future. Expand the timeline forward, aiming to explore potential connections with future beings. This exercise may be challenging as it requires a shift in thinking.

28.3.3.1 Imagining Future Individuals

Instructions: First - imagine people living in the future you may be connected to — perhaps a grandchild, their child, a relative yet to come, someone in your community who will live in the future, a future neighbour, or someone else you may never meet. Place them on your timeline, with dates. Then - choose one (or more) person from the future and, using collage, create a vivid representation of their attributes and traits. Combine old photos with magazine cut outs, use colours and textures to bring this person to life, creating their portrait if possible. Be playful. Experiment. If you prefer, draw, doodle or write. Reflect:

28.3.3.2 Key Questions

- What is the future person's name?
- What are their personality traits?
- What hobbies or interests do they have?
- Describe their daily routines, joys, challenges, and worldviews.
- If they could communicate with us, what message would they share?

28.3.3.3 Reflection Question

Consider the future birth date of that person from the future you are connected to. Calculate and mark a future point in time where these individuals will live, considering significant climate action deadlines. For example, if the person will live in 2100, reflect on how this relates to current strategies and deadlines for addressing climate change.

Once the first future person is envisioned, participants can add other elements or items from their imagined future world and explore connections with other future individuals.

28.3.4 Part C: The Future Us (30 min)

After populating the timeline with future people, attention shifts to the non-human world. Consider the following:

- Could future individuals live independently, and what would their relationship with nature be?
- Which other species and creatures would be essential to their existence?
- How balanced is the human–nonhuman interaction in their world?
- What environmental factors and interdependencies are crucial to their future?

Visual representations on the timeline should reflect these elements and their interconnections.

28.3.4.1 You Are My Voice (15–30 min)

To deepen reflection, participants choose a species from the future and describe the world from its perspective. They can collage or use natural elements to visually represent them, and ask:

- What does this species perceive and experience?
- If it could communicate, what message would it have for humanity?

Example: "I cannot speak for myself; you are my voice"—consider what the creature's message would convey.

28.3.4.2 Reflection and Discussion (30 min–1 h)

In pairs or small groups, share insights and visual representations of future people, species and/or other non-human entities. Discuss the significance of these representations and why they were chosen. Finally - reflect on their messages to our present selves - what would they say to us?

28.3.5 Closing (30 min)

To summarize the experience, invite participants to craft a concise message or personal pledge to the future people they "met", encapsulating reflections from the activity. Drawing inspiration from the pledge created by Allan Tough, encourage participants to express their pledge in their own language and style, even if it's just a short sentence (2002). Finally, following the activity, you can share findings and pledges created in a public exhibition or showcase, to inspire and engage others.

28.3.5.1 Additional Element

To enhance and extend the experience, provide blank postcard templates to participants, so they can create their pledge in the form of a postcard.

These postcards / pledges can then be "posted" in a celebratory manner by the group, kept as mementos, or shared with family and friends to extend the discussions and outcomes of the day (Table 28.1).

Table 28.1 Method overview

Main purpose
The activity aims to cultivate a profound personal connection to the future among participants. It seeks to develop skills in long-term thinking, future literacy, and ethical responsibility toward future generations and the environment.
Gained competences
Developing future literacy and long-term thinking skills. Enhancing art / collage-making and visual storytelling abilities. Cultivating empathy and perspective-taking across generations. Engaging in ethical reflection and embracing sustainability principles.
Educational setting
This activity is tailored for educational settings focused on sustainability, ethics, and future studies. It is suitable for workshops, seminars, or educational programs aimed at fostering forward-thinking mindsets.
Space requirements/restrictions
It requires a spacious environment that can comfortably accommodate up to 20 participants. Tables for collage-making and designated areas for group discussions are essential.
Resources and necessary materials
Magazines for cut-outs, natural elements (such as leaves and dry flowers—could be incorporated into the collage), scissors, glue or masking tape, and optional blank postcard templates or paper for creating personal pledges.
Number of participants
The activity is designed for groups of up to approximately 20 participants to facilitate meaningful interaction, collaboration, and in-depth discussions.
Facilitator competences and skills
Guiding group discussions and reflective activities. Knowledge of sustainability issues, futures literacy, and ethical considerations. Ability to encourage creativity in activities like collage-making and narrative construction.
Participants' skills/age/competences
Participants should ideally be adults or mature students capable of engaging deeply in reflective exercises and discussions on complex ethical and environmental topics. Proficiency in creative expression, empathy, and openness to long-term thinking is advantageous.
Duration
Approximately 2.5 h—duration may be adjusted according to specific requirements.

References

Clarkson L, Morrissette V, Régallet G (1992) Our responsibility to the seventh generation: indigenous peoples and sustainable development. International Institute for Sustainable Development, Winnipeg

Magnason AS (2021) On time and water. Open Letter, Rochester

Morton T (2013) Hyperobjects: philosophy and ecology after the end of the world. University of Minnesota Press, Minneapolis

Tough A (2002) What future generations need from us. Futures 34(2):137–144. https://doi.org/10.1016/S0016-3287(01)00089-0

Chapter 29
Imagination Activation Exercise: What If…?

Kate Rudd

29.1 Introduction

This is an engaging and powerful imagination activation exercise authored by Rob Hopkins (2020), co-founder of Transition Town Totnes and Transition Network (2020), which he often delivers during public talks or when running workshops.

It involves guiding a group of almost any size through an imaginary act of collective time travel to a future 10 years from now. The group "steps" into a time machine and "emerges" into a new world that—while not a utopia—is the result of doing everything possible to improve their community/bioregion/organization (or other specific area of focus). The group "walks" around this new world in their imagination and is prompted by the facilitator from time to time to help enrich their vision. Afterward, they share and discuss their experience with a partner and then among the wider group. The most popular ideas are recorded to create alignment on a vision and then used for a follow-up planning exercise, which helps the group create a roadmap for action.

This exercise is ideal for collective dreaming and co-visioning workshops as it provides a safe container in which to explore different positive visions of the future, arrive at an ideal common vision, align on desired outcomes, and co-create pathways for achieving them without the usual constraints.

K. Rudd (✉)
University of York, York, UK

J. Bentz, J. Ristić Trajković (eds.), *Imagining, Designing and Teaching Regenerative Futures: Art-Science Approaches and Inspirations From Around the World*, Science for Sustainable Societies,
https://doi.org/10.1007/978-981-96-9029-9_29

29.2 Transformative Potential

Unlocking the collective imagination in this way has significant transformative potential, as it frees people from the constraints of their current reality. It enables them to envisage a future that is not a linear progression of the present and where path dependency does not dictate. Paradigmatic assumptions can be momentarily suspended, meaning that the socioeconomic, political, and cultural models and systems we have collectively built on them can look different and produce different outcomes. These new outcomes can be used as starting points to reverse engineer the kinds of models and systems that might be capable of producing them. According to Rob:

> [The exercise] often moves people deeply. There are sometimes tears. Once the evening concludes, people head off out into the world, and that fabulous dream can either go on to inspire them to work towards that new 'North Star' in their lives, or it can just dwindle as the 'real world' takes over once more. But it is profound work. A lot of people say that it felt like something deep shifted in terms of their sense of what is possible in the future. (Hopkins 2020; para 5)

29.3 Application

This is an extremely versatile exercise that works in almost any group context (indoors or outdoors, seated or standing) with almost any kind of audience (business leaders, educators, students, policymakers, communities, landscape partners, etc.) or as part of hackathons, horizon scanning sessions, conferences, or artistic projects. No resources are necessary, but paper and pens for participants and/or a board to write on can be helpful.

I have used this exercise in professional settings and seen it work brilliantly in a workshop delivered by a colleague, Karen Winekker, for Siendo Naturaleza (2024), an organization engaging in holistic regeneration work in the Peruvian Amazon, as part of a process to visualize and discuss the changes the team wanted to see for their organization and their wider community. It enabled the team to reflect, dream, share their hopes and visions for the future, surface any frictions, and address them collaboratively to co-create a shared vision for the future. We drew on this work to help craft the organization's value proposition and regenerative business model and to develop a roadmap for its operationalization.

29.3.1 Workshop Overview

The following workshops are adapted from *The Future We Want: What Is? What If? What Next?* guide (2024).[1]

29.3.1.1 Part One: What if: Creating a Common Vision

Time: Approximately 60 min.
Aim: Use our collective imagination to bring to life the future that could still be possible.
Mindset: Expansive—cultivate a culture of "Yes, and" rather than "Yes, but." It is fine for the group to have multiple visions—look for synergies among the different visions.
Method:

- Explain the aim and mindset.
- Introduce the exercise, telling the group that in a moment you will attempt an act of collective time travel.
- Invite the group to get into pairs, and make themselves comfortable.
- Ask them to close their eyes, and to take a breath.
- Read the following text or a version adapted to your exercise/audience:

> We are about to take a journey through time: ten years into the future. The times we travel through will be times of the most profound and remarkable transitions in human history. Change that in [current year] feels unimaginable will build in positive and accelerating cascades. Institutions that felt so permanent in [current year] will crumble and fall, and new, infinitely better ones will bloom in their place. Those 10 years were the most thrilling time to be alive. They are times that those that came afterwards told great stories about and sang great songs about.
>
> I am turning on my Time Machine, and we are travelling forwards together. [Play/make the noise of a machine or engine starting].
>
> [Pause several seconds]
>
> Let's step out into this new world. It's not a utopia, but it is the result of everything that could have been done, being done. Keep your eyes closed as you begin to explore it in your imagination using all your senses. (Idem; p 19)

- Leave people sitting in silence for eight minutes to explore the future. (You can play relaxing music during this time.)

[1] Created by Transition Together with the support of Transition Network under the Creative Commons CC BY-SA 4.0 DEED licence. The original guide was written by Rob Hopkins and Michael Thomas. It was updated and revised by Chris McCartney and Daniel Balla. You are free to share and adapt these materials, with conditions of attribution and share alike. Find out more: https://creativecommons.org/licenses/by-sa/4.0/

- Offer an occasional prompt, with long pauses between, such as: *What do you see as you walk around? What sounds do you hear? What does it smell like, feel like, taste like? What are people doing for work and leisure?*
- Invite people to stay in the future mindset as they open their eyes and spend 12 min talking in pairs about the world they have imagined.
- Then take five minutes to write two delicious ideas each that they would like to share with the wider group.
- Invite people to share their ideas with the group.
- Choose the most interesting ones and write them on the board.

29.3.1.2 Part Two: From What if to What Next?

Time: 1.5 h.
Aim: Explore how to make the shared vision a reality.
Mindset: Try to release the reality of the present moment and your current mindset, which is the cause of so many of the problems we face. Instead, position yourself in the future, and let a better world emerge from that perspective. Participants may have different visions, but they can be resolved through dialogue and collaboration—think "both/and" rather than "either/or."
Method:

- Spend 15 min as a whole group deciding which of the themes listed on the board from the previous exercise to focus on. Which have the potential to become viable strong initiatives? And which ideas do the group have the most enthusiasm for making a reality?
- Split into smaller groups to work on the selected themes (one group per theme).
- Each group should spend about 15 min on each of the following questions to come up with some next steps/actions to help make the envisioned idea a reality. (Emphasize this is simply about next steps, rather than designing a whole project.)
  - Who do we need to collaborate with to make it happen?
  - What resources do we need to do this?
  - What could we do right now to get started?
- Come back together as a whole group and listen to each group in turn, explaining their next steps and outlining what they hope to have achieved by next year.

Close, and celebrate the group's work! (Table 29.1)

Table 29.1 Method overview

Main purpose
Unlocks collective imagination and fosters deep, transformative thinking by guiding participants to envision a radically improved future. It builds awareness, inspires behavior change, aligns participants around a shared vision, and empowers them to collaboratively plan actionable steps toward realizing that vision.
Gained competences
Collective imagination activation, co-visioning, collaboration, co-design, and creating a roadmap for future action.
Educational setting
Informal (workshops, seminars, public talks); adaptable for various contexts including hackathons, horizon scanning sessions, conferences, and artistic projects.
Space requirements/restrictions
Flexible: Works in almost any setting (indoors or outdoors, seated or standing).
Resources and necessary materials
Minimal: Paper and pens for participants, a board to write on (optional), relaxing music (optional).
Number of participants
Group work, adaptable for almost any group size.
Facilitator competences and skills
Experience in guiding group exercises, strong facilitation skills, and ability to prompt imagination and manage group dynamics.
Participants' skills/age/competences
Suitable for a wide range of participants, including communities, students, policymakers, researchers, designers, and practitioners.
Duration
Part One: 60 min; Part Two: 1.5 h; Total: Approximately 2.5 h.

References

Hopkins R (2020) Imagination taking power. https://www.robhopkins.net/2020/04/21/when-a-resilient-future-calls-by-to-see-if-you-want-to-come-out-to-play/. Accessed 15 June 2024

Siendo Naturaleza (2024) Siendo Naturaleza A life project of regenerative learning and living in the cloud forest of Peru. https://www.siendonaturaleza.com/. Accessed 15 June 2024

Transition Network (2020) Transition: bounce forward. What is, what if. What next? What if visioning exercise. https://transitionnetwork.org/wp-content/uploads/2020/11/What-If-Visioning-Exercise.pdf. Accessed 15 June 2024

Transition Together (2024) The future we want what is? What if? What next? Visioning exercise for groups and communities. https://transitiontogether.org.uk/wp-content/uploads/2023/01/Final-Visioning-Guide-Feb24.pdf. Accessed 26 Sept 2024

Chapter 30
Future Visioning as a Tool for Imagining Regenerative and Sustainable Futures

Anne Pender

30.1 Introduction

The multiple environmental, social, and economic crises we face today require urgent responses that not only seek to address immediate problems but also contribute to the essential transformation of the systems, structures, beliefs, and behaviors that have created and continue to underpin these crises (Pender 2023). Enabling such a "respond and realize" approach requires a clear vision of the kind of future we wish to generate for all life on earth, human and non-human alike.

Without such a vision to guide us, we are at risk of becoming stuck in mere "problem-solving" mode or of succumbing to dystopian and apocalyptic future narratives. Being able to generate our own personal vision of a sustainable and regenerative world helps to give our current work context and direction, enabling us to find opportunities, make decisions, and venture forth in a clearer and more purposeful way. When we can focus on our vision as a guide, we remain open to many paths for success, not just one. As poet and critic Rebecca Tamás (2023) puts it, "*we can commit to hope by imagining new futures and telling their stories, rather than revelling solely in dreams of the end.*"

One way to help us imagine such futures is to use future visioning. At its most basic, visioning is a process where we are guided through a series of statements or questions that enable us to paint a picture in our minds of what we would like our future world to look like. The approach I use in my teaching and consultancy is based on ideas developed by the late systems theorist and practitioner Donella Meadows, which are adaptable to any context or topic. Meadows used visioning in

A. Pender (✉)
School of Architecture, Planning and Environmental Policy, University College Dublin, Dublin, Ireland
e-mail: anne.pender@ucd.ie

J. Bentz, J. Ristić Trajković (eds.), *Imagining, Designing and Teaching Regenerative Futures: Art-Science Approaches and Inspirations From Around the World*, Science for Sustainable Societies,
https://doi.org/10.1007/978-981-96-9029-9_30

different aspects of her work, but the one I draw most on arose from a presentation entitled "Envisioning a Sustainable World," which she made in 1994 (Meadows 1996). The latter part of that presentation contains a series of questions designed to stimulate participants' imagination, drawing on all five senses—sight, sound, touch, smell, and taste—to enable a detailed vision to be built up in one's mind.

30.2 Transformative Potential

In the particular context of designing and teaching regenerative futures, we are challenged to support learners of all ages to develop the skills and capacities to think creatively about the future, as well as empower them to take action in support of their visions. As Toffler (1974) argues, "*All education springs from images of the future and all education creates images of the future… Unless we understand the powerful psychological role played by images of the future in motivating- or de-motivating – the learner, we cannot effectively overhaul our schools, colleges or universities, no matter what innovation we introduce*" (p. 3).

Future visioning is important for transformative and regenerative work in three key ways. First, it enables us to broaden our focus away from solely responding to problems "here and now" so that we can simultaneously be guided toward "realizing" the future we desire. This means valuing visioning as a tool for shaping our future, not just a pleasant abstract exercise.

Second, creating a future vision can allow us to step outside the limitations of the dominant paradigms and assumptions that constrain transformation efforts and enable us to move from "what is" to "what if," as the activist and writer Rob Hopkins puts it. Hopkins (2019) regards the practice of imagining a positive future as "*an act of immense courage, of resistance, of rebellion*" (p. 111).

Third, creating a personal vision of one's desired future can help to generate a sense of agency and empowerment to undertake and sustain the challenging work of realizing it. As we well know, the mental and emotional burdens that crises such as climate change place on us can overwhelm and sap us of the motivation to address them effectively. Having a clear positive vision to draw on can help to counteract this and enable us to believe that what we do really does matter.

30.3 Application

The Future Visioning tool can be used in a wide variety of educational settings, both formal and informal and with both children and adults. I have used it in a university setting with undergraduate students who were taking city planning and environmental policy-related modules, as well as in a summer school for PhD students working with digital and virtual reality technology.

With the undergraduate and postgraduate students, the visioning tool was used as part of a two-hour class on the topic of tools for environmental policy-making. For the PhD students, the tool was part of a half-day workshop on creative approaches to systems thinking. The time required for the exercise was the same in both cases, being approximately 45–50 min. This comprised approximately 5 min to introduce the concept of future visioning, 15 min to go through the exercise, 10 min for students to reflect on the exercise and note down the vision they had created, and 15–20 min for sharing and discussion.

The main preparation is to write the visioning script in advance. One example is Meadows' (1996) template, which uses a series of questions to guide listeners on a journey to imagine what their ideal sustainable world would look like. These include what would it look like to wake up in this future world, where would energy come from, what kind of work will be done in this world, how do people make decisions, and so on. She also encourages them to use all their senses in the creation of this vision—what can they see, touch, and smell in this future world, what sounds can they hear, etc. Once the script is ready, decide whether you will read the vision script aloud to participants or record it in advance and share the mp4 file with them beforehand so that they can listen to it individually through headphones during the session. I have found that delivering the vision script live can be a powerful way to help the group listen as a collective in the moment. However, a recording is useful if the exercise is to be part of an online session; also, participants may wish to engage in the visioning exercise again in their own time, and so having a recording available is useful for this.

The visioning exercise can be used with individuals or groups and works well with both. One advantage of using it with a group is that participants can share and discuss their visions after the exercise has been completed. This can generate valuable insights and inspiration, opening our eyes to the different values and priorities that others bring to their visions, as well as the elements in common that could form the basis for collaborative and inclusive action. It is also important to encourage participants to hold on to their visions and use them for ongoing inspiration and motivation in their work.

The futures visioning tool can also be used in a more in-depth way, by combining it with (a) an initial examination of what visions of the future have been used in the past to shape decision-making and (b) prototyping of the future visions created during the visioning exercise. One example of this is José Ramos' "FuturesLab" model (Ramos 2017).

In conclusion, the futures visioning tool is open to all who wish to harness the power of our imagination and creativity to support our work of generating a sustainable future for all. As Susan Griffin (1996) says, "*No one can stop us from imagining another kind of future, one which departs from the terrible cataclysm of violent conflict, of hateful divisions, poverty and suffering. Let us begin to imagine the worlds we would like to inhabit, the long lives we will share, and the many futures in our hands*." (Tables 30.1 and 30.2).

Table 30.1 Method overview

Main purpose
The main purpose of the method is to enable participants to harness their imagination and create a personal vision of a sustainable future, which can inspire and guide their own work.
Gained competences
Development of visualization and imaginative skills; use of creative thinking; development of experiential and transformative approaches to learning.
Educational setting
The method can be used in both formal and informal educational settings and with participants from all educational backgrounds.
Space requirements/restrictions
Can be conducted indoors or outdoors, ideally in a quiet place so that participants can listen to the visioning instructions.
Resources and necessary materials
A script to guide participants through the visioning exercise. If the script is to be pre-recorded (e.g., on an mp4 file), then participants will need headphones so they can listen to the recording.
Number of participants
Suitable for both individuals and groups of any number.
Facilitator competences and skills
Some practice in writing and delivering a vision script and a capacity to lead and hold the vision session space.
Participants' skills/age/competences
Suitable for participants of all skills, ages, and competences.
Duration
45–50 min.

Table 30.2 Steps in the future visioning process

STEP 1 Grounding	Participants begin by taking a few moments to pause, breathe deeply, and gather their presence for the exercise.
STEP 2 Imagining	Next, they are led through a series of guiding questions designed to stir their imagination into envisioning the tangible daily aspects of the future they wish to envision.
STEP 3 Consolidating	After completing the exercise, participants take some time to write down their vision, to use as a guide for their future decision-making.

References

Griffin S (1996) To love the marigold: the politics of imagination. Whole Earth Rev 1996:60–67. https://znetwork.org/znetarticle/to-love-the-marigold-hope-and-imagination-by-susan-griffin/

Hopkins R (2019) From what is to what if: unleashing the power of imagination to create the future we want. Chelsea Green Publishing, Vermont

Meadows D (1996) Envisioning a sustainable world. In: Costanza R, Segura O, Martinez-Alier J (eds) Getting Down to Earth, Practical Applications of Ecological Economics. The Third Biennial Meeting of the International Society for Ecological Economics, San Jose, Costa Rica, October 24 – 28. Island Press, Washington DC. https://donellameadows.org/archives/envisioning-a-sustainable-world/. Accessed 30 May 2024

Pender A (2023) From partial to integrated perspectives: how understanding worldviews can expand our capacity for transformative climate governance. Earth Syst Gov 16:10074. https://doi.org/10.1016/j.esg.2023.100174
Ramos J (2017) FuturesLab: anticipatory experimentation, social emergence and evolutionary change. J Futur Stud 22(2):107–118. https://doi.org/10.6531/JFS.2017.22(2).E107
Tamás R (2023) Why does imagining the end of the world feel easier than saving it?. Financial Times, 8 October 2022. https://www.ft.com/content/40a03001-7ca6-4653-9cbf-965f77d6128c. Accessed 30 May 2024
Toffler A (1974) Learning for tomorrow: the role of the future in education. Vintage Books, New York

Part IV
Enhancing Well-Being: Experiential Learning and Nature Connection

Regeneration begins with remembering our place within the living world. Methods presented in this part focus on experiential and embodied approaches, including perception walks, sensory journeys, deep-time storytelling, eco-somatic reflection, and mindfulness in nature, with the aim to reawaken our connection to nature, to ourselves, and to others. These methods invite learners to slow down, listen deeply, and engage with the world through the body, the senses, and feelings. In doing so, they activate emotional sensitivity and support an inner transformation. By fostering a sense of presence, empathy, and relational awareness, these practices contribute to a holistic understanding of well-being, not only as an individual element but also as part of a broader network of interdependencies. The methods invite us to reconsider how we perceive, understand, and design our spaces, exploring ways of how they can heal, connect, and regenerate us. They remind us about the importance of restoring our capacity to live in harmony with nature, to belong, to care, and to thrive together.

Chapter 31
Sensing the City

Julia Bentz, Jelena Ristić Trajković, and Kiat Ng

31.1 Introduction

Environmental education tends to be oriented toward the mindset that our cognitive processes are separate from our bodies and disconnected from the material and natural world. The commonly held belief of a body–mind duality has perpetuated a siloed way of thinking and a narrow way of approaching complex problems in a learning context that often failed to generate meaning, hope, and a sense of empowerment among learners.

There is an increasing agreement that cognitive, emotional, and embodied knowledge needs to be integrated into a learning process in order to support meaning-making, critical thinking, and the generation of a plurality of perspectives. Relatedly Pineau (1994) writes that "the active body learns in ways that are eminently more personal, applicable, critical and long-lasting than any other teaching method" (Pineau 1994 cited in Hamera 2006, p. 403).

Actively engaging the body and its wisdom in a learning context can be done in numerous ways. In our case, we used the senses as a medium to tap into embodied knowledge. The senses are seen as a powerful yet easily accessible gateway to connect to our bodies and the wisdom they carry (Serres, 2016; Oliveros, 2021). Beautifully said by John O'Donohue, "your senses are the guides to take you deep into the inner world of your heart. The greatest philosophers admit that to a large

J. Bentz (✉) · K. Ng
Center for Ecology, Evolution and Environmental Changes (CE3C), University of Lisbon, Lisbon, Portugal
e-mail: jhbentz@fc.ul.pf; kienng@ciencias.ulisboa.pt

J. Ristić Trajković
University of Belgrade – Faculty of Architecture, Belgrade, Serbia
e-mail: jelena.ristic@arh.bg.ac.rs

J. Bentz, J. Ristić Trajković (eds.), *Imagining, Designing and Teaching Regenerative Futures: Art-Science Approaches and Inspirations From Around the World*, Science for Sustainable Societies,
https://doi.org/10.1007/978-981-96-9029-9_31

degree all knowledge comes through the senses. The senses are our bridges to the world. Human skin is porous; the world flows through you. Your senses are large pores that let the world in" (O'Donohue 1997, p. 84).

Involving the body through the senses offers a means to draw on our personal and lived experiences and generate new insights on complex subjects such as regenerative urban futures (Haskell, 2022, Pallasmaa, 2012). The Sensing the City approach consists of guided tours that explore the city through the lens of one particular sense at a time. In our case, we offered three tours, one focusing on taste and smell, one on the visual, and one on the auditory sense. The tours intended to inspire alternative perspectives on the city, explore complexities, and evoke new imaginaries.

31.2 Transformative Potential

Engaging the body and tapping into embodied ways of knowing beholds transformative potential. Embodiment can be seen as the basis for how we make sense in the world. Experience, as phenomenologist Merleau-Ponty (2017) argues, exists between the mind and body. The body is the condition and context through which we have relations to objects and receive information. Based on this idea of the body, we can involve it to pose questions, connect with memories, sensations, and emotions, and explore theoretical concepts. Ultimately, it is through the senses and the body that we can see and experience ourselves and the world, and therefore, it is also through the body that we can potentially alter our ideas about the future. Education that offers spaces for creatively exploring embodied ways of knowing thus beholds regenerative and transformative potential for learners as well as educators (Leavy 2015; Wiebe and Snowber 2011).

31.3 Application

This Sensing the City approach can be used to explore cities and neighborhoods and specific topics such as regeneration, futures, or transformation. We here illustrate how we used it in an international PhD summer school focusing on urban futures. The summer school entitled "Urban Imaginary - Exploring our urban futures" was a 5-day course and took place in the National Museum of Science and Natural History and the Botanical Garden of Lisbon, Portugal, between 3 and 7 July 2023. It was conceptualized as a learning space that engaged cognitive and embodied knowledge and that nurtured both body and mind in a simultaneous and synergistic way. Applying a transdisciplinary approach allowed trainers and participants to navigate between the many polarities, contradictions, and challenges around life in cities and to explore new, regenerative imaginaries of the future (Bentz et al, 2024).

Apart from theoretical lectures and interactive sessions, it involved three parallel guided Sensing the City tours as well as other embodied and experiential learning sessions throughout the course. The chosen transdisciplinary art-science approach aimed to provide learners with a felt experience and conceptual understanding of transdisciplinarity and how to explore a real-world problem through a holistic approach.

To ground imaginations in spatial reality as well as bring forth practical future visions, participants were guided through Lisbon city focusing on auditory, visual, or taste and smell senses. The taste and smell tour took participants to sweet-smelling orange trees in busy streets and in shady parks, to family-grown coffee beans, and to a traditional tea shop. The visual tour called upon creative drawings of the urban buildings and contrasted the visual experiences with the lack of visual sense through blindfolding experiences. The audio tour went in search of the sounds of the city and stimulated deep listening through the sounds of the river and the urban lives.

The Sensing the City approach involves three steps, namely, the development of the itinerary, implementation of the tours, and reflections and lessons learned. In the following, we illustrate each step and provide guidance for applying it in an educational context.

31.3.1 Development of the Itinerary

For the development of the tours, it is recommended to collaborate with guides who know the city well and have expertise related to the specific sense, for example, visual artists or architects for the visual sense and perfumers or cooks for the taste and smell tour.

Together with the guides select 7–10 locations in the city. The whole tour should not be longer than 3 km. We selected locations and activities that offered opportunities to explore the city and its history in a particular sense. These involved, for example, parks with (exotic) trees and flowers, a pastry shop, a fish market (taste and smell tour), small alleys with traditional houses and viewpoints (visual tour), and noisy streets and silent churches (audio tour). We started the three (parallel) tours at the same location and ended them at a different and coinciding location and time in order to offer participants to exchange experiences after the tours.

31.3.2 Implementation of the Tours (Approximately 2–3 Hours)

Participants should be advised to bring adequate shoes as well as water and sun protection. Per tour, not more than 12 participants are recommended. To start the tours, the guides can explain the functioning of the particular sense, that is, how we hear, how we smell, and how we see. Each tour can have different activities such as drawing and walking blindfolded (visual tour), touching different surfaces (tactile tour), and tasting food (gustatory and olfactory tour) at the selected locations. Then the group walks to the next location. At the end, all tours can meet at a previously selected location to exchange experiences and reflect on lessons learned.

31.3.3 Reflections and Lessons Learned (Approximately 30 Min)

After the tours, we facilitated reflections using exploratory questions and prompts (Table 31.1). We also created a space where participants could share their experiences with participants who participated in a tour of a different sense (Table 31.2).

Table 31.1 Exploratory questions and prompts for the respective senses

Sense	Exploratory questions and prompts
Sight (Visual)	How do different architectural styles, shapes, and color schemes in various neighborhoods tell a story about the diversity of cultures within the city? What are the identities and values hidden behind the visual appearance of our urban environments? Where do you encounter traces of the diversity of human activity in the visual appearance of the city? What images or visual memories would you like to take to the future?
Taste (Gustatory) and smell (Olfactory)	Which smells do you find (un)pleasant in this environment? What smells and tastes do you associate with this city/environment/neighborhood? What memories and images are evoked by certain smells and tastes as you wander through the city? Which smells and tastes would you like to take to the future and which would you leave behind?
Hearing (Auditory)	What emotions arise for you while you are walking through the city soundscape? Are there particular sounds that represent the cultural identity of the city (e.g. street music, the ocean, a river)? How do the sounds of nature, such as birds, wind, or water, interact with the urban soundscape? Which sounds and rhythms would you like to take to the future?
Touch (Tactile)	How do different surfaces influence your experience of the city (cobble stones, tree bark, glass, marble, grass, pavement, etc.)? What memories and feelings arise when you touch different textures of the city? Which textures of the city are in opposition to your idea of regeneration? What kind of textures and haptic experiences would you like to take to the future?

Table 31.2 Method overview

Main purpose
(Re)connecting to our bodies, embodiment, awareness building through the senses, fostering a deeper understanding of the urban environment and intangible natural and cultural values, discovering hidden narratives, and gaining a new perspective on the city or a particular neighborhood
Gained competences
Mindfulness and embodied presence, sensory awareness, observational skills, empathy, exploratory thinking, and reflective listening
Educational setting
Applicable in various educational settings, both formal (practical courses, design studios) and informal (workshops, art and science labs, seminars, summer schools)
Space requirements/restrictions
Urban locations that offer adequate sensory inputs and triggers, such as urban and natural sounds (water, birds, wind), visual elements (architecture, artworks, nature), tactile experiences that engage the sense of touch (natural textures, urban textures), and scents (flowers, fresh produce, and local cuisine)
Resources and necessary materials
Art supplies for drawing (paints, markers, paper), cameras, audio recorders, and cloth for blind-folding
Number of participants
5–12 participants per group
Facilitator competences and skills
Facilitators should be familiar with the area and prepare tours and sensory explorations in advance
Participants skills/age/competences
Young people, elderly people, artists, researchers, students, and practitioners
Duration
90–180 min

References

Bentz J, Ristić Trajković J, Ng K (2024) Tomorrow's Odyssey: a time traveler's guide to our shared future(s). FCSH, Lisboa

Hamera JA (2006) The SAGE handbook of performance studies. SAGE, Thousand Oaks

Haskell D (2022) Sounds wild and broken: sonic marvels, evolution's creativity, and the crisis of sensory extinction. Viking, New York

Leavy P (2015) Method meets art: Arts-based research practice (2nd ed.). The Guilford Press

Merleau-Ponty M (2017) Phenomenology of perception. Andesite Press, London

O'Donohue J (1997) Anam Cara – spiritual wisdom from the celtic world. Bantam Press, London

Oliveros P (2021) Deep listening – a composer's sound practice. iUniverse, New York

Pallasmaa J (2012) The eyes of the skin: architecture and the senses. Wiley, New York

Serres M (2016) The five senses: a philosophy of mingled bodies. Bloomsbury, New York

Wiebe S, Snowber C (2011) The visceral imagination: a fertile space for non-textual knowing. J Curric Theorizing 27(2):101–113

Chapter 32
Deep Time Walk

Carolina Mello-Silva, Antje Disterheft, and Robert Woodford

32.1 Introduction

Deep Time Walk (DTW) is an educational activity based on embodied learning approaches combined with story-telling, enabling participants to experience a time dimension of Planet Earth's evolution and thereby providing new entry points for reflection on our relationship to the Earth. DTW is a space–time metaphor for the 4.6 billion years of the Earth's existence over a 4.6-km walk, with each meter of the walk equivalent to 1 million years. The most important evolutionary events are told as a story—later called *Earth Stations*—in which important scientific facts are presented and integrated into short reflective and/or playful activities. The objective is to offer a transformative walk that could provide students with an embodied experience and a new perspective on time, increasing the participants' sense of belonging to the web of life, and a greater understanding of the systemic nature of the climate

C. Mello-Silva
CENSE – Center for Environmental and Sustainability Research & CHANGE – Global Change and Sustainability Institute, NOVA School of Science and Technology, NOVA University Lisbon, Lisbon, Portugal
e-mail: cme.silva@campus.fct.unl.pt

A. Disterheft (✉)
CENSE – Center for Environmental and Sustainability Research & CHANGE – Global Change and Sustainability Institute, NOVA School of Science and Technology, NOVA University Lisbon, Lisbon, Portugal

Instituto de Ciências Sociais, Universidade de Lisboa, Lisbon, Portugal
e-mail: antje.disterheft@ics.ulisboa.pt

R. Woodford
Deep Time Walk C.I.C., Devon, UK
e-mail: robert@deeptimewalk.org; https://www.deeptimewalk.org

J. Bentz, J. Ristić Trajković (eds.), *Imagining, Designing and Teaching Regenerative Futures: Art-Science Approaches and Inspirations From Around the World*, Science for Sustainable Societies,
https://doi.org/10.1007/978-981-96-9029-9_32

and biodiversity crises, as well as increasing hope and active participation in solutions to the current anthropogenic predicament.

32.2 Transformative Potential

The DTW connects well with approaches in transformative learning and sustainability education that focus on relationships and interconnectedness, as it offers a hands-on, immersive learning experience. By using the power of stories, DTW strives for our abilities to connect with one another and to awaken a feeling of identification with the cycle of life and with the planet as a body that shelters our own among other species. Thereby, DTW responds to the need for new stories that foster our inner relationship with ourselves and the world and that address our mind and emotions to help overcome current patterns of unsustainability and separation (Wamsler 2022).

DTW is also linked to some of the concepts underpinning Deep Ecology, as the activities during the walk aim for deep experience, deep questioning, and deep commitment (Harding 2009, p. 57 ff.) and foster the understanding of a living planet. DTW offers a chance to reflect on our entanglement with all life and to contemplate the legacy we want to leave, inviting participants to rethink their role in the Earth's story and develop personal meaningful actions that can help regenerate the imbalances humanity has created. As a result, DTW can become a truly transformative and regenerative learning experience.

32.3 Application

DTW is a versatile activity that can be undertaken in a multitude of ways, depending on the facilitator guiding the walk. A facilitator can adapt the activity to different contexts (e.g., geographical and cultural) as long as they abide by the DTW Facilitator Compass (Deep Time Walk Project 2024)—the code of conduct that draws out a number of values and principles underpinning the walk, including alignment with the Earth Charter (The Earth Charter International 2023). A DTW can be carried out with different groups, in different types of places, with diverse focuses, and using the material in pluralistic ways.

DTW is literally a walk where the displacement in space is equivalent to the passage of time, with important suggested stops related to a certain relevant moment of Earth's story: the *Earth Stations*. The walk happens outdoors, promoting contact with the more-than-human world and experiential connectedness, with the facilitator choosing either a previously prepared path or designing a new route. One can use a Global Positioning System (GPS) or similar mapping tools to calculate the distance between the Earth Stations.

To start exploring the process of facilitating a DTW and access essential resources, it is recommended to register on the DTW Community Platform (https://community.deeptimewalk.org). DTW Project created a range of training materials that support the facilitator regarding both the content and the functioning of the activity. There are three substantial supporting resources of note: (1) Earth Stations—a resource guide for planning a DTW (Woodford and Braddock 2024b), which includes the dates of evolutionary events, important information associated with the referred point in deep time, a sample narrative and curiosities (information that might surprise or promote deeper reflections, as well as embodied activities and question prompts for group reflection and discussion); (2) Holding Space—a guide to organizing and facilitating a DTW (Woodford et al. 2024). This provides guidance on both the outer and inner aspects of hosting group dynamics and gives support for the logistics before and after the experience; and (3) Bedrock—A guide to the science that underpins a DTW (Woodford and Braddock 2024a). This provides further detail and review from over 300 peer-reviewed science articles, enabling facilitators to delve into further detail about the events explored at each Earth Station.

Based on the Earth Stations and the Holding Space training materials (Woodford and Braddock 2024a, b; Woodford et al. 2024), each facilitator can highlight what one considers more important within the 18 *Earth Stations* (stops). It is advised to use a multimodal approach to communicating the Earth Station knowledge, including elements such as images, objects, and music in order to more clearly illustrate scientific concepts that otherwise can be difficult to understand. Facilitators are also encouraged to use the ecosystems present around them within the walking route to support the learning experience.

32.3.1 Guidance Step by Step on How to Facilitate a DTW

32.3.1.1 Step 1: Familiarize Yourself with the DTW Concept

- Explore the DTW community network: Visit the community.deeptimewalk.org network to learn about its impact and download resources.
- Complete a facilitator training course: Register to participate in DTW training to gain insight and confidence in facilitation (optional).

32.3.1.2 Step 2: Gather and Study Core Materials

- Obtain facilitation materials: Download the documents “Earth Stations,” “Holding Space,” “Bedrock,” “TerraSheets,” and the Deep Time Cards. It is also possible to complement with other materials.
- Review the Earth’s history: using the resources provided, study major events in Earth’s 4.6-billion-year timeline, integrating elements from geology, biology, and other sciences.

32.3.1.3 Step 3: Practice with a Small Group

- Prepare a 4.6-km path: Delimitate a path using a tool such as plotaroute.com.
- Lead a practice walk with the TerraSheets: Start with a familiar group of friends or colleagues to refine your approach.
- Note: the facilitation materials provide concrete texts for scientific storytelling during the walk and include prompts of images and reflective questions at each Earth Station. For example, at Earth Station 3, after having walked approximately 500 m and having arrived 4100 million years ago, it is explained how the comets and asteroids of the Late Heavy Bombardment enriched the Earth with elements for life and how through condensation the Earth becomes like a ball of water—ocean earth. The facilitator can then invite for reflection with prompts like the following:

  Today, our precious oceans and freshwater rivers and lakes are widely polluted. Water is a limited resource—a sacred resource many would say—with the integrity of the water cycle in danger. **In pairs:** *what are some of the impacts that human activities are having on Earth's water today?* **Or:** *what is your relationship with water ? (TerraSheets, Earth Station 3).*

- Collect Feedback: Use feedback to adjust pacing, delivery, and clarity.

32.3.1.4 Step 4: Begin Public Walks

- Promote your walks: Announce your DTW sessions in your community or online.
- Facilitate with adaptability: Tailor each walk to the participant's needs, setting a tone of humility, curiosity, and exploration.

32.3.1.5 Step 5: Reflect and Improve

- Review and Enhance: After each walk, reflect on what worked and refine your approach.
- Continue learning: Keep deepening your knowledge of DTW and Earth's history to enrich future walks. The DTW is an ongoing journey of learning and discovery. With each walk, you and your participants deepen your connection to Earth's story (Table 32.1).

Table 32.1 Method overview

Main purpose
To provide a tangible and immersive way to understand the vast history of Earth, spanning 4.6 billion years; to foster a deep sense of connection with and awareness of Earth as a living system
Gained competences
Scientific knowledge (e.g., geology, biology, evolution, chemistry, and physics); environmental awareness; systemic and interdisciplinary thinking; reflective thinking
Educational setting
Formal and informal education, for example, in schools and universities; educational programs; community events; and companies
Space requirements/restrictions
Outdoor spaces and accessible paths; depend on suitable weather conditions
Resources and necessary materials
DTW cards; DTW Line; extra visual elements (all of them are recommended, but optional) (available at https://www.deeptimewalk.org/kit/, Deep Time Walk Project 2022). Resource guides: Earth Stations, Holding Space, Bedrock, TerraSheets (available at community.deeptimewalk.org)
Number of participants
Ideally min. 3, max. 15–20
Facilitator competences and skills
Participation in DTW facilitators training; group management skills; communication and facilitation skills; storytelling skills; knowledge in Earth Sciences (facilitators without scientific background can study using the materials offered in the training); and adaptability and flexibility
Participants skills/age/competences
Basic physical conditions to walk for 3–4 h
Duration
Min. 3 h, up to 4 h, or with pre- and post-workshops a whole day

References

Deep Time Walk Project (2022) Field Kit. https://www.deeptimewalk.org/kit/. Accessed 22 July 2025

Deep Time Walk Project (2024) Deep time walk facilitator compass. Deep Time Walk C.I.C. https://community.deeptimewalk.org. Accessed 22 July 2025

Harding S (2009) Animate earth: science, intuition, and Gaia. Chelsea Green Publishing, Vermont

The Earth Charter International (2023) The Earth Charter https://earthcharter.org/read-the-earth-charter/. Accessed 22 July 2025

Wamsler C (2022) What the Mind has to do with the Climate Crisis [Blog]. Mind & Life Institute. https://www.mindandlife.org/insight/what-the-mind-has-to-do. Accessed 22 July 2025

Woodford R, Braddock M (eds) (2024a) Bedrock. Deep Time Walk C.I.C, A guide to the science that underpins a Deep Time Walk, p 63

Woodford R, Braddock M (eds) (2024b) Earth stations. A practical guide for narrating a deep time walk. Deep Time Walk C.I.C, p 63

Woodford R, Maurel O, Gros L (eds) (2024) Holding space. Deep Time Walk C.I.C, A head-heart-body guide to facilitating a deep time walk experience, p 41

Chapter 33
Building an Umwelt Apparatus

Daniel Metcalfe

33.1 Introduction

The Umwelt Apparatus is a design exercise in which designers attempt to build a device that would help them experience the world from the perspective of different animal species. It is rooted in a tradition of empathic design, where designers use various techniques to get closer to the lives and experiences of existing or future users to gain a deeper understanding of their experience in the world. In the Umwelt Apparatus exercise, this approach is extended to learn about nonhuman species and the manifold ways in which they interact and experience the world, with the subsequent aim of including these perspectives in the practice of design.

The name of the exercise is derived from Jakob von Uexküll's concept of the Umwelt, introduced in a monograph from 1934 titled A Stroll through the Worlds of Animals and Men. In it, von Uexküll (1992) explores the possibility of experiencing the world from the points of view of species with entirely different sets of sensory organs than us humans and invites us to take part in a thought experiment of imagining the world through the sensory organs of a tick, a mollusk or a fly, for example.

Inspired by this work, the idea of the umwelt apparatus was born: a simulation apparatus that would help us experience the world from a selected animal's perspective—offering a glimpse into the unique ways in which the animal senses and interacts with the world. Such an apparatus may, for example, physically change our depth or field of view, limit some senses while enhancing others, or provide a protocol for communication informed by the ways other species communicate.

The exercise does not focus on creating an accurate, objective representation of the animal's experience of the world. Rather, it is about shifting our own perspective

D. Metcalfe (✉)
Technion- Israel Institute of Technology, Haifa, Israel
e-mail: daniel.met@technion.ac.il

J. Bentz, J. Ristić Trajković (eds.), *Imagining, Designing and Teaching Regenerative Futures: Art-Science Approaches and Inspirations From Around the World*, Science for Sustainable Societies,
https://doi.org/10.1007/978-981-96-9029-9_33

and observing what changes in us when we try to get closer to the animal's experience. Engaging in the design of such a device encourages us to learn about the animal holistically, put aside our preconceptions and beliefs, and observe the world with new eyes. It is often a humbling experience that can have a transformative effect on the way we perceive and relate to the animal. It reminds us that human perception is only one of many ways of being and engaging with the world.

33.2 Transformative Potential

A multitude of nonhuman species live in and around anthropogenic systems. Insects, reptiles, birds, and small mammals live above, under, and within our homes and infrastructures. Wild animals interact with expanding human-made systems on land, in the air, and sea, and some actively migrate into human-dominated habitats in search of food and shelter. In most cases, these systems are not designed with animal perspectives in mind and these interactions may lead to harm and conflicts.

In recent years, there has been a gradual shift in approach regarding urban wildlife. Ecologists, who have long ignored urban ecosystems, realize that cities are not only the problem but also part of the solution and that integrating nature in cities can benefit both people and biodiversity conservation (see Colléony and Shwartz 2019). The last three decades of urban ecology research have demonstrated that urban green spaces have the potential to support rich biodiversity that can be enhanced by implementing certain design and management solutions. In addition, cities designed for wildlife may help promote positive human–nature interactions, addressing what Robert Pyle referred to as the Extinction of Experience (1993)—the growing alienation of people from nature, with negative effects on, both, physical and mental well-being and people's motivation to protect nature.

To begin imagining and developing our surroundings as living habitats that support a multitude of life of diverse species, we need to develop design methods and approaches that help us study and represent more-than-human perspectives in the design process. The umwelt apparatus is one method that may help us shift our focus toward new possibilities of being in and experiencing the world, develop and evaluate our designs from multiple perspectives, and foster a sense of care and responsibility toward other species.

33.3 Application

It is important to properly frame the exercise from the beginning as an open and creative exploration. We are not seeking, nor can we create an objective representation of the animal's self-world. We aim to explore new ways of learning about the animal, try to get closer to its unique way of experiencing and interacting with the world, and, in the process, expand our own perception. We do this with the hope of unveiling some of our blind spots regarding the possible influences of our designs on other species.

With students, I typically dedicate a couple of weeks to the exercise as part of a wider course on more-than-human design for graduate design students. We start by reading von Uexküll's monograph, explaining the scope of the exercise, and looking at a few examples from previous years. The students are then instructed to choose an animal they would like to work with. We typically focus on wild animals that live in and around our immediate surroundings, so we have the chance to observe and study them first-hand.

Next, students collect information about the animal from both secondary and primary sources. They read about the animal, watch videos, and observe and document it in its natural surroundings. We focus on behavior and sensory abilities. Guiding questions to this part can be: What are the primary senses the animal uses to make sense of and navigate its surroundings? How do these senses manifest themselves and how are they different from the way humans perceive the world? What influences the animal's decision-making and behavior? How does it communicate? What is the animal's self-world comprised of? It is often helpful to consult experts on animal behavior at this stage. The students choose one aspect of the animal's perception or behavior to focus on that feels most significant or enigmatic to them.

Next, they are prompted to think creatively about how they might generate access to this aspect of the animal's self-world for themselves and others via guiding questions such as: What should shift in the way we perceive and interact with the world to get closer to the way the animal does? Which of our sensory abilities should we alter, enhance, or limit? Is there a way to add new sensing abilities? What tools might help with this task? These can be physical and technological tools such as filters, mirrors, computer simulations, augmented reality, etc., or imaginative ones, such as role-playing, games, and movement protocols. Next, students are advised to prototype and test different versions and approaches until they have one they are happy with. Finally, they build their umwelt apparatus and write down instructions for using it. The exercise concludes with a gathering, where students share their work and experience using the different apparatuses created by their peers.

The second part of the course is dedicated to developing a design intervention around a changing theme that takes into consideration multi-species points of view. At this stage, students are encouraged to keep using their apparatuses as part of the design process—to explore and observe different landscapes and artifacts with new perspectives, to enact different scenarios of interaction, to evaluate and compare design alternatives, and to identify potential risks or hazards.

More about this method, examples of apparatuses developed by students, and a discussion on how the exercise influenced their approach and relation to the studied animals can be found at Metcalfe (2023) (Table 33.1).

Table 33.1 Method overview

Main purpose
Empathy building/expanding perception/multispecies design
Gained competences
A deeper understanding of other species' "self-worlds"
Educational setting
Graduate and undergraduate design students
Space requirements/restrictions
Workshop space for developing the project and outdoor spaces for experiencing the apparatuses
Resources and necessary materials
Students need access to prototyping equipment and tools based on their projects
Number of participants
Can be individual or in small groups (two to three participants)
Facilitator competences and skills
I would not limit. If you feel like you want to give it a go—go for it. Technical design skills and knowledge in biology and animal behavior can help
Participants skills/age/competences
The exercise was conceived for design students; however, it can be relevant for other audiences as well (communities, artists, researchers, policymakers, etc.).
Duration
I do it over a course of 3–5 weeks as part of a larger course. It typically takes about an hour to explain and launch the exercise, 6–12 hours of individual/group work (with facilitators available for consultation), and 2–3 hours for presentations.

References

Colléony A, Shwartz A (2019) Beyond assuming co-benefits in nature-based solutions: a human-centered approach to optimize social and ecological outcomes for advancing sustainable urban planning. Sustainability (Switzerland) 11(18):4924. https://doi.org/10.3390/su11184924

Metcalfe D (2023) Empathic multispecies design – using empathy to design with animal perspectives in mind. In: Goldschmidt G, Tarazi E (eds) Expanding the frontiers of design. Routledge. https://doi.org/10.1201/b22630

Pyle RM (1993) The thunder tree: lessons from an urban wildland. Houghton Mifflin, Boston

von Uexküll J (1992) A stroll through the worlds of animals and men: a picture book of invisible worlds. Semiotica 89(4):319–391. https://doi.org/10.1515/semi.1992.89.4.319

Chapter 34
Eco-Sensory Learning Pathways for Neurodiverse Children

Eglantina Dervishi

34.1 Introduction

In recent years, the global discourse on education has evolved dramatically, placing an increasing emphasis on inclusivity, sustainability, and the recognition of diverse learning styles (Sobel 1996). Within this context, the intersection of environmental education and neurodiversity has gained prominence. Neurodiversity encompasses the idea that neurological differences, such as Autism Spectrum Disorder and Intellectual Disabilities, are natural variations of the human experience rather than deficits to be corrected (Armstrong 2019). Inclusive and equitable education that considers the needs and abilities of neuro-diverse individuals is not only a matter of social justice but also an essential component of building a sustainable and regenerative future.

One educational approach that has gained traction in this regard is Eco-Sensory Learning Pathways. Eco-Sensory Learning Pathways represents a holistic framework that recognizes the potential for sensory-rich experiences to bridge the gap between neuro-diverse learners and their environment (Alexander and Bissaker 2023). It leverages the principles of sensory integration therapy, ecological psychology, and sustainability education to provide a transformative educational experience. Through Eco-Sensory Learning Pathways, neuro-diverse children are not only encouraged to connect with nature but are also actively engaged in the process of environmental exploration, nurturing, and learning.

The Sensory Eco-Pathways for Inclusive Learning is an innovative educational approach designed to create inclusive and sustainable learning environments for neuro-diverse children, with a particular focus on those with autism spectrum

E. Dervishi (✉)
Department of Psychology and Pedagogy, Faculty of Social Sciences, University of Tirana, Tirana, Albania

J. Bentz, J. Ristić Trajković (eds.), *Imagining, Designing and Teaching Regenerative Futures: Art-Science Approaches and Inspirations From Around the World*, Science for Sustainable Societies,
https://doi.org/10.1007/978-981-96-9029-9_34

disorder and intellectual disabilities. This approach recognizes that neuro-diverse children have diverse sensory profiles and unique learning needs. The Sensory Eco-Pathways for Inclusive Learning leverages sensory-rich experiences and ecological sensitivity to foster holistic development, environmental awareness, and inclusivity.

34.2 Transformative Potential

Sensory-Centric Learning places sensory experiences at the core of the learning process. Educators and specialists identify the sensory preferences and sensitivities of each child. This understanding guides the creation of personalized learning plans, ensuring that the learning environment caters to the specific sensory needs of the child. Schools and educational institutions implementing the Sensory Eco-Pathways for Inclusive Learning create eco-friendly and sensory-rich learning spaces. These spaces may include natural elements such as sensory gardens, eco-conscious materials, and adaptive classroom furniture to support sensory experiences. The Sensory Eco-Pathways for Inclusive Learning emphasizes collaboration among educators, families, therapists, and students. Regular communication and collaborative efforts ensure that each child receives comprehensive support and that the learning experience extends beyond the classroom. Nature-based learning is an integral part of Sensory Eco-Pathways for Inclusive Learning. Children engage with the natural world through sensory exploration, fostering an appreciation for the environment and a sense of responsibility for its preservation.

Sensory Eco-Pathways for Inclusive Learning stands out for several reasons.

The Sensory Eco-Pathways for Inclusive Learning prioritizes the holistic development of neurodiverse children, encompassing their cognitive, emotional, social, and sensory well-being. By tailoring educational experiences to individual needs, it has the potential to regenerate self-esteem and self-confidence in these children (Walshe et al. 2022.) The approach promotes environmental awareness and responsibility. Fostering a deep appreciation for the environment has the potential to create environmentally conscious citizens who are more likely to engage in ecological conservation efforts. The Sensory Eco-Pathways for Inclusive Learning challenges stereotypes and misconceptions about neurodiversity, transforming societal attitudes toward these individuals. This transformation can lead to greater acceptance, reduced stigma, and increased opportunities for neurodiverse children. The individualized and sensory-friendly approach of Sensory Eco-Pathways for Inclusive Learning can transform traditional educational methods. It recognizes that neurodiverse children may have unique learning styles and needs, shifting the focus from conformity to celebrating diversity in learning.

34.3 Application

The Sensory Eco-Pathways for Inclusive Learning is useful in educational settings, including mainstream schools, special education programs, and other institutions focused on the holistic development of neurodiverse children. It can be applied in contexts where children with ASD and ID require specialized support to thrive academically, socially, and emotionally. Furthermore, the Sensory Eco-Pathways for Inclusive Learning has the potential to be integrated into educational policies and practices at regional, national, and international levels to make sensory-rich, inclusive education a standard practice.

Overall, Sensory Eco-Pathways for Inclusive Learning is a transformative and holistic approach to education that not only benefits neuro-diverse children but also has a positive impact on the environment and society as a whole.

By conscientiously acknowledging the profound influence of lived experiences, popular culture, and preconceptions on perceptions of neurodiversity, educators can adeptly engineer eco-sensory learning pathways that are attuned to the distinctive requirements and strengths of all neuro-diverse learners. This approach harmoniously aligns with the overarching objective of nurturing sustainability by ensuring that every child, regardless of their neurodiversity, is provided with the requisite opportunities to excel and make meaningful contributions toward a more sustainable future.

In our ever-evolving educational landscape, the significance of Sensory Learning cannot be overstated. The Eco-Sensory Learning Pathways program has successfully developed an evidence-based model of practice, tailored to address the socio-emotional, developmental, physical, and behavioral needs of children diagnosed with autism spectrum disorder and intellectual disabilities. Through Eco-Sensory Learning Pathways, we aim to empower children, regardless of their neurodiversity, to transcend the confines of traditional classrooms and embrace their local natural surroundings as a holistic medium for healing and regulation in their daily educational pursuits. Sensory Learning Theory underscores the fundamental idea that active engagement of our senses enhances the learning experience. This theory also acknowledges the diversity of sensory learning styles among individuals. It recognizes that students may exhibit a preference for one sensory modality over another, which can vary depending on the subject matter at hand. For instance, some students may excel in auditory learning, while others thrive in visual or kinesthetic learning environments. A proficient educator recognizes and accommodates these differences in sensory learning styles (Pawlik et al. 2023). This approach is known as multi-sensory instruction, whereby various sensory channels are employed to facilitate learning. This inclusive method ensures that every student is provided with the opportunity to learn in a way that suits their strengths and preferences. For instance, in the context of teaching reading, a skill crucial for a child's educational journey, a multi-sensory approach offers multiple options. It allows students to select the mode of learning that resonates most with them. Kinesthetic learners, for example, can benefit from air-writing letters, engaging their muscle memory effectively. The importance of sensory skills extends beyond the classroom (Spielmann et al. 2023). As we cultivate these skills, we forge well-established neural pathways in the brain, enhancing our ability to recall and utilize learned information. The brain, akin to a map, creates numerous routes to access knowledge, thereby deepening our understanding and memory retention. Numerous educational systems, such as Montessori, the Wilson Reading System, the Barton Reading Program, and the Multisensory Structural Language Education, have harnessed the power of sensory learning to successfully teach reading, and mathematics, and support individuals with learning disabilities. Research substantiates the effectiveness of sensory learning in education, and it is also shown to benefit brain health when acquiring new languages, linking auditory and visual cues for improved memory retention. This concept is further corroborated by the psychological concept of "crossmodal interaction," wherein our various senses collaborate to optimize our perception and cognitive processes, reinforcing the belief that the human brain has evolved to thrive in multisensory environments. Our senses serve as invaluable tools for gathering accurate information about our environment, with each sense offering distinct insights. Touch, for instance, provides information about temperature and texture, while hearing alerts us to auditory cues. These sensory inputs collaborate to build a comprehensive understanding of the world around us (Mohan and Tiwari 2023). The adage in neuroscience, "What fires together, wires together," highlights the strengthening of neural connections through repeated activation. This principle is crucial in a child's learning journey, where experimentation is key to fostering cognitive development. Sensory play, therefore, becomes an essential component of early childhood education. It goes beyond mere enjoyment and mess-making; it facilitates experiential learning. Children, through sensory play, fine-tune their thresholds for sensory stimulation, develop preferences, and discern safe from dangerous stimuli. Furthermore, it nurtures essential social skills such as cooperation and effective sharing with peers. The incorporation of multi-sensory instruction into learning activities at any age is an ideal approach. This, however, necessitates well-designed activities that align with desired learning outcomes and are suitable for the age group. Though it may require time and creativity, the rewards of enhanced student engagement and improved learning are well worth the effort.

Understanding a child's sensory sensitivities and preferences is the cornerstone of crafting personalized learning experiences. The success of Eco-Sensory Learning Pathways relies on the collaboration of educators, families, therapists, and students, ensuring a comprehensive support system. Schools should invest in sensory-friendly tools, including sensory kits, sensory gardens, and adaptable classroom materials, to create an environment rich in sensory experiences (Rocco et al. 2023). These kits contain a variety of sensory tools to assist students during their learning journey, such as fidget toys, noise-canceling headphones, and textured objects. Creating sensory gardens or outdoor spaces allows students to engage with nature, offering calming sensory experiences and opportunities for exploration. Technology can play a valuable role through personalized learning apps and software, adapting to each child's unique learning style and pace. Educators receive training to understand sensory needs and integrate sensory activities into the curriculum. Continuous professional development ensures effective implementation.

In summary, Eco-Sensory Learning Pathways holds the promise of a regenerative, transformative, and ecological approach to education for neuro-diverse children. Teachers can integrate environmental consciousness into the education of children with autism and intellectual disabilities (Bourgeois 2023). This approach benefits both the children and the world they inhabit, creating a positive impact on their lives and the environment. By celebrating sensory diversity, embracing personalized learning, and nurturing eco-friendly environments, Eco-Sensory Learning Pathways empowers children to excel academically, emotionally, and socially. Through collaboration, sensory-friendly resources, and ongoing training, Eco-Sensory Learning Pathways paves the way for a more inclusive and sustainable future in education. It stands as a beacon of hope, showcasing the potential for every child to thrive and make meaningful contributions to society when provided with the right support and environment. In the grander scheme, Eco-Sensory Learning Pathways is not just about protecting the planet but also about nurturing the learning journey of neuro-diverse children and encouraging them to be mindful caretakers of the Earth. This approach benefits both the children and the world they inhabit, creating a positive impact on their lives and the environment (Table 34.1).

Table 34.1 Method overview

Main purpose
Awareness building, behavior change, environmental consciousness, inclusive education
Gained competences
Sensory awareness; environmental responsibility; social skills; cognitive development; emotional regulation; and fine and gross motor skills
Educational setting
Formal—special education programs and inclusive classrooms (primary and secondary levels) Informal—workshops, outdoor education programs, and community programs
Space requirements/restrictions
Outdoor activities—sensory gardens, nature trails, and parks Indoor activities—sensory-friendly classrooms, and therapy rooms
Resources and necessary materials
Sensory tools—fidget toys, noise-canceling headphones, and weighted blankets Eco-friendly materials—natural elements (plants, soil) and recycled materials Educational materials—visual schedules and adaptive learning tools Technology—personalized learning apps and tablets
Number of participants
Group work—small groups of 5–10 students Individual work—personalized learning plans for each student Number of groups—multiple groups can participate simultaneously depending on space and resources.
Facilitator competences and skills
Background in special education, occupational therapy, or environmental education, experience working with neuro-diverse children, skills in sensory integration therapy, adaptive teaching methods, and environmental education

(continued)

Table 34.1 (continued)

Participants skills/age/competences
Parents, caregivers, neuro-diverse children (ages 5–18 years), educational administrators special education and environmental education researchers, educational program designers, occupational therapists, and special education teachers
Duration
Initial implementation—1–3 months for setup and initial training Ongoing application—daily activities integrated into the curriculum, with periodic assessments and adjustments

References

Alexander V, Bissaker K (2023) Same word, same picture, different responses: exploring teachers' and autistic adolescents' conceptions of autism. Educ Sci 13(7):734. https://doi.org/10.3390/educsci13070734

Armstrong T (2019) Neurodiversity in the classroom: strength-based strategies to help students with special needs succeed in school and life. ASCD, Alexandria

Bourgeois D (2023) Occupational therapists' role in educating caregivers on pediatric sensory interventions guided by neurodiversity principles [Doctoral project, University of St Augustine for Health Sciences]. SOAR, Student Capstone Papers Collection, St. Augustine, Florida, USA. https://doi.org/10.46409/sr.UJDX3116

Mohan HS, Tiwari S (2023) A survey on professionals' beliefs and practices related to emergent literacy skills in children with autism spectrum disorder. Speech, Lang Hear 26(4):1–11. https://doi.org/10.1080/2050571X.2023.2226532

Pawlik S, Rzeźnicka-Krupa J, Gierczyk M, Hornby G (2023) The voices of autistic adolescents on diversity, education and school learning in Poland. Educ Sci 13(4):368. https://doi.org/10.3390/educsci13040368

Rocco K, Drobnyk W, Bruce S, Soumerai SB (2023) Ayres sensory integration therapy for a child with Rett syndrome: a case report. Clin Med Insights Pediatr 17:1–12. https://doi.org/10.1177/11795565231188939

Sobel D (1996) Beyond ecophobia: reclaiming the heart in nature education. Orion Society

Spielmann V, Burke HK, McCulloch S, Mason A, Lane SJ (2023) Linking sensory integration and processing with mental health in autism: a retrospective review of survey data. Am J Occup Ther 77(2):7702185080. https://doi.org/10.5014/ajot.2023.050127

Walshe N, Moula Z, Lee E (2022) Eco-capabilities as a pathway to wellbeing and sustainability. Sustainability 14(6):3582. https://doi.org/10.3390/su14063582

Chapter 35
Developing the Explorers in the Garden: Exploring our Natural Self and Restoring our Bonds with the Land

Karel Sterckx

35.1 Introduction

Every person is born with natural gifts and strengths and an instinctive desire to develop and grow. There also exists a natural reciprocal relationship between humans and place. This can be clearly seen when we look at indigenous communities, who are still so close to nature and the land, that it is part of their identity (Cajete 1994). In Western society, many of us grew away from our natural gifts and strengths, and have forgotten about our natural bond with the land. However, these gifts and strengths are like dormant seeds, waiting for suitable conditions to germinate. Nature provides an inclusive learning environment for all kinds of learners and provides opportunities for active, experiential learning that contributes to remembering our natural selves.

Outdoor adventure activities provide a setting for life-changing experiences and personal development (Ewert and Davidson 2017). Participation in group activities in nature, with emphasis on nature connection, supports health, well-being, and resilience. Intentionally designed experiences offer opportunities for learning about regeneration and living systems thinking: how can we gain an understanding about our own gifts and purpose through outdoor adventures in which we explore the environment?

Exploring the garden is about exploring our own natural self and restoring our bonds with the land (Plotkin 2008) through outdoor activities, intentionally designed to offer the opportunity to learn through experience (Kolb 2015).

K. Sterckx (✉)
Field Research Leadership, Inari, Finland
e-mail: karel@karelsterckx.be

J. Bentz, J. Ristić Trajković (eds.), *Imagining, Designing and Teaching Regenerative Futures: Art-Science Approaches and Inspirations From Around the World*, Science for Sustainable Societies,
https://doi.org/10.1007/978-981-96-9029-9_35

35.2 Transformative Potential

This approach contributes to self-discovery and awareness. It provides an environment in which transformations are possible across scales: one's natural self can emerge on its own, in reciprocal relationships with other humans, the group, and the natural environment (Table 35.1).

Table 35.1 Method overview

Main purpose
Awareness building/interconnectedness human–nature/regenerative thinking development/ changing mental attitudes/experience regenerative thinking
Gained competences
Sense of natural self versus social self/a clear purpose/sense of community membership/ experienced life skills/the insight of the possibility to change worldviews and mindsets/ understanding of nature as a teacher and mirror/understanding the regenerative cycle of life/ understanding the possibility to create a regenerative culture/physical, mental, emotional, and spiritual well-being/healthy habits to nourish life/teamwork and one's role in a team/basic physical skills for survival/experienced endurance, and grit/awareness of Indigenous perspectives
Educational setting
Informal, as a living system: Facilitators create a safe space, context, and frame for participants. Participants contribute, experiment, and engage in the program. Together, they build the educational setting. Preparatory meetings and workshops: facilitators share information on how to prepare for the basecamp and assist participants in their preparation. Participants take ownership of their preparation. Basecamp: outdoor activities intentionally designed to experience awareness of natural self, interconnectedness with others and the environment, and regenerative reflection.
Space requirements/restrictions
During the preparation: indoors During the execution of the program: activities will be mainly outdoors. Sleeping, eating, showering, and so on can be done indoors. *Note:* *It is recommended to use an indoor facility as a base camp so that participants do not have to go too far out of their comfort zone to meet basic needs. For instance, for many people, sleeping in a tent is a serious stretch for their peace of mind or going to the toilet in nature.* *One or two nights in a tent are enough.*
Resources and necessary materials
Lashing rope (polypropylene), water filtering system, axe, knives, firesteels, matches, cooking pots, sporks, plates, field guides (animals, plants), paper, paint, sleeping bags and liners, sleeping pads, groups first aid kit, tents or tarps, backpacks, field guides, maps, compasses, a talking stick, orienteering control flags
Number of participants
Maximum 8 participants per pair of facilitators Maximum 24 participants per camp
Facilitator competences and skills

(continued)

Table 35.1 (continued)

Facilitators should be familiar with at least experiential learning and should have experiences in outdoor adventure education, leadership development, group dynamics, regenerative design, and nature-based human development. Ideally, they work in pairs. At least one per pair should have a valid certification in Wilderness Advanced First Aid (WAFA). More info on WAFA Courses: https://www.wmaeurope.com/wfa
Participants skills/age/competences
Teenagers/students/young adults/practitioners/teachers/youth workers/corporate team building groups
Duration
5 months divided into: Preparation: 2.5 months Basecamp: 8 days (1 week) Follow-up: 2 months

All species have their unique role and purpose in the complexity of nature. Within each species, each member also has its own life-nourishing role and purpose. Every member of a species is born to occupy a particular place in nature. This place is connected to the entire natural community, not only to the community of one's own species. The same goes for us humans: we all have our own unique role and purpose, interconnected with and taking care of all other forms of life in nature. That is one's natural self. Cultivating and building a life based on that natural self is a challenge in Western societies. Since the emphasis lies not on exploring and developing one's natural self, but on a societal self: the one a person becomes through socialization in child-raising practices, education methods, core values, ways of organizing communities, and the, often physical and mental, separation from the rest of nature. The natural self is, as it were, wrapped and covered with a social self. Rediscovering that natural self is a challenge for which social support from a group is very welcome. The aim is not to separate from society, but to contribute to society more consciously: self-aware, life-nourishing, and interconnected with the community and the environment.

The group is placed in a natural environment and is presented with challenges that require the use of every participant's potential, cooperation, perseverance, and connection with nature. This leads to awareness of one's physical, interpersonal, intrapersonal, and naturalist skills and of one's role in and contribution to the collective, which, in turn, plants the seeds of reorganization of meaning, potential, purpose, mindset, and worldview.

The whole group (facilitators and participants) functions as a living organization (Wolfe 2011), with clear roles rooted in reciprocal relationships between facilitators and participants. The role of the facilitators is to facilitate a safe space, context, and frame; to support from authentic leadership the thrivability of the group; and to be aware of what they are really teaching participants: it is not just about the content of the activities. We teach them who we are. Secure preparation, having the trust of the participants, and being able to manage their behavior are essential—last but not least, to stay at the periphery and assist the participants to explore their own potentials. The role of the participants is to be motivated to participate, contribute, and learn; to experiment with initiative, responsibility, and ownership; and to engage in reciprocal relationships with other humans, nature, and the environment.

The desired outcome is threefold: the first outcome is an increased self-awareness among participants; the second is an increased awareness about the interconnection between humans and the environment; and last, but not least, an increased attachment to the environment.

Through reflection, participants gain more clarity on how to apply regenerative thinking in their lives.

It is powerful to contribute to regenerative futures, precisely because outdoor adventures can create moments of awe (Long 2014). Moments that inspire and enlighten people and fuel their passion for life. We know that many will think back on their time with us, even so long after they have left us.

35.3 Application

There are three parts, each of equal importance: preparation, basecamp, and follow-up.

35.3.1 Preparation (2.5 Months)

Preparation of the 3-day group hike during the basecamp through online meetings: the aim is to facilitate space for participants to actively engage in and contribute to the design of the basecamp; exchange energy between facilitators and participants; and motivate participants. Facilitators share information on how to prepare for the basecamp and assist participants in their preparation. Participants take ownership of their preparation.

35.3.2 Basecamp: 8 Days of Adventure (1 Week)

A residence where participants can sleep and eat indoors, and where there are rooms to alternate indoor briefings and outdoor activities, is recommended. This helps to create a safe environment and to reach participants within their comfort zone. Self-discovery activities inherently cause some nervousness and one night in a tent can be challenging enough for many participants. It is not survival week (Table 35.2).

Table 35.2 Program of the basecamp week

Day	Focus	Activities
1	Survival skills and group culture	Fire-making, shelter building, and water fetching A circle to introduce the values, norms, and frame for interaction and behavior
2	Exploring the key life-giving elements (water, energy, food, soil, nutrients, organisms, and infrastructure)	Identification of the key life-giving elements in small groups through mapping and consulting field guides, followed by circling up in the big group for reflection on the relationships between these elements and how they affect them, the group, and the region
3	Individual connection with nature	Every participant hikes to an indicated place where they stay the night individually. During their stay, they are invited to start a personal dialogue with nature, to create their own rituals.
4	Returning and sharing experiences preparation of the group hike	Share stories about the individual journeys, and create art from the experiences After lunch, they start preparing for the 3-day group hike they designed during the preparation phase
5-6-7	Group hike	Participants take turns taking the lead. Facilitators facilitate the space for the group to develop itself
8	Wrapping up, closing the adventure, and returning home	Circle with reflection on the experiences and how to translate them into their personal situations

If there is a natural environment within walking distance, that is fine. This helps to encourage participants to also look for nature nearby at home. Of course, facilitators should visit the area beforehand and also view the camping options for the group hike. Maybe there are free bivouac areas. Or should a pasture be rented from a farmer?

One suggestion: Apart from the facilitators who execute the program, it is a huge help to have one extra person who is exclusively in charge of logistics.

35.3.3 Follow-Up (2 Months)

Two weeks after the basecamp, a follow-up questionnaire is sent to the participants. It is partly a participation satisfaction survey, partly a survey to be able to observe the degrees of impact of the program on changes in awareness about oneself and one's connection to the environment.

A follow-up moment is organized 1.5 months after the basecamp. All facilitators and participants are invited to the gathering. They determine whether the gathering is open to the general public, in whole or in part.

References

Cajete G (1994) Look to the mountain. An ecology of indigenous education. Kivaki Press, Skyland

Ewert A, Davidson C (2017) Behavior and group management in outdoor adventure education. Theory, research and practice. Routledge, London and New York

Kolb DA (2015) Experiential learning. Experience as the source of learning and development. Pearson Education, New Jersey

Long K (2014) An outdoor educator's guide to awe. Understanding high impact learning. The Outward Bound Trust, Cumbria

Plotkin B (2008) Nature and the human soul. Cultivating wholeness and community in a fragmented world. New World Library, California

Wolfe N (2011) The living organization. Transforming business to create extraordinary results. Quantum Leaders Publishing, Irvine

Chapter 36
Connecting to Self and to Nature

Ruth Förster and Anaïs Sägesser

36.1 Introduction

The practice of Connecting to self and to nature is based on checking in with ourselves and cultivating connection through our senses, from there having new insights and creating pathways toward sustainable, regenerative transformations. It aligns with principles of mindfulness, inter- and intrabeing, and ecological consciousness and allows for experiencing ourselves as nature (Abram 2011; Young 2016).

It starts with checking in with oneself on four Levels of Awareness, that is, physical, emotional, and mental (Halprin 2002), as well as a sensing of the wider field, and then continues with Sensory Awareness.

It addresses: How am I present right now? How am I in connection with myself, other beings, the environment, and nature via my senses?

Sensory Awareness allows us to connect to others, our environment, as well as to oneself via our five senses: sight (visual), sound (auditive), touch (haptic), smell (olfactory), or taste (gustatory). Basically, the information via sensory channels from inward or outward is evaluated in a complex process in our body, that is, in our brain and nervous system, largely unconsciously and rapidly. A result of these evaluations is emotions that ready us for action of the "internal and external sort" (Siegel 2020, p. 148), for example, as can be observed in approaching something pleasant and avoiding something unpleasant.

R. Förster (✉)
Dr. Ruth Förster, Training & Counseling, Zürich, Switzerland

A. Sägesser
scaling4good, Riehen, Switzerland
e-mail: anais.saegesser@scaling4good.com

J. Bentz, J. Ristić Trajković (eds.), *Imagining, Designing and Teaching Regenerative Futures: Art-Science Approaches and Inspirations From Around the World*, Science for Sustainable Societies,
https://doi.org/10.1007/978-981-96-9029-9_36

Through Sensory Awareness, we mindfully tune into the present moment and deepen our sensory experiences, without judgment or analysis. Checking "how am I present" via the four Levels of Awareness, one can then oscillate between turning inward and outward, focusing each time on only one distinct sensory channel or level (Rytz 2018), for example, focusing on listening to a black bird singing, perceiving sound, and melodies, then focusing on our physical level, sensing an increase of our heartbeat, an emotional arousal of joy, a mental image of the bird arising before our inner eye and then proceeding into sensing what else is present, then returning to listening to the blackbird's song, and noticing the nuances of the sound.

Such conscious practices may remind us of our inter- and intrarelatedness with other living beings that we are prone to perceive as other or foreign: animals, the environment, and nature (Abram 2011). They allow us to pause, be at peace, and stop automated, unconscious reaction patterns (in thinking, feeling, acting) and particularly stress reactions. We can thus influence the lenses through which we see the world and make meaning.

36.2 Transformative Potential

The transformative potential of this method Connecting to self and nature lies in supporting creative processes toward regenerative futures as follows:

- Connecting to oneself in the present moment supports self-regulation and regeneration:

- Tuning into Sensory Awareness and checking in on the four Levels of Awareness in the present moment, including a non-judgmental attitude, allows one to become aware of one's emotional–physical–mental states. This is particularly important for self-regulation, for example, regulating the further activation of automated stress reaction patterns (like fight, flight, or faint) and more generally our reactivity. When stuck in automated reaction patterns, one is unable to be creative, to regenerate, or to transform (Siegel 2020; Singer-Brodowski et al. 2022).
- Allowing us to experience ourselves as part of bigger systems and of nature, experiencing intraconnectedness and interdependence (Siegel 2020; Abram 2011):

- Through connecting to other beings, we open up a precious space in which we can listen, sense, and potentially perceive a response or resonance within us. We may realize that our previously held meaning perspective of separation from others and nature is no longer valid. A new meaning perspective, potentially one of deep intraconnectedness, can emerge, supporting more regenerative practices. Thus, we can act and particularly create from a place of (intra)connectedness including taking care of nature and of ourselves.

Accessing different ways of knowing, particularly experiential via senses, and intuitive ones (Abram 2011; Halprin 2002; Heron and Reason 2006), complements analytical knowledge.

36.3 Application

The method of Connecting to self and nature can be used to tune into a state of presence and groundedness individually or in a group. If used in a group, it makes sense that someone facilitates it. We apply it, to connect to something greater than us, to other beings, to nature, and to get access to different ways of knowing. This connection informs us in a way that increases the likelihood of creating regenerative solutions. For example, when we want to investigate the regeneration of land, we can first connect and ask ourselves: What do we sense? What may be needed or what may be asked for?

The Sensory Awareness part is a low-threshold exercise, accessible to most people without much framing and preparation. Please see an overview in Table 36.1.

36.3.1 Setting

- Site: outdoors, for example, in a park, the woods, or your garden. Ensure, as facilitator or practicing person, that it is a space where you or the group can be undisturbed for about 30 minutes and where it is generally safe to close one's eyes.

Table 36.1 Using four Levels of Awareness and Sensory Awareness

<table>
<tr><td colspan="2">Overall: How am I here/present right now?
Just notice without needing to change anything</td></tr>
<tr><td>Focus inward: guiding questions</td><td>Focus outward: guiding questions</td></tr>
<tr><td>1. Physical level
How am I here physically?
What is present in my body?
Any tensions, movement, warm or cold spots… ?</td><td rowspan="4">What I perceive via sensory channels:
sound (auditive), touch (haptic), smell (olfactory), taste (gustatory), or sight (visual)
What do I hear/smell/touch/taste/see?
How does this place sound/ smell/feel via touch/taste/look like?
How does XY (the being you are in connection with) sound/ smell/feel via touch/ taste/look like?</td></tr>
<tr><td>2. Emotional level
How am I here emotionally?
Which emotions are present?</td></tr>
<tr><td>3. Mental level
How am I here mentally?
Which mental images, thoughts or mental talks are present right now?</td></tr>
<tr><td>4. Level: sensing of the wider field
What else is maybe present?
Do I notice anything else in the field?</td></tr>
</table>

- Group size: individual or in a group of up to 30 people (make sure that everyone has a space and the facilitator can be heard). The ideal size for a fruitful debriefing in the whole group is 5–12 people.
- Overall time frame: 40–60 minutes depending on group size.
- Frequency: you can repeat connecting on a regular basis (multiple times per day in shorter sequences), or use it in specific situations.
- Facilitator: connect with yourself before starting and have some experience in holding connection, acting from a place of connection.

36.3.2 The Flow (30–40 Minutes)

In a comfortable space outdoors and preferably standing (although it is not a prerequisite).

Close your eyes (or simply lower your gaze).

Start with a check-in on four Levels of Awareness, inward (ca. 3–5 minutes): *How am I here right now*:

- Physically: for example, inner movement or tensions
- Emotionally: for example, joy or sadness
- Mentally: Which mental images, thoughts, or mental talks are evoked?
- Sensing the field: What else is here /there, which is maybe not easy to be named?

Then continue with: sound (auditive), touch (haptic), smell (olfactory), or taste (gustatory), and at the end, sight (visual) outward directed. Explore each specific sense, one at a time for 3–5 minutes, a total of 15–25 minutes.

Starting with the auditive is recommended, as it is easily relatable to most people. Ending with the visual is recommended, as it is so often predominant. The order of the other senses can be changed.

Guiding questions are: *What do I hear/smell...? How does this place sound, smell...?*

- Focus first on sounds far away and then nearby.
- Observe the sensory experiences without judgment. Immerse in the present moment.
- Whenever thoughts (images) arise, return to the sense that you are focusing on.
- Finish each sense by allowing yourself to be at ease, enjoy a moment of stillness, and move consciously to focus on the next sense.
- At the end, turn to the visual sense, slowly opening your eyes, and see where your sight feels drawn to. Explore visually from a distance 2–3 minutes.
- Then move towards what attracts you, this “plant, tree, object,” and make close contact engaging all your senses, one by one, totally for 5 minutes.

At the end of the explorations, check in on the four Levels of Awareness: *How am I here right now: physically, emotionally, mentally? Anything else I notice in the field? (3–5 minutes).*

Thank the land and place as you end your practice.

36.3.3 Debriefing (10–20 Minutes) (e.g., Generative Dialogue or Dialogue Circle)

Share and reflect on your experience either in a circle or in pairs or groups of three. If you did the practice on your own, you may now write down your experiences.

Guiding questions: *What needs to be shared/heard/spoken in this circle?*

What have I experienced in connecting to myself, to nature? How does this experience inform my design practice?

36.3.4 Follow-Up

1. Rooted in this experience, you may continue, for example, with a creative session or ask more specific questions connected to a design task from a place of being in connection.
2. This short and accessible practice is the basis for a deeper, longer practice "Connecting as nature" which allows one to experience oneself as nature (Sägesser and Förster 2025) (Table 36.2).

Table 36.2 Method overview

Main purpose
Cultivating a state of presence and groundedness individually or in a group as a basis for (transformative) perspective, behavior change, and creativity via sensory awareness
Gained competences
Connecting to oneself in the present moment via senses Experiences intraconnectedness with others and nature Self-regulation and regeneration Accessing different ways of knowing, particularly experiential via senses Creating from a place of presence resp. connection
Educational setting
It can be used in a formal setting, to tune into a state of presence and groundedness individually or in a group as a basis for (transformative) perspective, behavior change, and creativity. It can be used to introduce an individual, regular practice of connectedness in informal settings for the same purposes.
Space requirements/restrictions
Outside activities, for example, in a park, the woods, or a garden. A space where the group can be undisturbed for about 30 minutes and where it is generally safe to close one's eyes. If practicing on your own, you can choose any place where you can stand or sit undisturbed.
Resources and necessary materials
See above space requirements.

(continued)

Table 36.2 (continued)

Number of participants
Individual or group work: 1—max. 30 people; Ideal size for a fruitful debriefing: 5–12 people; Make sure that everyone has a space and the facilitator can be heard.
Facilitator competences and skills
Connect with yourself before starting, using the four Levels of Awareness, and have some experience in holding connection, acting from a place of connection. We suggest that you repeat connecting based on four Levels of Awareness on a regular basis in shorter sequences (ca. 10 minutes) (see Table 36.1).
Participants skills/age/competences
Community/students/policymakers/researchers/designers/practitioners, etc. Everyone who is involved in transformative, creative processes asking for presence and groundedness.
Duration
(a) 40–60 minutes: for the whole sequence in the group incl. introduction to four Levels of Awareness; (b) approx. 10 minutes: as a short form practiced on a regular basis for checking in on four Levels of Awareness.

References

Abram D (2011) Becoming animal: an earthly cosmology. Knopf Doubleday Publishing Group, New York

Halprin D (2002) The expressive body in life, art, and therapy: working with movement, metaphor and meaning. Jessica Kingsley Publishers, London

Heron J, Reason P (2006) The practice of co-operative inquiry: research 'with' rather than 'on' people. In: Reason P, Bradbury-Huang H (eds) Handbook of action research: concise paperback edition. SAGE Publications Ltd, pp 144–154

Rytz T (2018) Achtsam bei sich und in Kontakt: emotions regulation und Stressreduktion durch achtsame Wahrnehmung. Hogrefe AG, Göttingen

Sägesser A, Förster R (2025) Relating: connecting as nature a practice towards relational, collective leadership grounded in ecocentrism. In Egel E, Campo Suarez (eds) Inner development goals: stories of collective leadership in action. De Gruyter

Siegel DJ (2020) The developing mind: how relationships and the brain interact to shape who we are. Guilford Publications, New York

Singer-Brodowski M, Förster R, Eschenbacher S, Biberhofer P, Getzin S (2022) Facing crises of unsustainability: creating and holding safe enough spaces for transformative learning in higher education for sustainable development. Front Educ 7:787490. https://www.frontiersin.org/articles/10.3389/feduc.2022.787490

Young S (2016) What is mindfulness? WhatIsMindfulness_SY_Public_ver1.5.doc. https://www.shinzen.org/wp-content/uploads/2016/08/WhatIsMindfulness_SY_Public_ver1.5.pdf. Accessed 26 Dec 2024

Chapter 37
Walking as Relational Aesthetics

Susanne Bosch

37.1 Introduction

37.1.1 The Walk: Bodily and in Silence

This artistic excursion invites a group of participants for a nocturnal, silent, "light or torch-free" walk in the nearest forest. Group sizes from 2 to 15 are ideal for this artistic excursion into the nocturnal green space—and into the inner space of imagination that everyone associates with darkness and the nocturnal forest/nature. The route has a maximum length of 5 km. The group walks at a slow pace. The walk will last about 1.5–2 hours.

Darkness: With less visibility, the forest becomes more substantial and no longer sharply outlined. The group will from time to time stop and pause. The participants will walk one after the other, so no one is ever alone and yet everyone is for oneself.

Silence: Every participant is asked to be silent for the duration of the walk. This allows everyone to listen and hear better, both inwardly and outwardly. The participants are invited to check in once in a while with their inner choir and to go off their thoughts.

Paths: The group will walk on existing paths. They have been walked by many (humans and animals) in time. Some paths are less intensively used and slowly lose their contours.

The person in front: The person in front leads the way and sets the pace. The person will check from time to time if everyone is ok. All without words. The front person decides when the group stops and when the group moves on. The group agrees collectively on a codeword for emergency cases.

S. Bosch (✉)
Independent Artist, Berlin, Germany
e-mail: bosch@susannebosch.de

J. Bentz, J. Ristić Trajković (eds.), *Imagining, Designing and Teaching Regenerative Futures: Art-Science Approaches and Inspirations From Around the World*, Science for Sustainable Societies,
https://doi.org/10.1007/978-981-96-9029-9_37

Perception: Each participant uses his/her/their entire body as an organ of perception. Each participant is open to something he/she/they do not yet know: this experience in this forest in this darkness, at this moment, in these weather conditions. Everything counts: what one smells, feels, hears, and sees, what one suspects, thinks, and dreams, and how one moves. In the forest, one's bodily perception becomes essential. Only the mind sometimes wanders away from the body and returns again.

The group: Even if no one says anything, one will hear the others, their steps on the ground, how their clothes rustle, and how they breathe. One will smell the others. One is not alone and will feel that.

37.1.2 Spiral Dialogue

The group will talk about the "inner and outer walk" afterward in a spiral dialogue.

A spiral dialogue is a format of oral, verbal exchange sitting in a circle. The order of contributions is regulated by "passing on" to the person next to oneself. Everyone will have their turn and time; one can also just listen and be silent. Spiral dialogues are not about discussing, interrupting, or convincing someone. They are about listening, openness, and trustfully suspending prejudices for the duration of the dialogue. This type of dialogue has been developed and explored, for example, by David Bohm and Martin Buber. The spiral dialogue helps to connect and create collective meaning.

This method of walking together in darkness at places, where one tends to fail to walk alone, responds to various cultural assumptions that most (western) humans face. As a method, it allows for an embodied experience as an individual but also as a collective. Through verbal sharing of the experience afterward, the embodied experience is reflected and witnessed by others. It becomes a collective learning process. The reflection and learning addresses the various cultural assumptions Western societies have about darkness and forests by night. The method foregrounds the effect of these assumptions on one's daily behavior. After the experience, participants might develop a different approach and relationship to their environment. The experience might be strong enough to alter general cultural assumptions through one's own embodied knowledge.

This practice is useful in urban and rural contexts where participants aim to develop an extended understanding of the importance and value of night and darkness for all living beings as well as expand their freedom of movement in darkness. The effect of dark deprivation upon all living creatures as environmental protection and nature conservation is significant.

37.2 Transformative Potential

Walking with our bodies through space by night, activating all our senses as tools of perception for new knowledge, is a very simple yet powerful way to gain an embodied individual and collective understanding of a condition that the planet faces every day: The formation of day and night on earth is due to the rotation of the earth around its own axis. Night is the period of ambient darkness from sunset to sunrise during each 24-hour day when the sun is below the horizon at our part of the earth. The exact time when night begins and ends depends on the location and varies throughout the year, based on factors such as season and latitude.

Darkness: Too much light at night disrupts the natural rhythm of organisms. Medical experts warn that "light pollution" is making people and animals ill. Our planet shines brighter today than ever before. The amount of light it emits into the night doubles every 11 years. Two-thirds of Europeans and Americans no longer experience a real night; they have lost the darkness. All the lanterns and lamps that are supposed to keep people safe are becoming a threat. Due to the shadow of trees, the forest is darker than other places.

Forest: The forest is defined in many different ways. The ecological definition says: Although the collection of trees is an important unit in the forest community, it is primarily an ecosystem in its own right. Only when trees are so dense and numerous that a typical forest climate develops beneath them, it is called a forest. The temperatures in the forest are more balanced (cooler in summer and warmer in winter), it is less windy in the forest, more protected from light, and the humidity is slightly higher than in the surrounding area. According to §2 of the German Federal Forest Act, a forest also includes clear-cut or thinned ground areas, forest paths, forest division and security strips, forest glades and clearings, forest meadows, game grazing areas, wood storage areas, and other areas connected to and serving the forest.

A collective, playful exploration of darkness in a forest: The method touches upon theories of fear we feel at night as both natural and salutary, prompting in us an increased sense of aliveness by allowing us to face, through darkness, some intimation of our inevitable death. In darkness or dim light, people overcome mental boundaries more easily: We think more globally and creatively, pay less attention to social rules, and engage in riskier behavior, social science found out. Darkness swallows up the eyes of others. It gives us a feeling of anonymity and isolation. It gives us the courage to break norms and also to act more collectively, feeling ourselves as individuals more. The night is the other, a frighteningly insecure and promisingly free space, and it is the other side of our daytime mind. Darkness in a forest is also the ultimate imaginary place of lawlessness, as it is read as mysterious and powerful. When actually walking through a forest by night, one will discover the often in poetry described quietness and the essential nature of forest itself. Darkness in a forest is the ultimate imaginary place: Our imagination grows, and we think more creatively and act more daringly.

Sensing: Darkness allows us to sense, behave, and explore the world differently. Undefined collective, culturally transmitted fear often does not allow us to explore ourselves in dark and darkness itself. The potential of this method is to sense the essence of a forest through stepping by night into the territory where the body can sense the different climates beneath the trees. Our eyes will read the environment through different shades of gray and black with no clearly determined shapes. Yet, all participants will discover how safely one can navigate through the darkness and how joyful it is to explore the quality of this space. Our ability to read a dark forest space will become instantly clear.

Joy: Fear deprives most Western humans of the experience of walking in a forest at night. We often experience guided night walks as teenagers, where fear and horror are actively initiated. The potential of this method is to experience the essence of a forest in its peaceful quietness with its enveloping quality.

37.3 Application

Preparatory work: Send out an invitation for this experience. Participants should clearly express an interest in taking part. As an initiator, one needs to know the walking route and lead the group. The leader might try the route several times in several weather and light conditions as he/she/they will lead the group with one's inner compass. The person leading might have a tracking app recording the movement in the dark while walking, in case one feels insecure. It is recommended to inform local hunters that a group will be walking by night/in darkness in the forest. It is recommended to implement a starting and closing ritual to mark the time of silence.

Group size: 2–15

Overnight stay and start of the walk: The walk begins at a time and in a location when/where usually traveling is not easy. An overnight stay close by might be necessary to all be there in time and to be able to start collectively. Weather + clothing: People need to wear clothes according to the weather forecast and their urge to move. Participants are asked to wear shoes that they can walk well in (no sandals). Due to unevenness on the unpaved forest floor (stones, branches lying on the ground, tree roots, etc.), the walk is not completely barrier-free. The walk is largely weather-independent. Heavy rain and severe thunderstorms can lead to a postponement of the walk.

Mosquitoes + ticks: Mosquitoes and ticks may be an issue. It is recommended to bring whatever one wants to protect him/her/itself against mosquitoes.

Procedure and time: The route should have a maximum length of 5 km. The group will walk slowly because of the darkness. It is expected to take about 2 hours. There will be a post-walk talk, a duration of approximately 60–90 minutes. Ideally, the group will sit in a circle, round a fire, and celebrate the walk with hot or cold drinks and some snacks.

WC in the forest: If one needs to use the forest as a toilet, please do not leave any rubbish behind.

Further thoughts: The walk is not really suitable for small children or pets.

Do not take pictures in between, and do not use any mobile devices. Ask people not to smoke during the walk (Table 37.1).

Table 37.1 Method overview

Main purpose
Awareness building Embodied experience with self, other, and nature Individual learning experience and collective reflection
Gained competences
Trust building Being present Using many of the body senses Changing perception about darkness
Educational setting
Informal: walk and reflection dialogue
Space requirements/restrictions
Outside activity Forest Darkness/nighttime A place close by for reflection Ideally a space for a fire
Resources and necessary materials
Clothing according to weather conditions Good shoes Mosquitoes + ticks protection A one-night accommodation in case transport at night to the site is not possible
Number of participants
Group and individual work. Group size: 2–15
Facilitator competences and skills
Needs to know the site well and be able to walk through the forest by night without light Needs some experience in group facilitation
Participants skills/age/competences
Open for people of any age and background with basic fitness for a walk Not really suitable for small children or pets Not barrier-free
Duration
2-hour walk 1-hour dialogue/reflection

References

Between politics and aesthetics. https://susannebosch.de/portfolio_page/between-politics-and-aesthetics/ Accessed 14 June 2024
Between night and day. https://susannebosch.de/portfolio_page/between-night-and-day/ Accessed 14 June 2024
Crossing Jericho. https://susannebosch.de/portfolio_page/crossing-jericho/ Accessed 14 June 2024
Crossing Jericho. https://walkingart.interartive.org/2018/12/susanne-bosch Accessed 14 June 2024
Crossing Jericho. https://www.thereroutingproject.org/crossing-jericho-susanne-bosch Accessed 14 June 2024
Eklöf J (2023) The darkness manifesto, on light pollution, night ecology, and the ancient rhythms that sustain life. Scribner, New York

Chapter 38
Developing Daily Sustainable Habits

Valeriia Kostynets and Iuliia Kostynets

38.1 Introduction

The transformation of the educational space is closely connected with the ongoing rethinking of the mission of education as a whole, its goals, and ontological and axiological foundations in the global world (UNESCO 2024). Redesigning the educational space as one of the most important socializing institutions of society is becoming a fundamental scientific and methodological task that cannot be solved only within the framework of pedagogy.

In turn, educational environments are increasingly acquiring an interdisciplinary character, using a wide range of approaches, methods, and techniques from different areas.

The goal of daily sustainable habits is that students and university staff adopt and embrace the idea of a regenerative future and sustainable transformation. The main advantage of this method is the combination of practice-oriented outdoor classes with volunteer activities with the aim of not only preserving but also restoring the ecosystem. This practice is helpful both for the formation and development of daily sustainable habits and contributes to a sustainable transformation of the surrounding area.

V. Kostynets (✉)
Department of International Economic Relations, National Academy of Management, Kyiv, Ukraine

I. Kostynets
Department of Marketing, Economics, Management and Administration, National Academy of Management, Kyiv, Ukraine

J. Bentz, J. Ristić Trajković (eds.), *Imagining, Designing and Teaching Regenerative Futures: Art-Science Approaches and Inspirations From Around the World*, Science for Sustainable Societies,
https://doi.org/10.1007/978-981-96-9029-9_38

38.2 Transformative Potential

We regularly use and propose using open-air lectures for university students. Conducting classes outdoors does more than just change the environment—it can inspire new ways of both teaching and learning (Bodin 2018). Research confirms that schoolchildren who had their lessons outdoors once a week had lower levels of the stress hormone cortisol than children who only studied in traditional classrooms (Dettweiler et al. 2017). Moreover, there is evidence that high school students who had math and science lessons in the open-air format interacted more with their peers and remembered the material better and were able to reproduce it more successfully several months later than students who studied the same material but in classrooms (Fägerstam and Blom 2012). It shows that students whose lessons are held in an open-air format are more enthusiastic about what they are learning, and this enthusiasm remains with them when they return to the classroom. This is comparable to the results we have observed during our open-air lectures for students of one of the well-known Kyiv universities over the past 6 years (taking into account existing and possible organizational limitations).

38.3 Application

Conducting open-air lectures is recommended for students, but such classes can also be adapted for schoolchildren if desired. This form of conducting classes can be adapted for students of various educational programs. Conducting open-air lectures for students requires careful preparation and a thoughtful approach. Organizing open-air lectures involves several steps, listed below.

38.3.1 Step 1—Preparation and Planning

Selecting a place/space: find a comfortable and safe place outside. This could be a park on campus or any other public space. It is important to consider the noise level, the presence of green areas, shade, and accessibility for all students.

Weather assessment: weather is key when conducting open-air lectures. Pay attention to the weather forecast to avoid unpleasant surprises such as rain or too-hot weather.

Purpose of the lecture: determine which aspect of the lecture or discipline will be most effectively taught outdoors. This may be related to urban space, urban studies, ecology, tourism, and so on.

38.3.2 Step 2—Preparing and Informing Students

Notifying students: notify students in advance that the lecture will be held outside so that they can prepare. Explain the specifics of this activity and why it is being held. Remind students of possible weather conditions (e.g., bring hats and water if it is hot).

Instructions for participation: provide students with clear instructions on how to behave, including the importance of outdoor safety, and inform them of the necessary props to bring (camping mat or pad (or blankets), tablet, or laptop).

38.3.3 Step 3—Organizing the Space

Marking the space: divide the area for different groups of students (if the lecture will include discussions or group assignments). Provide comfortable seating or, if necessary, provide camping mats.

Preparing materials: prepare the necessary materials for the lecture: visual aids, handouts, and possible audio or video materials (use smartphones and laptops/ tablets).

38.3.4 Step 4—Delivering the Lecture

Introduction and establishing a connection with the environment: start the lecture with a short introduction and tell the students why the lecture is taking place open-air and how this relates to the topic. For example, if the topic is related to tourism, you can discuss how the urban environment affects the organization of city tours.

Using the environment as an educational tool: draw students' attention to the surrounding objects—architecture, plants, movement of people, and street infrastructure. This will help connect theoretical knowledge with real practice.

Dynamic delivery methods: the street can be very distracting, so use active teaching methods: discussions, surveys, group assignments, and practical observations.

Group discussions: divide students into small groups and ask them to discuss aspects of the urban environment and ecology that may be related to the lecture topic. You can use city maps or photographs for visualization.

Practical observations: ask students to note features of the urban environment that could be improved from an environmental point of view. This could be, for example, related to transport, public spaces, and street cleanliness.

Involvement of business representatives: if possible, invite business representatives to discuss the topic, which will make the lecture more lively and practical.

38.3.5 Step 5—Conclusion

Discussion of results: at the end of the main part of the lecture, discuss the results of the lecture with students. This can be either a free discussion or a structured analysis.

Summarizing: summarize the main points of the lecture and link them to theoretical knowledge.

Feedback: invite students to ask questions or share their impressions of the outdoor lecture. This will help to improve the format in the future.

38.3.6 Step 6—Evaluation and Reflection

Evaluation of the success of the lecture: after the open-air lecture, ask students to share their impressions, what they liked, and what could be improved. This can be done through questionnaires or group discussions.

Evaluation of the experience: conduct your own analysis of how the lecture went, whether there were any difficulties (e.g., noise and lack of materials), and what could be improved in the future.

When conducting lectures outdoors, it is important to remember that this form of learning is not only a way to diversify the learning process but also an opportunity to make it more lively and closely connected to the real world.

Such lectures have a positive impact on the formation of daily sustainable habits by the following:

"Clean the planet together." Conducting open-air classes for students helps them to develop eco-habits related to sorting garbage and keeping the environment clean. Thus, open-air classes are provided directly sitting on the lawns in specially designated and permitted places in the university campus or a park. If the space is contaminated with garbage, it is necessary to allocate 10 minutes to clean it up and only then begin the lecture. In this way, the eco-principle of joint cleaning of the planet is manifested and the habit of not leaving garbage behind is further developed.

Development of a positive attitude toward urbanism. Taking classes outdoors can have a positive impact on students' attitudes toward urbanism in several important ways:

Direct connection with the urban environment: when students spend time outdoors, they begin to see the city as a living, multifaceted organism. The feeling of space and interaction with elements of urban infrastructure helps them feel a connection with the environment and understand the importance of urban solutions.

Active involvement in real urban practice: outdoor classes can involve observing, analyzing, and discussing real urban solutions such as architecture, transportation, recreation areas, and public spaces and their impact on the comfort of residents. This practical interaction encourages students to think more deeply about urban problems and possible solutions.

Emotional involvement and experience: being in an urban environment, students begin to feel the "living" urban atmosphere. For example, walking through pedestrian areas or discussing the concepts of "green" technologies in the urban environment can evoke an emotional response, improving attitudes toward urbanism and stimulating interest in improving urban life.

Generally, outdoor activities can help students develop a deeper and more positive perception of the city as a living, developing organism that needs to be treated with care and attention, which, in turn, can be attributed to the development of daily sustainable habits.

As a conclusion in favor of this method of education, we will cite several quotes from Ukrainian students who participated in open-air classes:

Artem, educational program "Tourism Business Management": "Open air, beautiful places, the atmosphere is freer, simpler, you feel lighter and sometimes even feel as if you are resting and not studying, unity with nature. I don't see any disadvantages in such activities."

Zhanna, educational program "Tourism Business Management": "During open-air classes, the information received in the lecture was absorbed much better than during classes in the classroom, and the desire to conduct classes on clean lawns in the park fostered a culture of caring for nature."

Anna, *educational program "Hotel Business":* "These were the best classes we had during our entire studies!" (Table 38.1)

Table 38.1 Method overview

Main purpose
Awareness building and behavioral change
Gained competences
Professional competencies according to the topic; soft skills: environmental responsibility and sustainable behavior
Educational setting
Formal—practical-oriented lectures for undergraduate and graduate students Informal—seminars for the same students
Space requirements/restrictions
Outside activities
Resources and necessary materials
Camping mat or pad (or blankets), tablet, or laptop Location—campus or city park/garden
Number of participants
Group work—academic group, recommended up to 25 students
Facilitator competences and skills
Background in the professional field in accordance with the requirements of the educational program, environmental responsibility, and sustainable behavior
Participants skills/age/competences
No specific requirements
Duration
Initial implementation—1 month for setup and initial training Ongoing application—open-air activities integrated into the curriculum, with periodic assessments and adjustments Lesson duration—90 minutes (45 × 45) and brake up to 10 minutes

References

Bodin M (2018, May 1) Open-Air Learning. Retrieved from https://www.ascd.org/el/articles/open-air-learning. Accessed 26 Dec 2024

Dettweiler U, Becker C, Auestad BH, Simon P, Kirsch P (2017) Stress in school. Some empirical hints on the circadian cortisol rhythm of children in outdoor and indoor classes. Int J Environ Res Public Health 14(5):475. https://doi.org/10.3390/ijerph14050475

Fägerstam E, Blom J (2012) Learning biology and mathematics outdoors: effects and attitudes in a Swedish high school context. J Adventure Educ Outdoor Learning 13(1):56–75. https://doi.org/10.1080/14729679.2011.647432

UNESCO (2024) The Futures of Education. https://www.unesco.org/en/futures-education. Accessed 26 Dec 2024

Chapter 39
Perception Walks: Using Clothing as a Portal and Reflection as a Guide

Zoe Gilbertson, Niels Devisscher, and Rūta Žemčugovaitė

39.1 Introduction

Clothing is an accessible and immediate tool to connect human activity directly to nature—a portal to reflecting on the consequences of modernity and its potential impacts on other living beings. This practice uses clothing as a way for participants to connect more fully to place and nature. It combines walking, writing, and the arts of noticing to create a reflective process we refer to as a perception walk. It can be adapted to various educational contexts. The process was developed for students on a sustainable fashion master's degree, who found it a valuable contrast to their online study, allowing them to build personal resilience and develop greater empathy with nature over a timescale of several months.

Experimental and flexible, the practice invites participants to record how they feel, what they think, see, and sense all around them through recording thoughts immediately after a meditative period of walking in nature.

The practice can be guided by a facilitator as a one-off workshop or become a self-led independent repeated activity with guidance provided in advance. It can be used by any group of people, not just those with a direct interest in fashion. Each outing will be different because environments change with the seasons, as will internal states of being.

Educators may review the references given for this article to provide as supplementary readings before the exercise.

Z. Gilbertson (✉)
Falmouth University, Cornwall, UK
e-mail: zoe.gilbertson@falmouth.ac.uk

N. Devisscher · R. Žemčugovaitė
Sympoiesis.world, Glienicke/Nordbahn, Germany

J. Bentz, J. Ristić Trajković (eds.), *Imagining, Designing and Teaching Regenerative Futures: Art-Science Approaches and Inspirations From Around the World*, Science for Sustainable Societies,
https://doi.org/10.1007/978-981-96-9029-9_39

39.2 Transformative Potential

This method is a simple and effective practice developed to address the separation of humans from nature (Young J et al 2010). Using four frames—place, embodiment, materiality, and reflection—it leverages everyday experiences with clothing to cultivate awareness of our interconnectedness with the natural world. This encourages participants to reconsider their relationship with nature and recognize the broader implications of their choices, particularly regarding global fashion and its environmental impacts (Fletcher 2022).

Participants will use this method to deepen their connection to the environment and enhance their understanding of the complexities of global supply chains (Niessen 2020). Engaging in perception walks and reflective writing will help participants develop a richer, more nuanced perspective on how their actions impact the planet, promoting systems thinking and relational awareness. This process can facilitate more sustainable and regenerative practices in participants' daily lives by encouraging them to notice the often-overlooked details of their surroundings and reflect on the entanglement of human and non-human life.

The transformative power of this exercise lies in shifting focus from abstract environmental issues to tangible, personal experiences with clothing, place, seasonality, and time. By grounding reflections in sensory experiences and material realities, it cultivates a deeper, more embodied sense of ecological awareness and responsibility (Tsing 2010). The method asks us to consider how we can connect to other life forms through the beauty, functionality, and protection from the elements that clothing brings us and to develop gratitude for the earth's bounties that help us live comfortable lives. Here we can change our relationship from extractive to one of reverence and appreciation (Van Dooren et al 2016). Clothing is an accessible and immediate portal of relationality, a connection to our ancestry and the animacy of materials. Participants wearing clothing from synthetic fibers, such as polyester and nylon, may consider how these close-to-skin fabrics were once plants, algae, and bacteria, compressed over millions of years before turning into oil. We work with a multisensory, layered perception of space, reflecting how our bodies are located not only in the space itself but also within the clothing, which itself is grown from a place.

39.3 Application

To begin, decide on a place in nature to walk to each week. Choose a place that speaks to you, perhaps a park, forest, garden, or beach, although it must be a natural environment full of living, growing beings—the wilder, the better. If you are unable to walk, find a mode of transport that will enable you to reach a place close to nature and observe.

As you begin your walk, notice how you feel at this moment and be fully present in your surroundings. Reflect on the clothing you are wearing as you set off. How does it feel on your skin? Whether your clothes are made from natural or synthetic fibers or both, they have found their way to us from the ground, as plants or oil. If your clothing has synthetic fibers, it is made from oil and has been hundreds of millions of years in the making. If your clothing is natural, it was recently living.

Around you are dozens of plants, some of them are native, and some we call invasive. Consider how these plants arrived in your locality and how they are food for some animals and home for others. What benefits do these beings have for humans, be they medicinal or material; how have we exploited them; how can we care for them?

- As you walk, work through the following prompts, giving the time and space you need for each one. What is on your mind? What emotions/feelings/sensations do you feel? What parts of yourself do you want to bring to this walk?
- Notice how your clothes feel on your skin and your body. Do they feel comfortable or constricting? Are materials soft, itchy, or neutral? Can your skin breathe? How does the temperature or humidity feel?
- Imagine the journey your clothes have been on through time and space. Consider where they have come from. How far have they traveled? What living beings created them? What relationship do they have to the beings around you right now?
- Choose a plant (or other living being) to observe and connect with. What were our ancestral relationships with this non-human companion? What do you feel when connecting with this living being? Does it speak to you, what is it saying?

When you reach a beautiful place, allow yourself to become still and observe this place for at least 10 minutes. Sit down, stand, or lean against a tree, whatever allows you to get comfortable. Take in this space through all the senses, sight, smell, touch, and hearing. Allow yourself to feel into the place or a landscape. What is it telling you? What can you learn from it?

At the end of your time in stillness, record your immediate thoughts on a phone verbally or by taking notes on paper. You may take a photo. If this is weekly practice, then go to the same place to build a record over multiple weeks or months. Going at different times of the day, experiencing the change of seasons and different temperatures, and wearing a variety of clothing add to the depth of reflection and create a broader experience.

To close off the walk, thank all the beings that you have encountered along the way and return to your destination. You may wish to repeat the prompts on your way back.

39.3.1 Written Reflection

Upon returning, reflect on what you have experienced through journaling. You may add photos and drawings to this record each time you walk. Use these questions to get you started:

- How did you experience the walk and being in your clothing? What was your starting mood, and how did you feel at the end?
- Reflect on the walk prompts and your immediate notes/recording and refine your impressions. Consider if your clothes allow you to connect further with the place, or did they constrict you in any way?
- What plants (or other living beings) did you connect with? How did this affect the rest of your walk? Did your relationship with the place change after making this connection?
- How did the place look and feel to you this time? Has it changed since the last time you were there? What are you feeling about it now? (Table 39.1)

Table 39.1 Method overview

Main purpose
This practice uses clothing as a portal to connect more fully to place and nature. It combines walking, life writing, and the arts of noticing to create a reflective process referred to as a "perception walk." it can be adapted to multiple educational contexts. Experimental and flexible, the practice invites participants to record how they feel, what they think, see, and sense all around them through recording thoughts immediately after a meditative period of walking in nature.
Gained competences
Nature connection, systems thinking, integrative thinking, empathy building, and ecological awareness
Educational setting
Secondary education, undergraduate, and post-graduate. Online or in-person
Space requirements/restrictions
Access to wild space or nature for walking
Resources and necessary materials
Note-taking materials, phone, or camera
Number of participants
1–10
Facilitator competences and skills
Nature connection and journaling experience are advised but not necessary.
Participants skills/age/competences
From teenage upward. Could be adapted for younger ages by facilitators with experience. No skills or experience needed
Duration
30-minute activity that can be repeated weekly

References/Further Reading

Fletcher K (2022) Life writing as an ecological research method. Fash Prac 14(2):193–202. https://doi.org/10.1080/17569370.2021.1882760
Niessen S (2020) Fashion, its sacrifice zone, and sustainability. Fash Theory 24(6):859–877. https://doi.org/10.1080/1362704X.2020.1800984
Tsing A (2010) Arts of inclusion, or how to love a mushroom. Manoa 22(2):191–203. http://www.jstor.org/stable/41479491
Van Dooren T et al (2016) Multispecies studies: cultivating arts of attentiveness. Environ Humanit 8(1):1–23. https://doi.org/10.1215/22011919-3527695
Young J et al (2010) Coyote's guide to connecting with nature. OWLLink Media, Walnut

Chapter 40
The Permaculture Approach

Censu Caruana

40.1 Introduction

Permaculture can be considered synonymous with regeneration. In a world where so much damage has been done to the ecosystem that conservation and preservation are not enough, regeneration and restoration need to take the central stage.

Permaculture is a "set of principles that integrates land, resources, people, and the environment through mutually beneficial synergies. It aims to imitate and recreate natural ecosystems through the use of closed-loop and no-waste techniques" (PermaModule, n.d.). Unlike intensive agriculture, which often sacrifices biodiversity for high productivity, permaculture offers holistic alternatives rooted in ethics, specifically earth care, people care, and fair shares.

My teaching approach in permaculture is rooted in fostering observation, problem-solving, and critical thinking. I believe that true understanding comes from active engagement with the context. Therefore, I encourage my students to deeply observe natural systems, question existing practices, and collaboratively devise innovative solutions. This approach transforms students from passive recipients of knowledge to active participants in their learning journey. In a project setup, while foundational knowledge of permaculture principles is essential, I emphasize that much of the learning will occur organically through hands-on experience and exploration. Motivation and curiosity are the driving forces behind successful learning, far outweighing initial expertise. Students are encouraged to start with basic knowledge but are supported to dive deeper into the complexities of permaculture through continuous learning and adaptation. This methodology not only imparts technical

C. Caruana (✉)
Centre for Environmental Education and Research, University of Malta, Msida, Malta
e-mail: censu.caruana@um.edu.mt

J. Bentz, J. Ristić Trajković (eds.), *Imagining, Designing and Teaching Regenerative Futures: Art-Science Approaches and Inspirations From Around the World*, Science for Sustainable Societies,
https://doi.org/10.1007/978-981-96-9029-9_40

skills but also nurtures the potential for regenerative practices and transformative change. By integrating permaculture ethics and principles, students learn to envision and create sustainable systems, fostering ecological and social regeneration.

40.2 Transformative Potential

The power of permaculture lies in its potential to promote sustainable and regenerative futures. It fosters biodiversity, enhances soil health, conserves water, and reduces waste. Beyond environmental factors, there is enormous potential for the social sphere. In practice, it often unites people together through collaborative projects such as community gardens, urban gardens, and cooperative farms. It supports social equity by promoting inclusivity and access to fresh, clean, healthy, and fair food for community members while fostering self-reliance by providing the knowledge and skills to produce one's food sustainably.

The teaching of permaculture involves active learning, requiring students to engage through discussions, observations, group work, problem-solving tasks, case studies, and fieldwork. Learning design in permaculture necessitates taking responsibility for one's learning while recognizing the teacher's guidance in the process. In a world where many young people are suffering from nature disconnectedness, permaculture nurtures personal growth and well-being by re-connecting individuals to nature.

To successfully apply permaculture, it is important to consider time, commitment, and available resources. It can be applied both individually and in groups, but the community aspects can help enhance social cohesion, resilience, and friendship.

The transformative potential of permaculture also lies in its offering agribusinesses sustainable practices that improve the soil, reduce operational costs, and improve long-term yields, thus fostering profitability while minimizing environmental impact. It also opens up new opportunities for diversified product lines and niche markets, successfully aligning economic success with environmental sustainability.

40.3 Application

To apply the permaculture approach, one needs to start by understanding its core principles and design methods. Here, I illustrate its application through a project I conducted at the University of Malta.

40.3.1 Project Setup

At the University of Malta, I utilized a Virtual Learning Platform to deliver the permaculture curriculum. Each lesson, or volume, was released every week or 2 weeks. The platform included open forums for students to share their learning experiences and discuss tasks with each other. A piece of land on the university campus was designated as a simulation site for students to apply permaculture design principles.

The content of the permaculture curriculum can vary according to whether one is introducing the principles or whether one is doing a full immersion over several days. Typically, it encompasses an exploration of principles and practices designed to promote sustainable living and environmental stewardship. It focuses on the ethics and foundational principles behind permaculture, guiding students on the ethical considerations of earth care, people care, and fair shares. Being fundamentally about design, methods of design form a core part of such a curriculum, teaching students how to apply these principles in designing sustainable systems, often using a real piece of land as a simulation.

Due to the challenge of climate change, and one potential response to the challenge being sustainable ecosystem management, a permaculture curriculum typically tackles basic content on soil health, water management, and biogeographical regions. These are often developed through equipping students with a basic understanding of the scientific principles and natural laws that underpin the understanding of permaculture. Finally, pattern understanding illuminates the interconnectedness of natural systems, helping students gain competencies in system thinking, and empowering them to recognize and replicate beneficial patterns, thus also developing their anticipatory competencies in co-creating a new world they are able to imagine.

40.3.2 Assignments

For their assessed assignments, students are required to keep a reflective journal (30%) and produce a design for the identified piece of land (70%).

For the reflective diary, I ask my students to make an entry after every lesson and visit the piece of land. Students are encouraged to:

- Use the reflective diary to measure their own learning against the learning outcomes of the course, as well as to see and monitor their own development and how their ideas change with time. Through this reflective diary, they are able to capture their own personal development and growth.
- The reflective diary allows not only for a description of their learning but also for a critical reflection on their own learning. It provides a space for students to ask questions, express their doubts, and write about their hopes. It allows them to develop their own identity as change-makers.

- Students are encouraged to go beyond text and use any other medium they feel comfortable in, such as writing poetry, sketching, painting, and including photographs.

The final design assignment involves identifying the essential characteristics of the simulation site and applying permaculture design methods to create a sustainable system. If students or their families own a piece of land, they can submit a design for their own property instead of the simulation site. One notable example was a student from an educational background who designed a permaculture garden for the school where she taught.

40.3.3 Practical Implementation

Students are encouraged to visit the simulation site frequently to observe and take notes on its characteristics and microclimate. Typical questions posed through the Virtual Learning Platform include the following:

- What is the relevance of permaculture to today's environmental, social, and economic challenges?
- What is the potential of the application of permaculture ethics and principles to radical ecological and social transformation?
- Choose one of the Regenerative Agriculture and Permaculture Core Principles and Tenets that you consider particularly useful to the simulation field you have seen during the field visits. Why did you choose this principle? How can you apply it to the field in question?

40.3.4 Considerations for Implementation

Permaculture is about progress rather than perfection. The context and resources available determine the possibilities, but the most crucial step is to start. Safety is paramount, so it is essential to wear suitable outdoor clothing, gloves, and safety shoes and to closely follow protocols when using power tools. When resources allow, conducting a thorough site assessment, including soil analysis, climate study, and mapping of existing resources and features, is highly beneficial (Table 40.1).

Table 40.1 Method overview

Main purpose
Enhance agency in growing one's own food Design according to the basic principles of care for the planet, care for people and fair shares
Gained competences
Critical thinking, biomimicry, ethical consumerism, and living within the carrying capacity of the earth
Educational setting
Nonformal and informal
Space requirements/restrictions
Ideally an allotment or piece of land suitable for growing food
Resources and necessary materials
Gloves, garden tools, notebook
Number of participants
3–12 people
Facilitator competences and skills
Knowledge of permaculture design principles
Participants skills/age/competences
Anyone interested in agroecology and in growing one's own food Agri-business students Age 18+ years Subject to the necessary consent and safety procedures, can be adapted to younger persons as well.
Duration
Introductory course 1 day Various combination Fuller courses 3 days+ Ideally, time is blocked for fieldwork and practical design

Part V
Playful Learning and Interactive Digital Tools

Play invites us into learning with curiosity, creativity, and joy. This part brings together game-based methodologies, interactive digital tools, simulation activities, improvisational exercises, and role-play activities that translate systems thinking and sustainability concepts into lived participatory experiences. These approaches cultivate engagement, experimentation, and collaborative problem-solving. By balancing structure with openness, these methods create space for iterative learning, productive failure, and the co-creation of meaning, all vital for navigating the complexity and dynamism of real-world sustainability challenges. Within the framework of sustainability science, playful methods serve as powerful vehicles for adaptive learning. They help embody abstract ideas, reinforce interconnections, and foster collaboration across disciplines and perspectives. These activities can play an important role in building resilience and generating insights that are both critical and imaginative. In regenerative education, play becomes a practice of possibility while reminding us that profound change can emerge through creativity, participation, and collective joy.

Chapter 41
Web-Based Application to Foster Pro-environmental Behavior via Empathy

Augusta Gaspar and Ana Paula Claúdio

41.1 Introduction

Forests are disappearing at a faster pace than ever. As the problem stems not only from business interests but also from consumer's lack of knowledge and weak motivation, we put Emotion Psychology into practice, creating a web page that addresses the problems and solutions for the world's great forests through gamification—to motivate people across the world to make a difference, raising their awareness and empathy, as a means to change small habits that nonetheless have great environmental impact.

Gamification and specifically serious games are increasingly popular in education and the cognitive sciences—they have assisted Psychology over the years in Psychoeducation, promotion of mental health, prevention, and therapy—for example, in the treatment of phobias (Claudio et al. 2018) or prevention of bullying (Raminhos et al. 2016), among others. Our approach is to provide a digital, universal, carry-on, and fun tool that uses gamification strategies to entice and engage young people. This chapter describes the *SaveDforest* Project, an inter-university project, that entails contributions from Emotion Psychology and Informatic Engineering. Its main goal is to change players' everyday behavior in ways that contribute to saving forests. Another important goal is to build an understanding that one's actions in one part of the world impact ecosystems in distant parts of the

A. Gaspar (✉)
Universidade Católica Portuguesa, Católica Research Centre for Psychological, Family and Social Wellbeing, Lisbon, Portugal
e-mail: augusta.gaspar@ucp.pt

A. P. Claúdio
University of Lisbon, Faculty of Sciences, LASIGE, Lisbon, Portugal
e-mail: apclaudio@ciencias.ulisboa.pt

J. Bentz, J. Ristić Trajković (eds.), *Imagining, Designing and Teaching Regenerative Futures: Art-Science Approaches and Inspirations From Around the World*, Science for Sustainable Societies,
https://doi.org/10.1007/978-981-96-9029-9_41

world, which, in turn, may be vital to the health of the entire planet and critical to environmental sustainability. The project entails a webpage on forests, from which several tools can be accessed and it lodges a game with the same name saveDforest. The first version of the game—*saveDforest 1.0* —is focused on the Borneo forest and is ready and is currently under testing by users and experts and is freely available online. The game can be found here https://savedforest-temp-test-2.onrender.com/. The developer team envisages a continued improvement of the webpage and the incorporation of games that include other forests. The development of *saveDforest 2.0* is underway focusing on the Amazon forest. This approach aims to promote knowledge and empathy toward critical world forests—saving forests one by one.

41.2 Transformative Potential

Specificity and "baby steps" make a difference—our methodological approach to building SaveDforest departs from the notion that, to change behavior, we need not only to provide critical information and change perceptions of the environmental crisis but also to be specific on the goals to achieve and the concrete actions one might take to reduce/minimize anthropogenic harm to the environment. Recent research shows that behavior change strategies have to go beyond changing attitudes to provide facilitators of action (Albarracín et al. 2024). In that regard, saveDforest provides not only keys to concrete actions but also points to foreseeable solutions, in addition to frequent and constructive feedback. The specificity of the actions is linked to the specific forest one sets out to save—and the overarching concept of the webpage is saving the great forests one by one—Borneo, in version 1, Amazon in version 2, and in the future, we hope to embed other games targeting other forests.

Emotion drives intrinsic sustainable motivation and persistent behavior—games often resort to gamification strategies only to keep the player motivated, such as showing points, offering badges, and giving positive feedback. While this game design also does that, we also use content and media that have been chosen to elicit emotion.

Within the environment topic game scenario, there are some games available or partly available that were developed to contribute to environmental education, and at least one has been validated for this purpose (Eco, Fjællingsdal & Klöckner 2019). They accomplish some information acquisition on ecosystems, recycling, or sustainable agriculture in an enjoyable activity. However, they usually do not use emotional mediators that may produce persistent changes as empathy does, and resort to challenge, the motivation to play, external rewards, and visual immersion, as games do.

To change behavior beyond the game and in real life, one has to change motivation. The specific type of game we propose and designed is built upon the assumption that no behavior change is truly possible, and sustainable over time, if not driven by emotion. Emotion is the best predictor of intrinsic motivation, and ethical behavior has largely relied on affective empathy much more than on moral principles (Gaspar 2016). Thus, the saveDforest game is tailored to enhance two dimensions of empathy: (1) affective empathy, which entails an emotional response, feeling emotionally affected by others' experience, and (2) cognitive empathy, which entails the ability to understand the state and perspective of another being. Empathy, and especially a subdimension of affective empathy called empathic concern, changes people's actions in ways that strictly informative or hedonic approaches do not. Consistent with this, while playing the player will meet triggers of these two dimensions of empathy—empathic concern and cognitive empathy. These dimensions are measured before a player engages in the first game session, and again once the player has played at least three times. Additionally, throughout the game, and after each scene (scenarios with powerful information), the player rates how he/she feels in a validated widely used pictoric emotion scale that appears in a pop-out window.

The game links to documentaries and reports, and uses music and sound effects to produce impact—either driving attention when providing critical information or offering quizzes to test the player's knowledge.

Realism and links to the real world with a transformative potential—the game entails a storyboard and presents a series of scenarios created with real photos of Borneo presented on a full-size screen along with intermittent informative content, and it provides challenges and questions to the player. The player makes choices on optional answers, what to learn, and what to do. The written content in the storyboard is aimed at creating interest and cognitive empathy. In each scenario, several different options to move forward are provided to the player. Some choices are more valuable to earn points, as they will expose the player to videos or non-governmental organization (NGO) webpages with more intense content on forest animals such as orangutans or Bornean elephants, or local Bornean people, and their plight—which are intended to induce affective empathy. Other choices are more hedonic (e.g., touristic), and the player drives forward to what she/he wishes but earns no points. One option is to take a quiz—if one feels confident about his/her knowledge on the topic—which is a way of rewarding both cognitive empathy (understanding) and affective empathy (as one can take this option after choosing first to see a video) in an objective way. The quiz is presented with a background animation showing an orangutan frolicking in the forest, with a combination of forest sounds and other visual effects.

All in all, the game provides a journey through images, videos, and written content, along which the player is challenged with optional paths and quizzes on specific knowledge or on how one feels. Even making the least valuable choices (the easy paths, moving forward, or seeing pleasurable videos only), the player will finish the game learning a substantial amount of information on the Borneo forest, the plight of its living beings, taking along important tips on how to change some of his/her impacting consumer habits.

To most players, the game will hopefully attain the goal of promoting empathy toward forest animals and people who are affected by forest destruction. This is the first step toward transformation. The first step is understanding the potential regenerative power that lies in the adoption of consumer choices that do not contribute to further damage to the Borneo Forest—the game promotes this not only by raising empathy but by rewarding answers and choices that convey knowledge of the problem and pro-environmental behavior choices through active learning. Although this version of the game focuses on the Borneo forest, its animals, and the native people of Borneo, it also raises awareness of the global effects of forest destruction and encourages pro-environmental behavior in general.

41.3 Application

The *saveDforest* application project targets mostly young audiences but is suitable for any age group. It is an educational tool that can be easily used in a school setting, at home, in a library, or any other place where there is a computer with a fast internet connection.

The *saveDforest* 1.0 game has been designed for a target audience of adolescents and young adults and young adults. Notwithstanding, the game can be easily played by older children and preadolescents and remains challenging and enjoyable to other adults. So, educators can incorporate the game into school activities or homework. They should promote 30–60-minute sessions each time students log in, as the game entails large amounts of information acquisition having in mind that the project games are available in English only.

The game is ideal for triggering, supporting, and enriching discussions on forest protection, but also to monitor behavior changes—for example, asking players about any changes in their environment-related habits as a result of the experience of playing this game.

NGOs interested in forest protection can create links to the project webpage to further reinforce their mission. By lodging the games in a web portal that conveys relevant environmental information and that links to several NGOs, we provide a community tool solely based on academic knowledge and critical assessment, with no commercial or institutional interest attached—a public service. That is an important reason for educators and NGOs (Table 41.1).

Table 41.1 Method overview

Main purpose
Awareness building/motivation and behavior change
Gained competences
Knowledge acquisition on environmental issues and key stakeholders; practical consumer habits in everyday life that impact the Borneo forest
Educational setting
Informal (format—The game is introduced in a website, that can be presented in a brief workshop, conferences, or through a link from a stakeholder website—School, environmental organization, etc.)
Space requirements/restrictions
Indoor activity; the game is available in English and players need to have a strong domain of the English language.
Resources and necessary materials
Requires a computer (preferably) or, alternatively, a tablet, with access to the internet.
Number of participants
The game is played individually, but small groups (up to 12) may be organized to compete and discuss what they have learned from playing the game
Facilitator competences and skills
No specific skills are required, other than experience using the computer (preferably) or tablet and the keyboard
Participants skills/age/competences
Target audience—Students and young people starting in adolescence from 12 to 29 years old, but extensive to any age. Ideal for environmental education settings
Duration
The game should be played for a minimum of four sessions throughout the week. Each session should take 30–60 minutes. The best results should be obtained with further sessions and the entire game being played at least twice

References

Albarracín D, Fayaz-Farkhad B, Granados Samayoa JA (2024) Determinants of behaviour and their efficacy as targets of behavioural change interventions. Nat Rev Psychol 3:377–392. https://doi.org/10.1038/s44159-024-00305-0

Cláudio AP, Carmo MB, Gaspar A, Teixeira R (2018) Expressive avatars in psychological intervention and therapy. In: Ursyn A (ed) Interface support for creativity, productivity, and expression in computer graphics, IGI Global, Hershey, pp 27–48. https://doi.org/10.4018/978-1-5225-7371-5.ch002

Fjællingsdal KS, Klöckner CA (2019) Gaming green:the educational potential of eco—a digital simulated ecosystem. Front Psychol 10:2846. https://doi.org/10.3389/fpsyg.2019.02846

Gaspar A (2016) Morality and empathy vs empathy and morality: a quest for the source of goodness in phylogenetic and ontogenetic contexts. In: Silva S (ed) Morality and emotion: (un) conscious journey to being. Routledge, pp 62–82

Raminhos C, Cláudio AP, Carmo B, Gaspar A, Carvalhosa S, Candeias MJ (2016) A serious game-based solution to prevent bullying. IJPCC 12(2):194–215. https://doi.org/10.1108/IJPCC-04-2016-002

Chapter 42
VROOM: A Group Activity for Sharing Energy, Maintaining Focus, and Encouraging Play

Ben Stasny

42.1 Introduction

VROOM is an improvisational and embodied activity that requires participants to gather in a circle and pass an invisible *ball of energy* using a variety of phrases and gestures. There are three purposes for this activity—the first is to build energy among participants, the second is to maintain focus, and the third is to encourage play. VROOM is an impactful exercise to use when working with groups of people in educational, theatrical, and social justice contexts.

The concept of a *ball of energy* is a malleable, shape-shifting focal point for the participants. There are specific ways to pass the *ball of energy* around the circle. A list of phrases and their corresponding actions can be found in the application section. Depending on the group and the time restraints, the facilitator can introduce as many or as few of the actions as they desire.

42.2 Transformative Potential

VROOM is a combination of different improvisational activities typically utilized by theatre practitioners. It adheres to guidelines commonly found in improv. Some guidelines are universal such as deferring success to the group and engaging in non-judgmental participation, and others are more recognizable and specific to theatrical improv like saying "yes, and" or building a collective world. However, all guidelines can be found in a variety of situations where groups of people gather to share

B. Stasny (✉)
University of Colorado Boulder, Boulder, CO, USA
e-mail: Benjamin.Stasny@colorado.edu

J. Bentz, J. Ristić Trajković (eds.), *Imagining, Designing and Teaching Regenerative Futures: Art-Science Approaches and Inspirations From Around the World*, Science for Sustainable Societies,
https://doi.org/10.1007/978-981-96-9029-9_42

energy toward a goal, be it improvisational theatre, social justice organizing, or arts-based research and practice.

Social justice facilitator Adrienne Maree Brown (2021), in her book *Holding Change*, writes, “in front of you is a ball of energy. It is yours, an extension of you. It hovers in the air, shifting with your attention” (p. 6). Brown’s concept of “a ball of energy” is used in this context for social justice facilitation and can inspire a variety of disciplines in which group collaboration is necessary. Samuel Jaye Tanner (2021) is an improviser and educator who utilizes improvisation in pedagogical contexts geared toward literacy and social justice work. He writes, “the improvisational practices and ethos … offer a way to think about pedagogy as an ongoing rehearsal of the uprooting of reality, the cultivation of comovements by groups of people toward emergence, surprise, wonder—toward magic” (p. 13).

Not only is VROOM a useful tool for passing energy and maintaining focus but it is deeply silly and fun which can be important qualities to emphasize depending on the context. I consistently return to this question posed by brown (2021), “what kinds of humans do we want to be and become? What kind of structures and practices do we need to be in to generate that future?” (p. 117). I have found that embodied improvisational activities like VROOM help support structures and affirm practices that can (re)generate hope for the future through the act of play and being silly.

Because VROOM is a highly flexible and collaborative exercise, it allows for the addition of new phrases/actions specific to what sort of work the group is engaging in. For example, in the fall 2023 Improv course I taught at the University of Colorado Boulder, students participated in an exercise that required them to embody the top greenhouse gases (GHGs) in our atmosphere, not exactly *lighthearted* subject matter but used in the context of promoting silliness on a dense subject. One morning and without any prompting, a student integrated a few of the embodied elements of the GHG embodiment exercise into VROOM and everyone accepted this new addition without questioning or resisting. From then on, it remained an action for the students to utilize in VROOM. This example emphasizes a crucial tenet for healthy improvisation—saying “yes, and.” Kevin Leander (2023) is a scholar in literacy and improvisational pedagogy; he writes that “to speak of attunement between individuals is not to speak of a static state, but of constant modification and adjustment in time” (p. 257). The students in my improv class were attuned to each other’s energies and were able to modify VROOM without any planning or questions.

42.3 Application

VROOM can be taught in 5 min, depending on the size of the group, the allotted time, and how many actions the facilitator chooses to include. VROOM requires no physical materials or preparation aside from knowing the phrases and gestures in order to teach them. It works best with groups ranging in size from 5 to 20 people. If there are over 20 people in the group, I suggest teaching it as a full group and then

Table 42.1 How to play

Phrase	How to play
VROOM	Clap hands together and loudly say "vroom!" to the left or right, can only travel in the direction from which it came. The standard way of passing the ball of energy
Reversal	Make a sound similar to a car slamming on its brakes and move your hand in a counter-clockwise fashion; this action reverses the direction of the ball of energy
Oil slick	Move your body in a gyrating motion while speaking in a low voice and saying "oil slick." the ball of energy skips the person next to you and continues in the same direction
Speed bump	Yell this and all participants must jump into the air. A person who calls "speed bump" holds the ball of energy until they choose a new action
European intersection	Yell this, everyone randomly moves to a new position in the circle. A person who calls this holds onto the ball of energy and continues it after all have arrived in a new position
Stage presence	Name one person in the circle and announce "stage presence!" both people switch positions in a dramatic way, and the person who was named continues the ball of energy
Call and response	A collaborative call and response that is unique to the group of participants

splitting it into smaller groups when it comes time to do the exercise. For facilitators, it is important to know that this exercise is a living thing. It can and should constantly be adapted depending on how frequently the group is meeting. It can be played indoors or outdoors. Below is a brief list of standard phrases and actions (Table 42.1).

One thing to keep in mind when leading VROOM is that it can easily become too chaotic depending on the group's ability to maintain focus and discipline. While younger people can be more susceptible to less-controlled transfers of energy, I have observed this happening among a variety of ages and contexts. As a facilitator, I continuously remind them that it is as much an exercise for maintaining focus as it is for increasing the energy and playful atmosphere of a group.

There are any number of energy-sharing exercises and activities that are similar to VROOM. Facilitators might read this and think of exercises they themselves have led for groups of people that operate similarly. What makes VROOM unique for a variety of contexts is that it utilizes many different phrases/actions and can be adapted by any participant, as evident in the earlier example.

VROOM can be a powerful tool for encouraging creativity, imagination, group flow, and sharing energy. Moreover, its impact on future-building and encouraging people to come back to the table is powerful no matter the ages of the participants. When groups of people engage in embodied actions and activities, they tend to be more present, grounded, and collaborative in the space. This is crucial when conceiving of transdisciplinary solutions to current issues (Table 42.2).

Table 42.2 Method overview

Main purpose
Building energy among participants, maintaining focus, and encouraging play
Gained competences
Collaborative workshops, community engagement, art-based practices, imagination, embodiment practices, listening and giving focus to group participants, and sharing energy
Educational setting
None
Space requirements/restrictions
Indoor or outdoor friendly. No requirements other than enough space to form a circle with the participants
Resources and necessary materials
None
Number of participants
5–20 (can be done with more but will require splitting up the group)
Facilitator competences and skills
Creativity, imagination, some familiarity with game playing, managing the passing and sharing of improvisational energy, attunement to group dynamics, and commitment to improvisational guidelines
Participants skills/age/competences
Suits all skill levels and ages
Duration
Typical gameplay lasts between 5 and 10 min, usually as a warm-up before more intense community-building activities. But can also last as long as the group cares to play

References

Brown AM (2021) Holding change: the way of emergent strategy facilitation and mediation. AK Press, Chico

Leander K et al (2023) 'We got so much better at reading each other's energy': knowing, acting, and attuning as an improv ensemble. J Learn Sci 32:250. https://doi.org/10.1080/10508406.2022.2154157.257

Tanner SJ et al (2021) Ways with worlds: bringing improvisational theater into play with Reading. Read Res Q 56(2):1–16. https://doi.org/10.1002/rrq.304.13

Chapter 43
Dancing in the Ruins of the Future

Marcus Bussey

43.1 Introduction

The present has a strong hold on our consciousness. We find it very hard to envisage beyond the given realities we inhabit; thus, regenerative futures are usually built as an assemblage from the data (images, media, family and friends, etc.) we gain from our contexts. Such futures tend to be "Business as Usual" tweaked with familiar hierarchies, recognizable technologies, and values that do not rock the boat. But deeply regenerative futures need to rock the boat. This activity challenges us to rock the boat by getting people into their bodies and sowing the creative potential to "hack" the dominant narratives that silence alternatives.

First, this activity emerges from the reading of culture as an algorithm. Thus, to "hack" culture makes sense as culture functions algorithmically as a sense-making set of codes that institute any given "reality." Reality is simply, as Yuval Harari (2015) demonstrates, an imagined order. There is no law that says that this reality is the only one, or the best one. We are born into realities that we accept as given. But if you live long enough, you will see that realities are constantly changing and alive to both internal and external forces. Futurists call such forces drivers (Godet 2010). We forget that drivers are not always impersonal and that we as individuals and communities can work with drivers, even act as drivers of change. Regeneration thus becomes a dimension of human and natural systems. We inhabit both.

Second, this activity leverages the deep cultural conditioning that we embody. Rutger Bregman argues that humans have domesticated themselves. That we have

M. Bussey (✉)
University of the Sunshine Coast, Sippy Downs, Australia

J. Bentz, J. Ristić Trajković (eds.), *Imagining, Designing and Teaching Regenerative Futures: Art-Science Approaches and Inspirations From Around the World*, Science for Sustainable Societies,
https://doi.org/10.1007/978-981-96-9029-9_43

been conditioned to dominant hierarchies and toward collective behaviors (Bregman 2020). Taking this understanding as an invitation to "hack" our conditioning this activity gets us moving: *improvising, dancing,* and *playing*. Workshopping sustainable education, and sustainability in general, tends to be conceptual. In our heads—yet, to invite the body into this process can be liberatory and joyful. As adrienne maree brown argues so eloquently (2019), we cannot underestimate the power of joy to disrupt dominant conditioning. Regeneration is a return of life energy to entropic systems. Joy is a powerful component of this process.

Third, and this is where the paper's title comes from, culture is a cumulative data set of everything that has gone before: the past. It is also localized to sites defined by language, historical experience, geography, ethnicity, faith systems, and so on. Our experience of this "past" is always partial. It can be understood as a set of ruins that we move through unconsciously as we live our lives. So, we in the West inhabit the ruins of a Judeo-Christian world, set now in a largely materialist capitalist, technologically powerful, and colonial world. We have inherited a hybrid of religious, spiritual, and also materialist imagined orders that Joseph Henrich insightfully describes as making us WEIRD people (2020), an acronym for Western, Industrial, Educated, Rich, and Democratic. Yet these ruins are not simply in the past or around us, they extend into our futures. We do, indeed, inhabit the ruins of the future.

43.2 Transformative Potential

Regeneration springs from the natural processes of healing, connecting, reconnecting, and experimenting with the materials at hand. New configurations—like a kaleidoscope playfully tumbling and patterning—spring into being. In a recent video philosopher (and Chess master), Jonathan Rowsom describes pathways beyond the "metacrisis." His analysis is hopeful and liberating, but he does not point to the embodied nature of conditioning that a crisis culture deploys to keep us all off balance. I am using the term "dance" and the phrase "dancing in the ruins" to indicate that our embodied rebellion against conditioning has transformative and regenerative potential. It releases all that joyful pent-up energy that living beings possess but which is often denied in the world of contained, conditioned, behavior.

Ruins are site-specific. Some are very local, whilst others can be planetary. This *cosmolocal* dimension is all around us and experienced as material, cultural, and ecological flows. Ruins have structure and form and in futures parlance are often projected as images of the future. The metacrisis offers us an image that disempowers, as does the term *Anthropocene.* In my work, it is essential that we rebel against images that foreclose on our futures as individuals, communities, and relationships with the more-than-human. To dance in the ruins transforms our relationship with these forms. Celeste Nazeli Snowber, a site-specific poet and embodied academic, describes aspects of this:

> I am after connections between humus, humility, and humour, where everything arising is connected to the body and the earth. We are storied human beings, and our stories are found in one another… Each moment is a place to return to what it means to be deeply alive. All life is material for creating. When we expose our selves to aliveness, we become alive. (2022, p. 13)

Pedagogically moving the body spawns connections. It releases us from constraints. It provokes us to step out into worlds of wonder and possibility beyond the stereotypes of body, being, problem-solving, and surviving that dominate current discourse. Dance and movement are surprisingly liberating and sometimes confronting (Bussey 2020). I use movement to shake up the psyches and souls of those I work with. To regenerate is like rewilding the soul, the story, the community, and the vision. So, if you as a pedagogue focused on regenerative futures want to unleash the human potential to go beyond, innovate, reimagine, and reconnect with meaning and the living world, then dancing in the ruins will take you there. Trust me LOL.

And if you cannot do that, here are some wise words from InterPlay co-founder and exponent of body wisdom, Cynthia Winton-Henry:

> In a world that has nearly lost the birthright of dance, my gift for it seems unexplainable and random, especially as I age. Yet, as I research art, dance history, theology, social justice, leadership, education, and spirituality across contexts, I've discovered the kinaesthetic, spiritual intelligence that belongs to all people, all ages, all of the time. (2020, p. 348)

Ruins are creative, even transformative spaces to dance in. They stretch into the past and into the future. To dance in them is to both sacralize them and also tap into sources of energy, imagination, and power that have the potential to regenerate our world. So, let us dance.

43.3 Application

We met up in Lisbon, in July 2023—a group of 48 eager graduate students and 10 teachers. Our task was to work with the concept of Urban Imaginaries. Every day I took staff and students through an embodied "wake up" that pointed (in a loose way) to themes we were to deal with that day. So, over the 5 days of the course, we built on the tools of InterPlay and developed a shared embodied "grammar." The urban imaginaries of all were activated, fed, and nurtured toward regenerative possibilities in our personal and professional practices.

Cities are alive and grow out of the ruins of the past and embody in the present all the signs of the ruined present, and old buildings are reconfigured, destroyed, or repurposed, and new ones emerge. The future also contains these ruins as a natural outgrowth of current actions and our imaginative engagement with possibilities presented by the urban environment. Here is a rough template for how I led people to dance in the ruins of the future.

Time: 45 min × 5 days
Site: Botanical Gardens in Central Lisbon, Portugal under a delightful large fig tree
Group: Any size from 2 to 50 or even more
Equipment: Large Bluetooth speaker and my smartphone (with suitable playlists)
Music drawn from multiple genres focused on raising or lowering energy, creating atmosphere and emotional arousal (Examples: *Wake Me Up* by Simply Three; *I Giorni* by Ludovico Einaudi; *The Spheres* by Ola Gjeilo; *Krigsgaldr* by Heilung; and *Lamentation for a Lost Life* by Max Richter)
Preparation: Two YouTube clips Why Movement and Urban Futures Imaginaries
Instructions: This is always improvised but runs something like the following:

> "Please stand in a circle" (we always start in a circle)
> "This session is designed to loosen up and clear our heads." Then I begin with a laugh… "Who here can walk?" They all laugh, and some put up their hands. (Of course, in some settings we may have folk in wheelchairs or with other physical issues … for them the question is who is/how you mobile?)
> "I am about to play a fun music track and I invite you to walk in any direction you wish. When I call out an instruction such as 'Stop'; 'Turn at right angles,' 'Walk backwards'; 'Run' etc please do so. Oh, and if you bump someone please say: 'Thank you' (that always gets a laugh)"

We do this for up to 5 min and then we stop, and I invite people to pair up with someone they do not know and share how they found this activity—one at a time, that is, one speaking and one listening. I use words like witness for this. Witnessing is key to awakening our perceptual capacity and being still and just absorbing the communication of the speaker.

I then repeat the activity adding a narrative that we are walking through a city, witnessing, noticing, sensing, and opening our embodied beings to the aliveness of the urban. I introduce the concept of "easy focus" where we allow the entire experience in a perceptual and embodied way (see: Porter 2020, p. 335). Again, we stop and pair and share. This activity is called "Walk, Stop, Run" (Winton-Henry 2009, p. 155) and begins each day but is modulated to fit with specific questions for each day (these are my questions for the 5 days, I do not share them rather they are my compass):

- How do we negotiate the Urban?
- How can we find stillness and solitude when we need it?
- What ruins of the urban do we inhabit?
- What ruins of the urban inhabit us?
- Honoring the Pasts, Presents, and Futures of urban imaginaries

Two key elements are added across the 5 days.

1. One-hand dancing—this is a beautiful form of movement that invites each individual to focus inward and express that inner universe with only a single hand. Dance becomes distilled in movements that flow and express a feeling of being alive but is beyond language and even "purpose" as in the control of the mind. See Winton-Henry (2009, p. 15).
2. Dancing on Behalf of—I introduce this activity on Day 4. On Day 5, we use it to close the morning exercise. To Dance on Behalf of requires two people or groups. One speaks of their concern, vision, hope, calling, fear, or whatever, and the other acts as a witness. When the speaking is done, the witnesses dance to a piece of music reflecting in some way the feelings, insights, and emotions evoked by the speakers. This can be cathartic and liberating for both the speaker and listener. It was profound for me to lead and witness this conclusion. The air was still and thick with a sense of "Wow!" when the activity concluded. For more details, see Winton-Henry (2009, pp. 101–102).

43.4 Conclusion

Dancing in the Ruins is an open-ended perpetually surprising and life-affirming practice. Regenerative futures emerge at the interface of the ruins we as individuals and collectivies inhabit. To hack our cultures is a necessity and a key ingredient to releasing alternatives and empowering individuals to reclaim their right to cultural agency. Society, culture, and civilization have grown up around our bodies, yet we forget how intimately these processes and our own psyches are embodied. Bringing the body back into rethinking our world is therefore a powerful way of regenerating all our futures.

Excellent resource: Cynthia Winton-Henry (2023) The Art of Ensoulment: A Play Book on How to Create from Body and Soul (Platypus Publishing, Fair Oaks, CA) (Table 43.1).

Table 43.1 Method overview

Main purpose
Open participants to joy. Moving the work away from the head and into the body. Freedom emerges and creativity is linked to action
Gained competences
Co-creative purpose and reflexivity. Trust in the heart and increased optimism. Deepened appreciation of the weaving of pasts, presents, and futures into our consciousness and experience of "reality"
Educational setting
Formal and informal settings both within education and also in institutions and communities
Space requirements/restrictions
Outside activities/indoor activities/online
Resources and necessary materials
Music, space
Number of participants
Group or individual work, group size, number of groups; from one person to community
Facilitator competences and skills
Knowledge of InterPLay or related embodied process work
Participants skills/age/competences
Community/students/policymakers/researchers/designers/practitioners, and so on; all people of all ages and walks of life
Duration
Short sessions such as 45 min or can be done intensively over a number of days

References

Bregman R (2020) Humankind: a hopeful history. Bloomsbury Publishing, New York
Brown AM (2019) Pleasure activism: the politics of feeling good. AK Press, Chico
Bussey M (2020) The Heirs of Tiresias: grace, the uncanny and transformative action. In: Bussey M, Mozzini-Alister C (eds) Phenomenologies of grace. Springer, pp 19–40
Godet M (2010) Future memories. J Technol Forecast Soc Change 77(9):1457–1463
Harari YN (2015) Sapiens: a brief history of humankind. Harper, New York
Henrich J (2020) The WEIRDest people in the world: how the west became psychologically peculiar and particularly prosperous. Penguin UK
Porter P (2020) Grace moves: what wing it! Performance ensemble taught me about the relational nature of grace. In: Bussey M, Mozzini-Alister C (eds) Phenomenologies of grace. Springer, pp 331–343
Snowber CN (2022) Dance, place, and poetics: site-specific performance as a portal to knowing. Palgrave Macmillan, New York
Winton-Henry C (2009) Dance: the sacred art. Skylight Paths, Woodstock
Winton-Henry C (2020) Grace operatives: how body wisdom changed the world. In: Bussey M, Mozzini-Alister C (eds) Phenomenologies of grace. Springer, pp 345–357

Chapter 44
AROSA: Argumentative Role-Play

Gloria Gaviria and Susanne Bosch

44.1 Introduction

> People only play where they are human in the full sense of the word, and they are only fully human where they play—Friedrich von Schiller

With the game AROSA-Argumentative Role Play of the Involved, we developed an interventionist-participatory art project in 2019. The basic idea is to develop a form of artistic practice critical of capitalism for and with people in a neighborhood. At its core, the game invites people to develop a vision of possible collaborative and co-creative ways of shaping their living environment together by changing roles and getting agency of the argumentation.

AROSA is a participatory method that invites people to consider everyone's voices from a new perspective. The game aims to support participation and decision-making processes. The game principles focus on listening and finding arguments from a new perspective or role taken with the attitude of creating awareness of new issues and new ideas for new visions. The result is a discussion and dialogue process that enables participants to experience inspiring, surprising processes and ideas, as well as to experience new alliances and collaborations and to make decisions together for the better future development of a cause.

The aim of a round of playing is to negotiate an agreement together. Co-intelligence and co-creativity are increasingly desired and required in urban processes. People often experience powerlessness and a lack of understanding of their position. In this game, decisions are learned from a different perspective. The ability to negotiate is

G. Gaviria
Urban Planer, Urban Manager and Architect, Berlin, Germany

S. Bosch (✉)
Independent Artist, Berlin, Germany
e-mail: bosch@susannebosch.de

J. Bentz, J. Ristić Trajković (eds.), *Imagining, Designing and Teaching Regenerative Futures: Art-Science Approaches and Inspirations From Around the World*, Science for Sustainable Societies,
https://doi.org/10.1007/978-981-96-9029-9_44

something that is tested performatively from the role of both, the powerful and the powerless.

44.2 Transformative Potential

The game is a solution-orientated building block of grassroots democratic decision-making and urban development, in which concrete measures and strategies can be developed, as well as empathy for the arguments of the actors involved. The game is an alternative tool for discussing arguments in a didactic way. It can be employed as a participatory method or as a participatory tool, enabling all stakeholders in the city's development to engage in discourse on pertinent issues, enhance the decision-making process, and collectively address the community's needs.

The advantage of role-playing is that it involves roles that are identified in advance through on-site research. These will be recognized by the players during the process. On the other hand, these roles are embodied by other local people. This enables a new discussion about the issue.

The learning effect occurs at the moment when the participants explore other perspectives and experience facts that they have not been directly confronted with before. Participants learn by putting themselves in the shoes of others, recognizing how they have traveled this path and what challenges lie ahead. Practice is the main component of the game and provides an opportunity to reflect on and confront our listening, communicating, arguing, and mediating skills. The game is flexible because it changes, evolves, and adapts to the context. Not all conflicts in urban development are the same, even if they look similar. The main difference lies in the diversity of the scenario's protagonists.

To summarize in brief:

- There is often a general lack of understanding about the process itself.
- There is often a feeling of a lack of transparency and therefore incomprehensibility about the motives behind the decision.
- People can get to know different positions "from inside" through embodying this role.
- People can understand decisions from a different perspective.
- People can experience power*/powerlessness in their roles.
- By listening to all arguments and opinions, people will be able to make better decisions.

Our approach is the joint development of living spaces through dialogue, co-intelligence, and co-creativity. This results in a competence of many for the context, which is desirable in these processes and necessary due to the complexities. However, the ability to negotiate is something that has to be practiced, from the role of both the powerful and the powerless. The AROSA game helps in practicing (1) the inner attitude, (2) the language (technique of arguments, negotiation techniques, and forms of communication), and (3) the process expertise. Reflection questions supporting the game can be: What do I want/can I negotiate? What does the other person have that I don't? What do I have that the other person wants? What do I not want to give up under any circumstances? What is not so important if I don't get it? Should I cooperate? How and with whom can I cooperate? Is it better to work alone or in a group to achieve a goal? How can I get to be listened to? Do I take the time to listen to others? How can I keep myself motivated and motivate others?

The advantage of the game is that it is fun, enjoyable, and an open offer. It enables a broader field of action (further than everyday life often allows).

Taking on a role makes it possible not only to live out convictions and ideas but also to overcome them when it comes to finding jointly constructive solutions. It is about recognizing individual and collective potential. Playing with role changes makes it possible to reshape a situation through expression and arguments.

The game offers the opportunity to experience power in a different way: power in the sense of molding as well as power in the sense of being able, being able to act differently in a situation.

44.3 Application

AROSA is a participatory and interdisciplinary art form designed to change the state of society. It is a card game with action instruction cards for 9–12 people and a moderator role. The game lasts3,5 h, with the players reflecting on the experience in the last half hour. The starting point is developed according to the context: It is a concrete local urban planning scenario. Alternatively, topics such as mobility, public space, housing, infrastructure, social issues, education, climate, and so on can be discussed.

44.3.1 Who Plays, When, and Where

The game is played in real contexts in which urban situations are discussed, such as round tables in neighborhoods. The game can also be performed in groups that are institutionally connected and working on a topic.

44.3.2 Preparation

Research a specific problem or topic in a real context such as a neighborhood or an initiative. It is good if the topic is actually widely and controversially discussed, such as the current mobility transition. Write up a scenario. Identify 9–12 stakeholders in the topic/problem: What position do they represent, who are they, and what are their arguments. Compile their arguments. Anonymize these in a "role" in the game. Make cards for all the roles and their potential actions.

It is essential to ensure that the arguments presented in the card games are crafted in a manner that encourages or triggers the involvement and participation of other roles in the game. The arguments may be formulated in a variety of ways, including those that are provocative, collaborative, collective, or individual. The objective of such formulations is to elicit automatic reactions or the activation of other participants to achieve collaborations, establish strategic alliances, set up common goals, and ascertain the interests of all active members in the development process.

It is recommended that card games are named with active verbs to ensure that participants are aware of how to use the action cards and what they are meant to achieve.

Finally, action cards for the moderator should be developed without description and with simple words or active verbs such as cooperation, participation, setting priorities, democratic process, listening, activation, and so on.

Additional elements such as maps or photos of the situation can be used to better understand and illustrate the problem or the development of new plans.

44.3.3 Game Steps

44.3.3.1 Introduction (5 Min)

Introduction of the facilitators and the intention of the game.

- To jointly develop a good solution for the topic/problem with a view to the different needs of the people.
- Using a game as an instrument of the participatory process to discuss real arguments and ideas and to think about further arguments.
- The aim is to hear arguments from the different roles in the scenario, discuss them, and initiate/prevent actions or find common solutions.
- It is not about winning or losing, but rather about community development through joint dialogue.

44.3.3.2 Game Description (10 Min)

The objective of the game is to find out what the reason for today's discussion is and what the rules of the game are.

The scenario is explained. Questions are also clarified before the game begins.

- Introduce an imagined context for the role game to take place, like a citizens' assembly, for example, where the present group is asked to approve an application for the neighborhood.

- Each player takes on a role. This role is described in a few words on a card. Each player voluntarily chooses a role that does not correspond to his/her real-life role. The roles of the game are introduced.
- A maximum of 12 people with different roles will interact with each other. A moderator will accompany the game and intervene from time to time, either to ensure that everyone has their say or to interject information.
- Every player receives four different action cards + a joker. These action cards invite players to act out an attitude, for example, to argue, question, discuss, support, reflect, and confront. This means that every time you speak up, one of these cards should be actively used. To do this, place this card face up in the center of the table. Address your questions or invitations to respond directly to the colleagues at the table. Invite them to respond or react. Each card can only be used once.
- All players can activate their joker once at any time. The function of this card is to help express an idea that you consider important for the course of the game. This is a powerful card that you should use skillfully.
- After receiving and reading one's action cards, each player takes a moment to develop their character and possibly a game strategy.
- The players start to play.
- It is possible during the game to create strategies in small section groups with the objective of achieving joint synergies, listening to the contributions of others, and presenting arguments collectively to the group. This allows for direct exchange with a role that, in real life, could be challenging to achieve or a role that is typically contacted to reinforce and complement arguments.
- The moderator can intervene with an action card in accordance with the observations made during the game and can steer the game in different directions.
- Reflection phase: For this phase, it is important that we return to our real persona. We will reflect on the experience so far and bring a new perspective, present new arguments, and talk about the lessons learned. We evaluate ideas and experiences together so that we get a comprehensive perspective on the possibilities.

44.3.3.3 Procedure of the Game

- Round of introductions of all people present, distribute to game tables (6–9 people) (15 min)
- Read out the game scenario (5 min)
- Select roles (5 min)
- Introduce the roles drawn (15 min)
- Read and understand action cards and jokers (5 min)
- Play (90 min)
- Break (15 min)
- Reflection (45 min) (Table 44.1)

Table 44.1 Method overview

Main purpose
Developing through role-play a vision of possible collaborative and co-creative ways of shaping their living environment Getting agency of the argumentation Negotiating an agreement together
Gained competences
Participating in decision-making processes Listening and finding arguments from a new perspective Negotiating through testing the argument performatively
Educational setting
Informal: Role play with ground rules A facilitator to hold the space
Space requirements/restrictions
Table for the map Chairs for everyone
Resources and necessary materials
A researched, site-specific scenario that holds tensions, possibly conflict A set of card games with four action cards for each player plus jokers A set of 9–12 characters and a description of each character Possibly a geographical map as table clothes positioning the scenario in a context
Number of participants
9–12 people and a moderator role
Facilitator competences and skills
The moderator frames the scenario in a fictional group meeting (like a council meeting or a citizen assembly, etc.) the moderator keeps the time and has the permission to also insert qualities that move the players into a different mindset. The moderator calls for a reflection break in between and facilitates the exchange. The reflection is based on an exchange of observations only
Participants skills/age/competences
The game can be performed in groups that work together on a topic, for example, a neighborhood or an initiative. It can also be performed if people are institutionally connected and work together on a topic
Duration
The game lasts 3.5 h, with players reflecting on the experience of the last 45 min or 1 h

Further Readings

"Kiezblocks"—Argumentatives Rollenspiel. https://www.rifs-potsdam.de/en/events/kiezblocks-argumentatives-rollenspiel. Accessed 14 June 2024

Arosa, https://susannebosch.de/portfolio_page/arosa-argumentatives-rollenspiel-der-agierenden/. Accessed 14 June 2024

https://www.instagram.com/arosa_rollenspiel/. Accessed 14 June 2024

Chapter 45
Circular Families Game

Marianna Marchesi

45.1 Introduction

The Circular Families Game is a card game that applies game elements in real-world scenarios to facilitate understanding of the circular economy in cities. Games for sustainability or eco-games are a type of game for learning (Spanellis A et al., 2022). They create attractive and enjoyable educational scenarios to intrinsically motivate individuals to engage with materials related to sustainability (Stanitsas et al. 2018; Pereira Oliveira et al. 2021).

The circular economy is a regenerative and restorative approach that aims to eliminate waste and pollution, extend the lifespan of products and materials, and regenerate nature to enable us to prosper sustainably. A circular economy fosters eliminating waste and pollution through practices such as encouraging active and public mobility. It helps to keep products and materials in use as long as possible by promoting practices like second-hand product selling. Nature is regenerated by initiatives like food growing. Moreover, circular initiatives such as repair centers and food hubs create new jobs, improve access to healthy foods, and connect people locally. Circular Families Game aims to support leaders within urban communities in fostering awareness of the circular economy in cities and promoting behavioral change within their communities. The game can be employed by the public sector (such as city councils) and the third sector (such as non-profit organizations, charities, housing associations, social enterprises, and community groups). Students in planning and urban design who wish to engage in design for a circular economy can also benefit. The game creates an imaginative space for playing and learning to understand the circular economy in cities collaboratively. Players are engaged in

M. Marchesi (✉)
Welsh School of Architecture, Cardiff University, Cardiff, UK
e-mail: marchesim@cardiff.ac.uk

J. Bentz, J. Ristić Trajković (eds.), *Imagining, Designing and Teaching Regenerative Futures: Art-Science Approaches and Inspirations From Around the World*, Science for Sustainable Societies,
https://doi.org/10.1007/978-981-96-9029-9_45

urban scenarios, citizen roles, and initiatives for a circular economy through game mechanisms. Initially, players are introduced to seven circular citizen characters—do-it-together citizen, sharing citizen, strategic citizen, zero-waste citizen, sensor citizen, do-it-yourself citizen, and service-oriented citizen—who performs a different production or consumption practice for a circular economy (Marchesi M, Tweed C, 2022), and asked to guess the type they are or would like to become.

Then, participants are engaged in playing the Circular Families card game to test their hypothesis. They play to form a circular family in a resource sector by closing the resource loop. Each turn a player must decide on a circular action which could be "take," "make," "use," or "recover" to close the loop. By playing, they explore individual and community-led practices for a circular economy in different resource sectors: food, product, building, green infrastructure, and mobility. These initiatives benefit individuals, communities, and neighborhoods by keeping materials in use, reducing waste, and regenerating nature with impacts on the environment, economy, and society. In terms of environmental impact, the game can help players explore suitable opportunities in their community network to reduce waste generation and raw material consumption through interventions such as sharing, reusing, and repurposing materials and products such as tool libraries and systems for donating unwanted goods. In terms of economic impact, the game can support players to explore circular skills and new and specialized jobs such as local exchange and repair centers and urban gardens for food system innovations. In terms of social impact, the game can support players in discovering opportunities to enhance social cohesion through community-led opportunities for making, reusing, sharing, and repairing as well as through organic producer networks linking communities with peripheral farms and alternative markets.

While playing, they are encouraged to collaboratively interact and actively learn in a safe environment about the circular economy, the circular citizens, and their initiatives to eliminate waste, circulate materials and products for as long as possible, and regenerate nature. The game elements include strategies, progression towards goals, feedback, competition, and rewards. These elements are combined through game rules and integrated with pedagogical patterns to ensure learning.

45.2 Transformative Potential

The Circular Families Game can be an effective active learning method for sustainable development education as it provides players with a better understanding of the circular economy and practices that they can pursue in communities and neighborhoods for a circular economy to strive sustainably. It further promotes motivation, engagement, and satisfaction which together can improve interest in the learning process and performance results (Pereira Oliveira et al. 2021). Circular Families Game creates an effective learning environment and redefines the learner experience through gamification. Playing the Circular Families Game, participants are engaged with circular economy concepts implemented in real-world situations within a gamified context and facilitated in learning about circular practices in cities through seven circular citizen characters while having fun. The game also offers a debriefing discussion afterward to encourage players to reflect upon their newly gained insights, deepening their learning experience. While simply possessing knowledge is not enough to support behavioral change, enhanced knowledge does have a favorable impact on attitude, and attitude plays a significant role in shaping behaviors (Chappin et al. 2017). Therefore, increasing knowledge about circular practices at the individual and community levels can directly affect environmental attitudes and indirectly promote behavioral change. It can increase players' understanding of the circular economy and their familiarity with practices to contribute to ecological thought and encourage behavior change.

45.3 Application

The Circular Families Game (Marchesi 2022) (Table 45.1) is designed for adults and teenagers (15+) and can be played in a group of two to six players using a card deck or multiple groups using multiple card decks. The game is available in two languages (English and Welsh) with the translation of the citizen identity cards into 23 European languages. The game brochure is freely downloadable from the CircuBED project webpage (https://www.cardiff.ac.uk/research/explore/find-a-project/view/2587142-circular-families-game) for home printing and making. It includes a do-it-yourself envelope to accommodate the cards, game instructions, and rules after having played. A video is also provided on the project webpage to facilitate game rules briefing. For each playgroup, a table and six chairs are necessary for playing. If the game rules are briefed by the video, a screen connected to a computer, loudspeakers, and an internet connection are also needed. The game session requires between 45 minutes and 1 hour if the session ends when a player builds the first family. It requires between 1 hour and 1 hour and a half if the session ends when the players finish all the cards in their hands.

Initially, the game facilitator introduces participants to the 7 circular citizen characters and asks participants to guess the circular citizen type they are, or they would like to become. Then, participants are engaged in playing the Circular Families card game to test their hypothesis. Before starting the game session, the game rules are introduced. Players aim to collect all four members from one of 13 "circular family" groups representing five resource sectors in cities: construction, manufacturing, food, green infrastructure, and transport. To form a circular family, each turn a player must decide on a circular action which could be "take," "make," "use," or "recover" to close the loop in a resource sector. Each action corresponds to an initiative for closed resource loops and zero waste implemented by one of the circular citizens in a resource sector. The player who collects the first circular family or most families wins. While playing, participants get familiar with the seven circular citizens and look into their circular interventions. After having played, they select the preferred intervention among all initiatives on the cards and identify the circular citizen in charge to verify their initial guess. Finally, players can discover features,

Table 45.1 Method overview

Main purpose
Awareness building (directly) and behavior change (indirectly)
Gained competences
Understanding of the circular economy approach applied to neighborhoods and urban communities
Educational setting
Informal (card game session)
Space requirements/restrictions
Outside activity/indoor activity
Resources and necessary materials
1 circular families game deck, one table, and two to six chairs; If the game rules are briefed by the video, a screen connected to a computer, loudspeakers, and an internet connection
Number of participants
Two to six participants
Facilitator competences and skills
Basic understanding of the circular economy approach and the seven circular citizen characters, and knowledge on the card game rules
Participants skills/age/competences
Community/students/designers
Duration
45 minutes to 1 hour if the session ends when a player builds the first family 1 hour to 1 hour and 30 minutes if the session ends when the players finish all the cards in their hands

practices, challenges, and impacts of their preferred character by looking at the circular citizen identity cards included in the card deck and then reflect on their own circular practices (Table 45.1).

References

Chappin EJL, Bijvoet X, Oei A (2017) Teaching sustainability to a broad audience through an entertainment game—the effect of Catan: Oil Springs. J Clean Prod 156:556–568. https://doi.org/10.1016/j.jclepro.2017.04.069

Marchesi M (2022) Circular Families Game. https://www.cardiff.ac.uk/research/explore/find-a-project/view/2587142-circular-families-game. Accessed 15 Nov 2024

Pereira OR, Gomes de Souza C, da Cunha Reis A, Medeiros de Souza W (2021) Gamification in E-learning and sustainability: a theoretical framework. Sustainability 13(21):11945. https://doi.org/10.3390/su132111945

Stanitsas M, Kirytopoulos K, Vareilles E (2018) Facilitating sustainability transition through serious games: a systematic literature review. J Clean Prod 208:924–936. https://doi.org/10.1016/j.jclepro.2018.10.157

Marchesi M, Tweed C (2021) Social innovation for a circular economy in social housing. Sustain Cities Soc 71:102925. https://doi.org/10.1016/j.scs.2021.102925

Spanellis A, McMenemy E, Borzenkova G, Cole D (2022) Guidelines for design and development of gamification apps and games for sustainability. Heriot-Watt University, Edinburgh

Chapter 46
Climate Community Street Play

Myriel Milićević, Jasmin Jossin, Desiree Foerster, and Ruttikorn Vuttikorn

46.1 Introduction

It is a 40 °C day. Where do you run to? Jump ahead and land on both feet. Right foot stands on sunny asphalt in painful 56 °C. Left foot in the shade of a tree at 34 °C. You jump ahead but get trapped inside a dome of hot air, finding yourself on a heat island. It's a mono-species place, a human figure says. You too are of the human species but cannot take it there. You follow a tree portal to escape. A forest spreads out in front of you, where you find rescue. Birds, pollinators, foragers, amphibia, and many other species are already there. Jump once more, spiral in the air by 180°, and land on both feet to read what is written upside down: Turn surroundings into multi-species places.

With Climate Community Street Play, we bring urban climate issues to the streets—to battle and navigate together the challenges of global warming in local neighborhoods. In reference to Donna Haraway (2016), one could say we are playing with the trouble. By climate community, we understand a support network of people, plants, animals, and other beings within shared and challenged environments. The concept was developed by Myriel Milićević and Ruttikorn Vuttikorn and was brought to life with students, youth, practitioners, and scholars of diverse

M. Milićević (✉) · J. Jossin
Department of Design, University of Applied Sciences, Potsdam, Germany
e-mail: myriel.milicevic@fh-potsdam.de; jasmin.jossin@fh-potsdam.de

D. Foerster
Department of Cinema and Media Studies, University of Chicago, Chicago, IL, USA
e-mail: dfoerster@uchicago.edu

R. Vuttikorn
Club Creative Co., Ltd., Bangkok, Thailand
e-mail: ruttikorn@ruttikorn.com

J. Bentz, J. Ristić Trajković (eds.), *Imagining, Designing and Teaching Regenerative Futures: Art-Science Approaches and Inspirations From Around the World*, Science for Sustainable Societies,
https://doi.org/10.1007/978-981-96-9029-9_46

backgrounds. We began by considering the socio-ecological dimensions of affected urban communities, along with the global tradition of street play—a form of public interaction familiar to people around the world. Building on this, we developed a series of street games that engage players with local risks and raise awareness of existing or potential alliances within their communities. Iterations have been developed for cities in Germany and Thailand. Although these places are facing different challenges, they all tell stories of people and other species coping with the impacts of the climate crisis. A game narrative, paired with engaging physical actions, can deeply immerse players in the experience, helping them understand the game's message and relate it to real-world scenarios. This immersive approach can be particularly effective in conveying complex issues such as climate risks, vulnerabilities, and community capacities.

To explore the risks and real or imagined adaptation opportunities of local climate communities, the participants first go on guided walks with environmental scientists to observe a neighborhood. They collect knowledge, stories, and experiences in conversations with local residents, teachers, gardeners, or shopkeepers, through their own sensory perceptions, and through exercises in adopting the perspectives of other species. Understanding, mapping, and visualizing the spatial, social, and natural relationships with interdisciplinary and multi-species perspectives creates the basis for developing the formats of the games that are later taken to the streets, where the players can engage with different scenarios—in our case from fighting together for street trees in droughts, turning heat islands into cool spaces, rescuing animals in floods, connecting with personal guardian species, or finding a networked path to biodiverse spots as a pollinator. The games intend to bring together players of all ages to encourage conversations and instigate possible actions for the local community. Playing and discussing opens a space for greater awareness for the interconnectedness of humans and other entities living in these climate conditions together, and for imagining speculative perspectives of future neighborhoods.

46.2 Transformative Potential

Global warming is confronting people with floodings, cloudbursts, storms, or prolonged droughts. Quieter and steadily intensifying effects such as heat stress are also forcing communities to prepare for the consequences of rising temperatures. Safeguarding livelihoods in a warming climate requires more than just adapting the built infrastructure, economic systems and healthcare systems. Community resilience must also be strengthened at the social and personal level through the cultivation of informal exchanges, local communication, and support networks. This was, for example, evident in the 1995 Chicago heat wave (Klinenberg 2002), where making it through the heat wave safely depended on social connections and access to communal places.

The concept of Climate Community Street Play extends the social support network beyond the human residents, highlighting the regenerative potential of supporting these communities for the human and the non-human world: A tree or ivy cools a street or a building and provides shade to people and other species as well as food and shelter for insects, bats, and birds, who again feed on mosquitoes that might carry diseases. Moreover, urban green spaces impact human health and may also increase social well-being through fostering social interaction and longer-term relations (e.g., Hornberg et al. 2016; Huang and Lin 2023; Nguyen et al. 2021).

The transformative potential of Climate Community Street Play is also of an educational kind: When entering a game, you might turn into an elderly person, a young bat, or become a ferocious heat wave. Imagination and play can put you into another state of mind where different rules apply. In general, you may learn better with higher bodily engagement (Skulmowski and Rey 2018). Playfulness and materiality in an adult learning setting may offer additional benefits (Jørgensen et al. 2022).

With this project, we contribute to a more nuanced perspective on the potential of imaginative and playful engagements with the climate crisis. As Roy Bendor stated (2018), current approaches to climate change are characterized by a crisis of imagination. What we therefore need are "worldmaking interactions ... that aim to promote the public's own ability to imagine alternative futures" (Bendor 2018, 132). The street games are such worldmaking interactions. The often ambiguous and complex processes associated with a heating climate are made experiential through role play, which can also open up new social imaginaries for future climate communities. These imaginaries will need to go beyond an anthropocentric perspective and consider the complex ways nonhuman animals and vegetal life are implicated and impacted by global warming as well.

46.3 Application

The process is made up of three parts:

First, we identify, research, and map vulnerable multi-species communities at the neighborhood level. How does ivy provide shelter for both people and other species? Why are bats and swifts at risk when raising their young on a south-facing façade? How is a shop owner connected to the birds that uplift people's moods? Why is a delivery service worker more at risk during floods? And how are frogs affected by canal walls built to prevent flooding?

Second, we design new formats for street games to address the specific challenges faced by urban climate communities or adapt games to suit local conditions. A game format based on districts with varying heat island effects due to sealed surfaces can, for instance, be mapped to other areas in different cities. Vulnerable groups, such as the elderly, young trees, or amphibians, could be affected anywhere, while some groups may face greater challenges in specific regions—such as particular species or people living in particular working or housing conditions. Both global and local challenges, along with opportunities for action, can form the basis for developing game concepts.

Third, we play the games and, where appropriate, evaluate them.

If you are interested in applying the approach, we recommend finding a public event where the street games can be played. If you want to adapt already existing games, a growing archive of guidelines for games, as well as examples of the research process will be made available on the project website.

Instead of just replaying games, a more effective approach is to adapt and localize them to tackle the specific challenges and conditions of the area. Who lives there—both people and other species? What are their needs, and how might they support others? During a neighborhood walk like the one described earlier, participants can step into the shoes of different species. Depending on the geographical context, the 'Animal Perspectives' tool available at xtopien.org/en-toolbox/ may be helpful for achieving this unfamiliar perspective-taking. A survey form and map can aid the entire observational process, helping participants observe and document aspects such as the spatial and temporal features of the area they are exploring, the resources available, the characteristics of the micro-climate, the presence of small or large water bodies, and even the emotions and sensations they experience in the space. Through a guided research process like this, participants can uncover local risks, opportunities, connections, and stories that can shape the content of an existing game format or inspire the creation of entirely new games. Table 46.1 provides additional information to help you plan your own application effectively.

When these games are eventually brought to the streets, there is an exciting moment where participants share the rules, context, and stories behind the games with the players. In doing so, they pass on their knowledge and take on the role of facilitators themselves.

Just like street games are changing from place to place and handed down from generation to generation, these games should be shared, transformed, and played in the streets of climate communities all around. If you want to become part of this community and inspire others to play and take action, your games can be featured at http://climatecommuniti.es.

Climate Community Street Play was first developed and iterated with students of Design and of Urban Futures at the University of Applied Sciences Potsdam in the spring of 2022. This was followed by workshops in Thailand with youth of the Klong Toey district and students at KMUTT's School of Architecture and Design in Bangkok, as well as high school students of Panyaden International School and students of Mass Communication at Chiang Mai University. In Spring 2024, another iteration was carried out by students of the international MSc Integrated Urbanism and Sustainable Design at the University of Stuttgart.

Table 46.1 Method overview

Main purpose
Awareness building/social interaction/worldmaking/creating local, more-than-human social support networks
Gained competences
Imagination, collaboration, and strategy. Knowledge on local climate impacts, possible measures for protection, and relation to biodiversity
Educational setting
Development: workshops and seminars. Play: street, public spaces, and schoolyards
Space requirements/restrictions
Safe and large outdoor space
Resources and necessary materials
Chalk/chalk spray, possibly stencils and other props depending on the game.
Number of participants
Development: 7–20 participants Play: 1-XX depending on the game and amount of games played at the same time
Facilitator competences and skills
Inviting, sense of humor, and knowledgeable of game rules and content. Possibly reflecting with participants after playing
Participants skills/age/competences
Anyone who enjoys playing and physical activity
Duration
Complete process of research and game development: 2–3 weeks. Implementation of street games: 1 hour–half a day. Playtime: several minutes per game round

References

Bendor R (2018) Interactive media for sustainability. Springer

Haraway DJ (2016) Staying with the trouble: making kin in the Chthulucene, Experimental futures: technological lives, scientific arts, anthropological voices. Duke University Press, Durham, NC

Hornberg C, Beyer R, Classen T, Herbst T, Hofmann M, Honold J, van der Meer E, Wissel S, Wüstemann H (2016) Stadtnatur fördert die Gesundheit. In: Kowarik I, Bartz R, Brenck M (eds) Naturkapital Deutschland—TEEB DE. Ökosystemleistungen in der Stadt – Gesundheit schützen und Lebensqualität erhöhen, Technische Universität Berlin, Helmholtz-Zentrum für Umweltforschung—UFZ, pp 98–125

Huang W, Lin G (2023) The relationship between urban green space and social health of individuals: a scoping review. Urban For Urban Green 85:127969. https://doi.org/10.1016/j.ufug.2023.127969

Jørgensen HH, Schrøder V, Skovbjerg HM (2022) Playful learning, space and materiality: an integrative literature review. Scand J Educ Res 67(3):419–432. https://doi.org/10.1080/00313831.2021.2021443

Klinenberg E (2002) Heat wave: a social autopsy of disaster in Chicago. University of Chicago Press, Chicago, IL

Nguyen PY, Astell-Burt T, Rahimi-Ardabili H, Feng X (2021) Green space quality and health: a systematic review. Int J Environ Res Public Health 18(21):11028. https://doi.org/10.3390/ijerph182111028

Skulmowski A, Rey GD (2018) Embodied learning: introducing a taxonomy based on bodily engagement and task integration. Cogn Res Princ Implic 3:6. https://doi.org/10.1186/s41235-018-0092-9

Chapter 47
Regen-D: Template-Based Games for Sustainability and Regeneration

Ulrike Zeshan

47.1 Introduction

Regen-D is a format for creating and playing serious games, and its name comes from regenerative design. Serious games are games for purposes beyond mere entertainment, such as collaboration or education. These games are powerful educational and facilitation tools because players are active in a group as they engage in dialogue with each other and interact with a game environment.

The aim of the Regen-D game series is to present case studies in sustainability and regeneration. In relation to sustainability, the Regen-D remit defines regeneration as going a step further, that is, not only to make sure that future generations and the natural world can be sustained but also to actively repair the damage that has already been done while cultivating new relationships with nature and society.

Each Regen-D game is a separate case study based on the generic Regen-D template. The template allows for an unlimited number of games to be created easily. This makes developing games feasible and accessible for sustainability and regeneration initiatives that would like to use the creative potential of games but do not have specific game design expertise.

The core component of the game is a collection of up to 12 wooden disks with QR codes linked to multimedia files, each of which represents an aspect of the initiative being presented. Players view each multimedia file on a screen, then flip the disk to uncover placeholder pictures on the reverse side, and successively assemble them into a diagram. In addition, three types of tasks are activated via action disks: *evaluating* an element, *linking* two elements, and introducing a new element with a

U. Zeshan (✉)
University of Central Lancashire, Preston, UK

Ishara Press CIC, Silverdale, UK
e-mail: uzeshan1@uclan.ac.uk

J. Bentz, J. Ristić Trajković (eds.), *Imagining, Designing and Teaching Regenerative Futures: Art-Science Approaches and Inspirations From Around the World*, Science for Sustainable Societies,
https://doi.org/10.1007/978-981-96-9029-9_47

joker. The type of disk to be played next is chance selected by spinning the arrow on a selector wheel. Each time a disk is added to the evolving diagram, there is a group discussion to understand the new element and its relationship with the other disks.

This game choreography leads players through the multimedia materials to discover and reconstruct the case study. However, the potential of Regen-D games is not only due to the immersive experience for groups in educational contexts, but it is also a way for projects in sustainability and regeneration to share what they do in an innovative way. The time it takes to design and produce a new Regen-D game is drastically reduced because we can rely on a tried-and-tested template. All that is needed for a new game is to swap the content of the multimedia files and tweak the game rules.

47.2 Transformative Potential

Regen-D games have the potential for sustainability education, for grassroots actors to support their work, and for people to become authors of their own games. The multiple uses are a particular strength of this game format.

The Regen-D series is published by the social enterprise publisher Ishara Press at https://ishara.uk/regen-d. Each game has an open-source e-book, from which the game components can be printed and self-assembled. The hardcopy games are available from Ishara Press but are not open source.

The examples below come from the game "Water management systems at Shikha Ecovillage," which is based in rural eastern India (Panda et al. 2023), and from the game "Local Loop: Collaborative finance for the circular economy," which is based in north-west England (Woodroof and Green 2024).

As a learning experience, a Regen-D game prompts the players to actively explore the topic of the game instead of being presented with preconfigured content. This leads to a deeper engagement with the case study, as there is ample room for group discussion. Assembling the visual diagram turn-by-turn makes learning a slower-placed, tangible, and engaging experience. The game fosters an inclusive and non-threatening learning environment where the turn-taking ensures active participation.

The Regen-D design is motivated by the belief that learning at a purely intellectual level is not good enough if we truly want to orient people toward a regenerative future. We must involve other human capacities, such as imagination, empathy, intuition, coordination with others, and the will to act. As a serious game, Regen-D has the potential to mobilize such holistic capacities.

Besides educational uses, Regen-D is designed to support grassroots regenerative initiatives, which often start with a small core group of innovators, early adopters, or frontrunners. When additional people subsequently join the initiative, there is an "onboarding" process, which means introducing them to the new context. Regen-D is a creative alternative to conventional onboarding.

By playing through a case study, new team members can learn about its interconnections and complexities. The "Local Loop" initiative needs a sufficient number of local businesses to be mobilized in a network for mutual credit clearing. The underlying financial mechanisms and algorithm are somewhat complex, so onboarding via playing the Regen-D game, with the information broken up into manageable chunks in multimedia form, is more accessible than reading a document or watching a presentation.

On-going grassroots initiatives often find it challenging to document and disseminate their work. This was the case for Shikha Ecovillage in India, which was developed as an off-grid campus with a boarding school for deaf children and chemical-free horticulture (see https://shicol.in). Local people were not interested in producing written texts or data tables. Instead, creating a Regen-D game motivated the local team to contribute the necessary multimedia materials.

Shikha Ecovillage also illustrates how Regen-D games allow people to become published authors who would not otherwise have this opportunity. The game content is a multimedia file collection of photos, videos, maps, diagrams, and other mixed media, with no need to write in an academic or formal style. In addition, the game was published bilingually with an English version and a version in the local languages Sambalpuri and Odia. This context has enabled one of the local team members to become a first-time author.

Bi- and multilingual publishing in local languages also means that Regen-D games are accessible to local people, not just to a distant English-speaking elite. This democratization of publishing, in terms of both authors and audience, is an important design feature of the Regen-D series (Zeshan, forthcoming).

47.3 Application

Regen-D games are played either internally with existing and aspiring members of an initiative, for example, for onboarding, or externally as an educational and dissemination activity, for example in Education for Sustainable Development classes. Playing through a case study takes ca. 90 min. If game materials are self-assembled instead of using a hardcopy game set, this takes about an hour. The QR-code links lead to open-source multimedia files, and a sufficiently large screen is needed as per group size to view them, using a handheld QR scanner or a smartphone. Teachers or facilitators may lead through the game, but groups can also play autonomously.

Regen-D is played in groups of 4–16 players. If there are more than eight players, they should form an inner and an outer circle. Whenever someone in the inner circle has played a turn, they exchange places with a player from the outer circle. Groups larger than 16 need several parallel groups or a larger outer circle. In the latter case, not everyone can play actively, but the whole group may still participate in discussions.

Teachers and facilitators can rely on the open source e-books for preparation. Along with the print-and-play materials for self-assembling the game, each book includes directions for use and game rules. Where available, the e-book also describes the case study coherently and comprehensively, with links for additional information and resources. Learners can read the book after playing the game, if that is considered useful.

The potential to easily create new case studies in the Regen-D format is just as important as playing them. Because of the time and expertise involved in game design and development, it is often not feasible for people without relevant experience to develop a well-constructed game from scratch. A workable prototype like the Regen-D template breaks down this barrier and allows people to become game authors without prior experience.

Regen-D case studies must fit the remit of the series, which includes fully implemented initiatives, work in progress, and design proposals in sustainability/regeneration. Game designers at Ishara Press guide the authors through the process and technical details of creating a well-balanced file collection and e-book. This takes 4–8 weeks, which may also suit lecturers to undertake as a creative learning activity with university students over a semester (Table 47.1).

Both the multimedia materials and the e-book are peer-reviewed for quality by the international editorial board, which includes both academics and practitioners. E-books with a comprehensive description of the case study are published and indexed via the open-source OAPEN platform at https://oapen.org in addition to the Ishara Press website.

Table 47.1 Method overview

Main purpose
Learning from case studies; creating new case studies from the game template
Gained competences
Experience a serious game; learn about regenerative initiatives; actively engage with regenerative thinking; multimedia publishing.
Educational setting
Informal (workshops, game sessions, and classrooms)
Space requirements/restrictions
An empty table with circular seating for four to eight people, or empty floor space if sitting on the floor
Resources and necessary materials
Sufficiently large screen in relation to group size; selector wheel/game spinner and game disks in two sizes (download and self-assembly available)
Number of participants
4–16 players; split into several groups for more than 16 players
Facilitator competences and skills
Background in serious games is useful but not mandatory.
Participants skills/age/competences
Secondary school and older; no specific skill and competence needed
Duration
90 min to play the game 4–8 weeks to develop a new game

References

Panda S, Panda S, Zeshan U (2023) *Water management systems at Shikha Ecovillage. Ishara Press, Lancaster, UK.* https://ishara.uk/regen-d-no-1/. Accessed 21 June 2024

Woodroof T, Green D (2024) *Local loop: collaborative finance for the circular economy. Ishara Press, Lancaster UK.* https://ishara.uk/local-loop-collaborative-finance-for-the-circular-economy/. Accessed 21 June 2024

Zeshan U (forthcoming) Multilingualism, multimodality and gamification: pathways to democratising educational spaces. In: Bagga-Gupta S (ed) *Palgrave handbook of decolonising the educational and language sciences. Palgrave Macmillan, London, UK*

Chapter 48
Creating Social Change Through Urban Gaming

Curt Gervich

48.1 Introduction

Urban gaming represents a new form of media and entertainment in which game designers create place-based activities that allow participants to experience cities through play. Scavenger hunts, outdoor puzzle races, augmented reality, and orienteering are common examples of urban games. These activities gamify the city by presenting participants with challenges and competitions to complete while moving about the city, searching for secret locations and objects, taking part in contests, cracking codes, and deciphering riddles. Some urban games are played without the help of mobile phones and apps, while others are driven by technology. This chapter provides a methodology and other support to urban changemakers designing their own games.

Activists in some cities are revolutionizing urban gaming to harness the platform for radical social change. These game designers are creating games that confront hegemonic power structures of oppression and violence and advance sustainability and resilience. Their games are unique to the cities and neighborhoods in which they occur and offer a multidimensional platform for reclaiming and retelling historical narratives of the city; reimagining the future of the city; empowering residents with the skills necessary for making positive change in their communities; and nurturing sense of place. When playing these games, urban gamers must apply geospatial thinking to navigate the city and understand its spatial dimensions; consider the city's evolution from past to present and imagine what it may become in the future; and think critically about social, political, and economic dynamics.

C. Gervich (✉)
Center for Earth and Environmental Science SUNY Plattsburgh, New York, NY, USA
e-mail: cgerv001@plattsburgh.edu

J. Bentz, J. Ristić Trajković (eds.), *Imagining, Designing and Teaching Regenerative Futures: Art-Science Approaches and Inspirations From Around the World*, Science for Sustainable Societies,
https://doi.org/10.1007/978-981-96-9029-9_48

Julia Byrne's *Martin Luther King Jr. in Atlanta: Unwrap his Life and Legacy* offers a good example of an urban game oriented toward social change. Byrne's game is a puzzle quest, hosted on the Questo App. Puzzle quests are outdoor escape games during which participants solve fictionalized mysteries by deciphering clues hidden in the urban environment. Many varieties of puzzle quests exist, but the typical format involves a mystery or action-adventure narrative that unfolds while solvers walk to a series of designated locations in sequence, solving puzzles as they go.

Julia Byrne's *Martin Luther King Jr. in Atlanta: Unwrap his Life and Legacy* is situated in the Sweet Auburn neighborhood of Atlanta, Georgia, in the United States. The Sweet Auburn district is a majority-black community that developed in the post-slavery period of the late 1800s. It grew to become a stylish neighborhood for affluent black families in the early 1900s. It is where Martin Luther King Jr. was born and raised. Julia Byrne's puzzle quest highlights locations significant to the neighborhood's development and Martin Luther King Jr.'s ascendance to the leadership of America's Civil Rights Movement. While playing the game, solvers are confronted by the same moral dilemma that defined Martin Luther King Jr.'s advocacy: remaining committed to nonviolent civil disobedience in the face of violence from law enforcement agencies and white southerners seeking to maintain the systems of oppression at work in the Jim Crow south. Byrne's game encourages players to imagine an alternative future where the power structures of Jim Crow segregation cease to exist. Or, as King put it, where people are "not judged by the color of their skin but by the content of their character (King's "I have a Dream Speech", August 28, 1963).

Residents and visitors to the Sweet Auburn neighborhood frequently pass the locations included in Byrne's puzzle quest, yet often without seeing the details and history embedded in them. This is not unusual. The details of cities often blend into the background like white noise, but the reality is that humans design cities with intention. We build the physical elements of cities through blood, sweat, and tears and co-construct their social dimensions through behaviors and interactions. Cities are multi-layered physical and metaphysical creations, and urban games can help people experience this multidimensionality.

48.2 Transformative Potential

Working toward resilient urban communities takes particular modes of thinking. Wiek et al. (2015) outlines six competencies important for sustainability changemakers. They are: systems thinking, futures thinking, strategic thinking, normative thinking, integrative thinking, and collaborative thinking. Designing and playing radical urban games can help develop and practice these modes of thinking.

Systems thinking describes the ability to see cities as living organisms that are greater than the sum of their individual parts. Systems thinking involves seeing cause-and-effect relationships, and feedback loops, among the built and non-built environments, and the social, economic, and political sectors of the city. For urban changemakers, seeing systems allows for the development of holistic approaches to resolving challenges and avoiding unintended consequences.

Futures thinking and normative thinking go hand-in-hand. Imagining a future free of oppression, environmental degradation, poverty, and inequity is an essential skill for changemakers. Futures thinking means being able to draw upon observations of the past to shape future scenarios. Normative thinking means envisioning the types of interactions and behaviors that will be normalized once a better future is achieved. These modes of thinking provide a north star that orients activists to the future they intend to create. Without it, building traction with supporters is difficult and progress may be haphazard.

Creating change requires working across the artificial boundaries that define the social, environmental, economic, and political work of many organizations. Bridging boundaries to build coalitions, attack problems from different angles, and nurture a common working approach among members of a movement requires the integration of diverse ways of knowing and operating and the development of shared strategies and plans. Strategic and integrative thinking describes the processes by which consensus visions for the future become reality.

Deconstructing systems of oppression and violence against people and nature requires collaboration. The challenges of urban transformation are simply too great for any single person or organization to resolve alone. Collaboration is not just a way of doing but also a way of thinking. Collaborative thinking requires thoughtful communication, patience, sharing, compromise, synchronization, and empathy, among other characteristics.

Radical urban gaming offers a platform for developing the skills Wiek describes as necessary for advancing sustainability. Urban games assist players in seeing the invisible systems that contribute to the formation of the city as a social and physical space. These games also are best played in groups, and group problem-solving improves collaboration and integrative and strategic thinking. The radical nature of the games being produced by activist designers is inherently founded upon futures and normative thinking, and as a result, players soak up these competencies through the gaming process.

48.3 Application

You do not need to be a professional game designer to build urban games in your community. Before designing your first game, I suggest playing a few to get a general feel for how they work. You can search for nearby games on all the common search engines or by using one of the apps listed in the resource section below. Use the following steps as a guide to creating your first urban game.

1. *Develop the theme of your game.* Think deeply about the systems that shape your city. Establish a theme that represents your objectives for systemic change.
2. *Develop the storyline for your game.* Great stories have common ingredients: characters, conflicts, and resolutions. Who do you want your players to think about while playing your game, and what conflicts are these characters up against? What resolutions illustrate the future you imagine for your city? Create a story that builds on these elements.

3. *Connect your game to a place.* Identify a neighborhood important for understanding the theme and story at the heart of your quest and locate your game in this location. Use competition, challenges, riddles, tasks, and other activities to facilitate thoughtful engagement with this place.
4. *Develop gamified activities.* During your game, you will want players to observe and interact with their surroundings. Look for interesting signs, architecture, historical markers, and monuments that are relevant to the theme of your game. Create riddles, trivia questions, scavenger hunts, and other activities that encourage engagement with these items.
5. *Playtest your game.* The beauty of urban games is that they are enjoyable *and* powerful. This harmony can be lost if navigating the city, following the story, or playing the activities is too frustrating. Feedback from playtesters is critical to enhancing the playability of your game.

The following resources will help you build your first urban game for social change.

- *Questo* is a puzzle quest app that offers a user-friendly game design portal that leads you through the game development process.
- The *Goosechase* app allows users to create place-based scavenger hunts with a variety of activities (Table 48.1).

Table 48.1 Method overview

Main purpose
Radical urban gaming uses gamification strategies to encourage learning about historical narratives in cities, observe systems of oppression, violence, and degradation, strategize how to build systems of equity, resilience, and sustainability, and reimagine the future of cities to allow all residents to achieve their personal, family, and community aspirations.
Gained competences
Systems thinking, futures thinking, normative thinking, integrative thinking, strategic thinking, and collaborative thinking
Educational setting
Urban and community centers
Space requirements/restrictions
Outdoor activities
Resources and necessary materials
Some urban games incorporate smartphone apps.
Number of participants
5–500
Facilitator competences and skills
Creativity, research skills, and some familiarity with game-playing
Participants skills/age/competences
Urban gaming suits all skill levels and ages.
Duration
Building an urban game requires 5–30 h depending on the complexity of the game. Basic games can be created quickly and simply.

Reference

Wiek A, Bernstein M, Foley R, Cohen M, Forrest N, Kuzdas C, Kay B, Withycombe KL (2015) Operationalising competencies in higher education for sustainable development. In: Barth M, Michelsen G, Rieckmann M, Thomas I (eds) Handbook of higher education for sustainable development. Routledge, London, pp 241–260

Chapter 49
A Listening Exercise in Care and Forgiveness

Florian Ruland

49.1 Introduction

Many of us living in the privileged part of the world carry a feeling of guilt for not taking a more active role in halting or reversing the global ecological crisis. This feeling, when nurtured and maintained, is not motivating but actually trapping us. Feeling bad about oneself is not a great place to start when trying to do anything, much less when imagining tackling an unprecedented global challenge. Conversely, if we manage to lift the weight from our shoulders temporarily, we suddenly find ourselves free to hope and act.

The goal of the Listening Exercise in Care and Forgiveness is to achieve this feeling of relief in five steps: After a short discussion about everyone's expectations, the size of the problem is reduced and focused on one species. The global ecological crisis is impossible to fathom, but we can project our feelings of loss and discomfort on one species (animal, plant, or fungus) that is especially dear to us. Literature and other resources are provided to conduct a short personal research on the species. Third, the weight of the guilt is reduced by opening up about it to a listening exercise partner. This is an important step to (i) acknowledge the feelings of guilt by speaking out about them and (ii) connect to the exercise partner by sharing this personal information. Then, the partner shows us the gift of forgiveness from the position of the focal species. This means that the act of opening up is rewarded and with a feeling of gratitude and connection comes hope. Finally, experiences are shared in the group with a highlight on the emotional process and what is needed now.

F. Ruland (✉)
Freie Universität Berlin, Berlin, Germany

IGB Berlin, Berlin, Germany

Náttúrustofa Vesturlands, Stykkishólmur, Iceland

J. Bentz, J. Ristić Trajković (eds.), *Imagining, Designing and Teaching Regenerative Futures: Art-Science Approaches and Inspirations From Around the World*, Science for Sustainable Societies,
https://doi.org/10.1007/978-981-96-9029-9_49

49.2 Transformative Potential

The act of opening up to the exercise partner and being rewarded with forgiveness yields a regenerative potential. Unpleasant and uncomfortable feelings that the participant quietly carried are now released and their spell is broken by the response. A general transformative potential lies in this exercise as it creates a bond of trust between participants and underlines the position of the human species within the global ecological network.

A problem with the most dire consequences is that humans see themselves as outside of an "innocent" and "threatened" "nature." Many people compare the human species on this planet to a disease or a curse for the beauty and diversity of life that is gradually destroyed by human actions. This conclusion may be relatable with observations of massive habitat destruction, pollution, and overexploitation of natural resources. However, in it lies limited potential for positive personal or political action as the restoration imperative would be to disappear completely as a species. To act responsibly on earth, we must feel welcome as fellow inhabitants. Creating a biodiverse urban garden, for example, will be all the more enjoyable if we allow ourselves to sit in it and observe the life around us.

49.3 Application

You can apply this method in a group seminar or simply as a personal exercise with a friend. In the following description, the two people applying the method are called participant A and participant B. The overall time frame of the exercise is approximately 60 min, no preparation is necessary. No material is required except a laptop/phone with internet access and optionally a way to take notes. The steps of the method are (1) preparation, (2) research, (3) release, (4) forgiveness, and (5) sharing. Steps 3 and 4 are each repeated once.

1. *Preparation (5 min)*

Clarify what are the expectations of participants. It may be that there are reservations of participants toward the role play to the point of fearing involuntary humor. Conversely, strong emotional reactions could be expected from the verbal sharing of deep feelings. These scenarios should be addressed in advance to allow everyone to participate in this exercise to their full potential.

2. *Research (20–30 min)*

The suggested resource would be the IUCN Red List of Threatened Species (iucnredlist.org). Both participants are invited to browse simultaneously for a target taxon (for example: turtles) and find the species that speaks to them. Information on threat status, habitat, and history of distribution and threats is available online. This research connects them with the living reality of how it is to be a member of this species today. Participants are invited to imagine life as a member of this species across time. There is time for this exercise in the seminar, but, alternatively, it can be an assignment before the session.

3. *Release (2 min)*

After participants have become familiar with the species, they form teams of two people each. Participant A will take the role of the other participant's target species and only listen for 2 min. Participant B will speak freely and openly about his/her feelings of pity, guilt, or whatever else might come up with respect to their target species. It is directly addressed to the target species. The target species does not interrupt the speech.

4. *Forgiveness (2 min)*

After the 2 min, the participant in the role of the target species will respond in the kindest way possible. He/she will not lash out or increase the guilt. There are many things the species might say that are a relief and still reasonable. For example: What would the person apologizing really have been able to do for a bat species in a tropical archipelago? Did they even know about the species' existence? It is important to hear that there is a human who cares. It feels good to hear someone apologize, even though damage has been done. What had been said by the first participant can be taken into account to make them feel heard in their apology. After this first round, either can the roles be reversed directly or participants get together in different pairs. It just has to be made sure that the one apologizing and the target species are switched, whoever just apologized will be the target species for someone who just forgave.

5. *Sharing (10 min)*

In the end, there should be room for feedback and reflection on how this exercise worked or did not work and how participants felt across the different steps of the exercise (Table 49.1). What worked well for them? What would they need—if anything—to feel more comfortable now?

Table 49.1 Method overview

Main purpose
Awareness, empathy, and courage
Gained competences
Knowledge about an endangered species
Educational setting
Can be part of a formal or informal course.
Space requirements/restrictions
Indoor or outdoor activity
Resources and necessary materials
No materials necessary
Number of participants
Participants work in teams of 2 each, and a total of 10 groups is manageable.
Facilitator competences and skills
Some experience as a workshop facilitator is beneficial.
Participants skills/age/competences
All ages and backgrounds
Duration
45 min

Part VI
Complexity and Integrative Thinking

To live wisely in a complex world, we must learn to act with awareness of interconnection and to learn from diverse and sometimes conflicting perspectives. This part presents methods that cultivate these capacities, including systems mapping, multilevel analysis, integrative design frameworks, and more-than-human future practices. These approaches support learners in tracing relationships, recognizing interdependencies, and developing a deep understanding of socio-ecological systems. They reflect a commitment to engage holistically with complexity, rather than reducing it, and to foster integrative thinking that connects disciplines, scales, and ways of knowing. More than analytical tools, these methods are imaginative instruments as they help us design interventions that reach beyond the surface and address the deeper patterns that sustain life. In this way, they support the weaving of regenerative responses into the complex environmental systems, nurturing the capacity not only to understand the world but also to participate in its transformation.

Chapter 50
Reframing Perspectives: A Systems Mapping Exploration

Karla Santos Zambrano

50.1 Introduction

Reframing Perspectives is an explorative system mapping exercise based on the *praxeomorphic way* as identified by sociologist Zygmunt Bauman (2012). According to Bauman, the *praxeomorphic way* is the way in which people understand the world around them, where a person's worldview is informed by the know-how of the day, what people can do, and how they do it. From an etymological perspective, *praxeo/praxis* relates to practice, habit, or custom, leading to an actionable outcome (**action**). *Morphic* relates to shape or form, pertaining to **structure**. Praxeomorphic then refers to the structure(s) that shape our actions or practices. These practices can be formed and informed by how we understand the world around us.

How can we create solutions for a problem we have not experienced ourselves? The Reframing Perspectives method can help us understand complex systems, such as the energy system, through the eyes of individual members of society or from the viewpoint of society as a whole and provide insights into the lived experience of particular phenomena, that is, social justice within the energy transition.

The method contemplates an exploration of a "complex system" involving social-related issues, such as poverty, justice, resilience, social transformation, or other topics comprising societal involvement, such as climate action, sustainable futures, and others. Reframing Perspectives asks us to compartmentalize a chosen

The original version of the chapter has been revised. A correction to this chapter can be found at https://doi.org/10.1007/978-981-96-9029-9_67

K. Santos Zambrano (✉)
Department of Sociology and Criminology, School of Society, Politics and Ethics, University College Cork, Cork, Ireland
e-mail: ksantos@ucc.ie

J. Bentz, J. Ristić Trajković (eds.), *Imagining, Designing and Teaching Regenerative Futures: Art-Science Approaches and Inspirations From Around the World*, Science for Sustainable Societies,
https://doi.org/10.1007/978-981-96-9029-9_50

issue within a given system (e.g., resilience in the energy system) into distinct parts or categories through group collaboration and knowledge co-production (Bolger et al. 2021; OECD 2020) and with the help of a mapping exercise. The categories are based on the *praxeomorphic way*: (1) know-how, (2) what people can do, and (3) how people do it. By know-how, we refer to current practical skills, knowledge, and experiences available to people. What people can do refers to abilities, something a person is able to do. Finally, how do people do it alludes to behaviors, processes, societally-influenced actions, and cooperative approaches, including why people do what they do, for example, putting on an extra sweater instead of turning up the heater could be a learned behavior from family members or a measure to save money.

Understanding the reasons behind people's thinking and processing of information, how they perceive the world, can help us identify gaps or more specific issues within a given system (such as transport, energy, and planning) and provide potential solutions for improvements. However, it is important that the participants have some experience in the issue or topic to explore, such as including low-income households in a study about poverty or wheelchair users in a study about urban realm improvement.

This method can support policymakers who are striving to produce effective policies yet may not fully comprehend the actual concerns or issues their policies are trying to address.

50.2 Transformative Potential

The main idea behind the Reframing Perspectives method is to understand the way others see the world, which could allow for a more comprehensive and inclusive discussion. Based on people's capabilities, the approach looks to deeply understand the reasons behind behaviors and customs (Ariely 2008) so that once identified, these can be acted upon—or integrated into a framework that enables approaching problems differently. In doing so, we might shift the way in which we communicate or frame any issue. The emphasis of the method is on the understanding of the context in which the problem exists and framing it from the perspective of people experiencing that problem through collaborative work and co-creation of ideas and solutions, for example, framing energy justice through the lens of low-income households or viewing playgrounds from a child's perspective.

Regarding regenerative futures, there is a need to further engage with others and unite in making this world more liveable and congenial. Reframing Perspectives attempts to understand and build a perspective that we might not entirely engage with, thus enabling us to understand the wider context and potential solutions to broader systemic issues. It is through the interactions of the system in which the problem is located that we can identify more specific problems and affected groups. Understanding the way the world is perceived by people experiencing a certain problem can be of great value for furthering policies and assisting in community building.

It is worth identifying which perspective is missing in any topic we want to explore. This is particularly important for policymaking and community engagement to help identify current problems or concerns. "Reframing Perspectives" aims to contribute to how we understand the context in which we operate, including the people we engage with.

50.3 Application

Reframing Perspectives is applied as a collaborative workshop that seeks to explore a previously defined topic, such as: How can we improve resilience in the energy system? or How can we make the public realm more accessible to wheelchair users?. This requires preparation from the facilitator.

The duration of the workshop is usually between 2 and 3 h, but this can change depending on the level of depth desired for the exploration of the chosen topic. The categories that divide the system are as follows: *(1) know-how, (2) what people can do, and (3) how do people do it*. For the workshop, the headlines for the categories can be adapted to explore *(1) Technologies/Skills*, *(2) Capabilities*, and *(3) Practices*. The workshop is designed to study the topic through those three categories to gain a comprehensive view of the system.

Step-by-Step Guide

1. Choose a topic of interest to explore, such as *Improving resilience in the Energy System*.

2. Assemble participants with experience or interest in the workshop topic, for example, people vulnerable to changes in the energy system and/or social workers if the issue relates to poverty or energy poverty, or residents of a particular street and/or council planners if the issue involves urban renewal. *Recommended*: Group consisting of 9–12 participants per facilitator for maximum engagement.
3. Gather materials for the workshop. This includes flipcharts, paper, markers, pens, sticky notes, and others, to write and/or visualize what participants say—visual facilitation is encouraged.
4. Prepare for the workshop—the facilitator should understand that the topic is to be divided into three categories: *(1) Technologies/Skills* that are available for people to use, *(2) Capabilities* that people have and count on, and *(3) Practices* or current behaviors associated with the topic.
5. During the workshop, introduce and navigate the topic, briefly explaining the expected outcome, that is, building a mind map that provides an overview and a comprehensive understanding of the topic.
6. Divide the main group into smaller groups of 2 or 3, depending on the total number of participants.
7. In a flipchart, write the category as a headline (as many as necessary for each group) and ask the participants to discuss the chosen topic from the perspective of each category. Participants can write their thoughts directly on the flipchart or on sticky notes before sticking them to the flipchart. For example: 3.

Practices: *List behaviors or practices that can improve people's resilience to changes in the energy system, such as increased electricity prices. Have you done something differently at home, like wearing more clothing to avoid turning on the heating?*

8. Once the small groups have filled out the categories, bring them back to the main group and provide a break of 10–15 min if necessary.
9. Ask the main group to share insights on each category and build a mind map with the answers provided by the group. This mind map provides a comprehensive awareness of the chosen topic and an overview of the parts involved in the system.
10. If potential calls to action are identified, they should be linked to the relevant category. The identified ideas or solutions can be included in a report for further reflection.
11. *Optional*: Take photos of the materials produced in the workshop (flipcharts, drawings, and mind-maps) and use the contents for further analysis or reflective discussion in a report.

It is important to note that the perspective that is being mapped can be unexpected, thus allowing for a perspective to be reframed from what was originally thought. Additionally, facilitators can use other engagement methods to provide a deeper insight into the topic being explored or other types of material outcomes.

Example: Fig. 50.1 contains an example of an application that seeks to gain insights into building resilience in the energy transition. By applying this method as a system mapping, we can use the categories: Technologies (i.e., know-how), Capabilities (i.e., what people can do), and Practices (i.e., how they do it), to build a mind map that allows to have a comprehensive overview of the parts contained within the energy transition that can help us build resilience. After setting up the group and finalizing a brief discussion, the group provides suggestions or items about the topic in relation to each category, for example: *What are the technologies available today that pertain to this issue? Solar panel installation can help alleviate reliance on fossil fuels.* This is done with each category to build a final mind map including all the categories. Based on the perspectives arising from the exploration, we can find and list further calls to action or solutions, and get an overview of the system as a whole (Table 50.1).

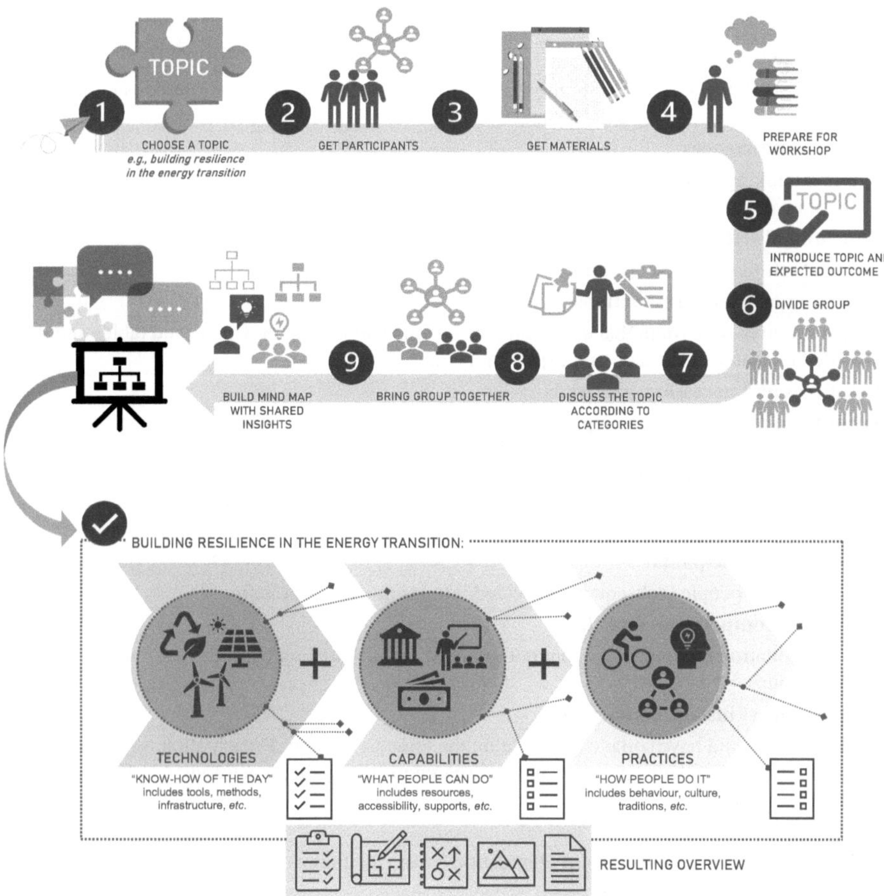

Fig. 50.1 The method applied to an energy system topic. (Author's original illustration based on Microsoft ClipArt)

Table 50.1 Method overview

Main purpose
Awareness building/system exploration/identification of stakeholders and solutions/narrative shift
Gained competences
Systems thinking and mapping of complex scenarios, new perspectives on a specific issue, and collaborative work.
Educational setting
Informal setting—workshop format.
Space requirements/restrictions
Indoor activity—one room that can fit all participants and facilitator. Chairs and tables should be provided to enable discussion and exploration among groups.
Resources and necessary materials
What to do? Pre-workshop preparation for facilitator (topic chosen before the workshop); no previous reading or preparation required for participants. **What is needed?** Materials for the workshop: paper, flipcharts or board, post-it/sticky notes, markers, pens, pencils—sticky tape and scissors or any reusable adhesive to stick papers on chart or board if required.
Number of participants
Group of 9–12 people (subsequently divided into groups of 2 s or 3 s)
Facilitator competences and skills
Visual facilitation recommended, but not required; good visual and oral expression; good at leading groups.
Participants skills/age/competences
Participants should have some experience on the topic to explore. Participation is welcomed from community, students, policy makers, researchers, designers, practitioners, and general public.
Duration
2–3 h (not including pre-workshop preparation by the facilitator)

References

Ariely D (2008) Predictably irrational: the hidden forces that shape our decisions. HarperCollins, New York

Bauman Z (2012) Liquid modernity. Polity Press, Cambridge

Bolger P, Brereton P, Grant O, Torney D (2021) Better together: knowledge co-production for a sustainable society. Royal Irish Academy, Dublin

OECD (2020) Addressing societal challenges using transdisciplinary research, OECD Science, Technology and Industry Policy Papers 88. OECD Publishing, Paris

Chapter 51
Systems Sensing Approach

Luea Ritter, Justus Wachs, and Nancy Zamierwoski

51.1 Introduction

Systems sensing is a broad term for somatic approaches that utilize felt senses to expand awareness, providing deeper insights into complex interrelationships within systems, such as organizations or networks. It connects with our inherent capacities for intuiting, perceiving, relating, and sense-making through diverse and subtle ways of knowing.

This approach is informed by systems thinking (Meadows 2008), a field that studies complex systems with their dynamic and interdependent relationships. Systems sensing adds a dimensionality of felt senses to systemic analyses. It includes somatic sensations, intuition, and emotions as signals exploring a system's interactive relationships, quality, and resonances (Ritter and Zamierowski 2021, p. 101–115).

The explorative sensing journey is one method within systems sensing. It is suitable for individuals and teams to explore shared topics and issues. It can be used within a larger transformative change process or as an intervention during challenges. The method's power lies in surfacing hidden relational, emotional, and

L. Ritter (✉)
Collective Transitions, Muttenz, Baselland, Switzerland
e-mail: luea@collectivetransitions.com

J. Wachs
McGill University, Montreal, QC, Canada
e-mail: johann-justus.wachs@mail.mcgill.ca

N. Zamierwoski
Collective Transitions, Muttenz, Baselland, Switzerland Transformative Studies, California Institute of Integral Studies, San Francisco, CA, USA
e-mail: nzamierowski@mymail.ciis.edu

J. Bentz, J. Ristić Trajković (eds.), *Imagining, Designing and Teaching Regenerative Futures: Art-Science Approaches and Inspirations From Around the World*, Science for Sustainable Societies,
https://doi.org/10.1007/978-981-96-9029-9_51

socio-cultural dynamics, including beliefs, assumptions, and mental structures. By widening the breadth and depth of awareness of underlying dynamics, teams can generate shared understanding and greater compassion, helping align resources and move forward with clarity.

This practice includes three steps: (1) developing a clear guiding question and intention and selecting essential elements, (2) the embodied practice of sensing into the elements, and (3) debriefing the experience as a group. We guide you through these steps using a case example in the final section of this chapter.

51.2 Transformative Potential

We recognize that the future does not lie ahead of us but exists right here in our current ways of being, thinking, and acting across perceived divides. The patterns that are present now determine how the future can unfold. These patterns may include past experiences that can continue influencing our current reality. Individuals or groups experiencing fragmentation, separation, or confusion can find relief by addressing existing and often deeply ingrained patterns. Catalyzing regenerative change on individual and collective levels requires methods that can bring these patterns into conscious awareness. Surfacing and intentionally working with these patterns can increase a sense of common ground, clarity, and alignment.

In our experience working within multiple sectors, including non-governmental organizations, activist groups, and organizations, systems sensing methods are invaluable in surfacing conscious and unconscious patterns so that they can be addressed and released. These methods uncover underlying beliefs, assumptions, and mental structures that may contribute to challenges or conflicts. Beyond just thinking about these challenges, systems sensing methods draw on diverse ways of knowing, including our bodies, emotions, and intuition, as instruments of perception. In our experience, a wider range and diversity of signals can support organizations in navigating the uncertainty inherent in tackling complex challenges. It can also help deepen their sense of embeddedness and belonging in wider systems, which can catalyze more meaningful collaborations and innovative solutions.

51.3 Application

This section discusses a case study as a practical application of an explorative sensing journey. We facilitated the exercise for a worker-owned consultants cooperative. The cooperative had reached an impasse in their search for value-aligned models to structure their work reconciling the diverse interests of individual members with their collective purpose. Our intervention aimed to facilitate movement toward greater collaboration. For brevity, we only discuss the general outline of the explorative sensing journey. To learn more or receive training in facilitating these methods, readers are invited to contact the authors.

51.3.1 Step 1: Crafting the Guiding Question and Elements (1 h)

In our first contact with the cooperative, we met several team members to understand the organization's challenges and aspirations. From this shared understanding, we collaboratively (1) crafted a guiding question and (2) selected elements for the sensing journey.

1. *Guiding question:* A guiding question defines the scope of the exploration and the underlying purpose of the practice. It should be open and welcoming to multiple responses. Instead of having a fixed answer, it should invite the exploration of diverse perspectives and interpretations.

For the co-op, we co-designed the following guiding question: "What is important to know and to nurture now toward a shared vision as a co-op and a regenerative future with mutual benefit for all?"

2. *Elements:* During the exploration, participants are invited to visit and embody different "elements" that are relevant to the given system; these may include people, places, emotions, qualities, or essential components of a system.

The team chose six elements: shared vision, personal motivation, unique gifts, contribution, pain, and an individual resource.

51.3.2 Step 2: The Explorative Sensing Journey (1–2 h)

To prepare, ensure that participants have materials to take notes and a comfortable space to sit, stand, or move around. The session may begin with a grounding exercise and a moment to bring the guiding question into awareness. Then, everyone finds a place that functions as a "neutral resting space" to come back at any point.

The facilitator guides participants from element to element at a slow pace, inviting them to "visit" an element and get to know its perspective. The invitation is to notice the qualities of the experience in each moment. For example, what is it like to transition to the element? Once "inside," what emotions or sensations are present in the body? What other signals can participants perceive? After visiting each element, the participants step back into neutral and note emerging insights.

In our session, the team explored the elements of *shared vision, personal motivation, unique gifts, contribution, pain,* and *an individual resource*. When visiting the "unique gifts," for example, participants were invited to experience the quality of these gifts, what made them uniquely positioned to do their work individually, and what connected them to the greater whole.

When complete, we guided participants in shedding the elements' roles and consciously closing the practice. We ended the session with a mindful moment and shared insights that surfaced. For example, the participants noted a strong sense of interdependence and connection and the clarity they could find what they needed to resource their collaboration within the collective.

51.3.3 Step 3: Reflection and Sense-Making (1–2 h)

We suggest integrating the experience for a few days before hosting a reflective session with others, beginning the group's sense-making process.

We suggest starting with check-in questions such as:

- "How are you coming to this session, and what is a sensation that stayed with you from the practice?" Moreover, the group can explore reflection questions such as:
- What are your main insights and what surprised you?
- What are important takeaways for us and our shared work?
- What could a gentle next step look like supporting integrating these insights?
- What calls for more exploration or is not yet clear?

Sometimes, witnessing and hearing each other's experiences might be "all" that's needed to foster a deeper sense of shared purpose and compassion for the complementary perspectives. Furthermore, it can support the surfacing patterns best addressed collectively.

Key insights included that the exercise allowed for (1) a powerful level of emotional honesty, (2) an understanding of the interdependence of relationships and a stock-taking with where individuals had contributed to collective challenges, and (3) new realizations of the gifts they contribute that had not been appreciated before. During the integration session, they experienced how much emotional charge they could hold safely now among themselves; this sense of capacity felt like a gift.

51.3.4 Conclusion

This contribution introduces systems sensing and offers the explorative sensing journey as a tool leaders and organizations can use to catalyze regenerative change, individually and collectively (Table 51.1). We acknowledge that working toward

Table 51.1 Method overview

Main purpose
Expanding awareness/working within complexity/including wider ways of knowing/collective sense-making
Gained competences
Individual: presence/compassion for other perspectives/accessing deeper insights into complex interrelationships within systems/deep listening to one's inherent capacities for intuiting, perceiving, relating, and sense-making through diverse and subtle ways of knowing As a group: deep listening to each other/being present with what is/holding emotional charge as a group/sense-making
Educational setting
Team workshop
Space requirements/restrictions
Outside, indoor, or online activity
Resources and necessary materials
Notebook and pens
Number of participants
Group and individual work/group size open to the context/one group
Facilitator competences and skills
Systems thinking knowledge, understanding of inquiry design, practice experience, ability to create a participatory environment, calm abiding presence, active listening, ability to be with tension and conflict, attunement with the group, and ability to guide the group through a contemplative practice
Participants skills/age/competences
Open to all people from all kinds of disciplines, backgrounds, professions, and ages/competence to be open, curious, and relating
Duration
3–5 h

regenerative futures requires a commitment to continuously face and be with unresolved and entrenched habits of thinking, collaborating, and working. At the same time, this path is filled with unacknowledged gifts and dormant capacities. Our experience has shown us that systems sensing can be an effective approach to releasing patterns and unleashing untapped potential (Wachs and Ritter 2023).

References

Ritter L, Zamierowski N (2021) Systems sensing and systemic constellation for organizational transformation: building collective capacity for navigating complexity. JASC 1(2):101–115

Wachs J, Ritter L (2023) Systems sensing for transformations organizations: lessons from the transformations conference 2023. SIJ 22

Meadows, D. H. (2008). Thinking in systems: A primer. Chelsea Green Publishing

Chapter 52
Exploring the Regenerative Potential of the Micro-Level

Milena Grbić

52.1 Introduction

When it comes to climate change, many people find themselves uncertain about what actions to take or may believe that their individual efforts are insufficient to address such a critical issue. However, believing that regeneration is what life has always done, there is a hypothesis that suggests that at the household or individual level (micro-level), there are conscious or subconscious actions that can still contribute to the fight against climate change.

This methodology aims to shed light on these potential actions as part of a broader strategy for regeneration. In other words, the intention is to explore and discover the actions of individuals who, through conscious or unconscious gestures, participate in the process of regeneration, so that at the level of an individual household, they can be supported and strengthened, focusing on more complex and long-term goals in the process of regeneration.

The methodology's intention is to encourage individuals (especially students engaged in disciplines related to spatial design) to explore the micro-level aspects of the urban environment, tapping into their intuitive understanding of how spaces can contribute to regeneration. It draws inspiration from Immanuel Kant's philosophy, which suggests that our minds comprehend only what falls within our empirical grasp, in rational concepts, but by treating the surrounding space as a tangible empirical world, one can uncover a deeper meaning. This approach incorporates personal perception to deconstruct conventional norms related to the concept of regeneration.

M. Grbić (✉)
University of Novi Sad – Faculty of Civil Engineering, Subotica, Serbia
e-mail: mgrbic@gf.uns.ac.rs

J. Bentz, J. Ristić Trajković (eds.), *Imagining, Designing and Teaching Regenerative Futures: Art-Science Approaches and Inspirations From Around the World*, Science for Sustainable Societies,
https://doi.org/10.1007/978-981-96-9029-9_52

The observer is the one who conducts the research, the observed is a specific segment of urban space. What is essential is for the mind to remain open to every stimulus. It is from this intuitive basis that creative power emerges.

52.2 Transformative Potential

The insights generated through this methodology are highly diverse and uniquely illuminate the behavior and decisions of individuals and their impact on the regeneration process at the micro or individual level (e.g., materialization of a house, ways of terrace or yard usage, shades, small vegetable gardens, rainwater usage, and presence of outdoor toilets).

At its core, this method explores the subtle interactions between what we observe and how we interpret it. On a micro level, everything becomes a realm of pure signals, filled with clues and hints that can inspire future actions. By noticing and understanding these signals, we gain numerous insights and directions for future decisions, many of which may be unexpected—such as recognizing an expression on someone's face during a conversation that shifts the course of the interaction in an unforeseen direction. This methodology is crucial because it helps us identify hidden opportunities and make informed decisions, even in seemingly unpredictable situations. The greatest value of this process lies precisely in the journey from discovery to decision-making.

Simultaneously, this methodology underscores the importance of acknowledging and embracing the broader context on the path to regeneration. Rather than crafting concepts and strategies in isolation, it advocates for an integrative approach that considers the context in its broadest sense. This inclusive perspective can serve to empower individuals, reinforcing the notion that they are already making meaningful contributions toward regeneration. It emphasizes that substantial deeds are not the sole requirement; even seemingly small gestures hold significant value and contribute meaningfully. And all of this can be included and stimulated through specific design practices if understood correctly.

The application of this methodology can be effectively integrated into educational curricula as an exercise, offering an excellent means of assessing its efficacy and fostering a deeper understanding of its principles among students.

52.3 Application

This method combines systematic observation with thoughtful interpretation, enabling meaningful comparisons and actionable insights for spatial regeneration. This task is a structured observation exercise requiring students to deeply analyze a specific spatial area and its interactions through systematic categorization.

52.3.1 Steps for Completing the Task

52.3.1.1 Step 1: Select a Specific Spatial Domain

- Choose a city block containing either single-family houses or apartment buildings.
- Define the boundaries of the domain, understanding that the entire space is observed subjectively.
- Represent the spatial domain using the largest circle in the diagram to graphically delineate the observed area.

52.3.1.2 Step 2: Observe and Record Phenomena

- Identify physical phenomena within the domain that exhibit regenerative potential. Phenomena represent operational domains in a broader sense of functioning. They are physical locations where the potential for regeneration has been revealed. What is important is to carefully examine the sphere of influence of each phenomenon and determine whether these influences extend to other phenomena, working together or operating independently. Pay attention to their physical presence, boundaries, and domains of influence (e.g., shading a garden, reflective roof surfaces, air-purifying plants, and community gardens).
- Represent phenomena in the diagram with three concentric circles: Innermost: The phenomenon's size relative to the domain; Middle: Nature of its boundary; and Outer: Its sphere of influence or impact.

52.3.1.3 Step 3: Identify and Examine Institutions

- Recognize existing institutional elements within the area that regulate behavior or facilitate regeneration (e.g., recycling bins, public transport stops, and water systems).
- Observe the frequency and manner of use and assess their impact.
- Represent institutions using circles with varying shades to indicate the intensity of their regulatory influence: Light shades: Low influence; Darker shades: Strong influence. Adjust the circle size to reflect their authority within the observed space.

52.3.1.4 Step 4: Analyze Agents

- Observe individuals (agents) using the space.
- Note down their movements, interactions with phenomena, and use of institutions.
- Investigate unique behavioral patterns and predictability of actions.

- Pay particular attention to: agent–institution relationships—how agents use or are influenced by institutions; agent–phenomena interactions—how agents modify or adapt to the space.
- Represent agents with the smallest circles in the diagram.

52.3.1.5 Step 5: Identify Desires and Aspirations

- Observe the intentions, needs, or preferences of agents, particularly related to regenerative potential (e.g., the desire to plant more trees, use public spaces, or install renewable energy systems, etc.).
- Use a combination of direct observation (e.g., behaviors or gestures) and indirect methods (e.g., dialogue).
- Represent desires and aspirations with crosses in the diagram.

52.3.2 How to Use Diagrams Effectively

The diagram is a relational database capturing the relationships between spatial domains, phenomena, institutions, agents, and desires. Create layers in the diagram to depict different categories and use unique identifiers for every observed occurrence. An example of a diagram is provided in the illustration, and the method of categorizing the observed data should be done according to the following classification (Fig. 52.1):

- Base layer: The spatial domain.
- Phenomena layer: Physical elements and their interactions.
- Institution layer: Regulatory structures and their influence.
- Agent layer: Movement and interaction of individuals.
- Dynamic elements layer: Aspirations and their impact on regenerative processes.

52.3.3 Considerations for Observation

- Time on-site: Spend ample time in the space to notice patterns and subtle interactions.
- Tools for observation: Use a notebook for detailed notes, diagrams, and optionally, photographs for reference.
- Behavioral nuances: Observe the time of day and environmental conditions, as these may influence behaviors and patterns (Table 52.1 and Fig. 52.1).

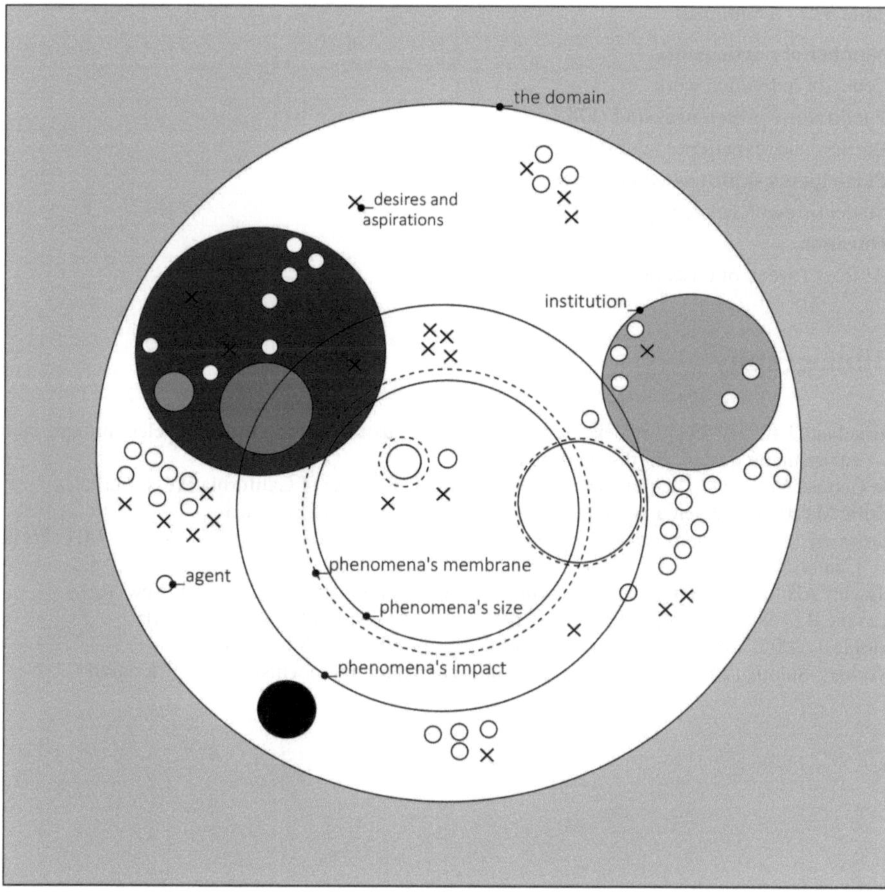

Fig. 52.1 Diagram example

Table 52.1 Method overview

Main purpose
Behavior change
Gained competences
Encouraging individuals engaged in disciplines related to spatial design that it is possible to contribute to the process of regeneration on multiple levels through the aspect of design
Educational setting
Informal (format—workshops)
Space requirements/restrictions
Outside activities
Resources and necessary materials
Paper and pencil

(continued)

Table 52.1 (continued)

Number of participants
Group or individual work
Facilitator competences and skills
Background, experience, and skills
Participants skills/age/competences
Students/researchers/designers/practitioners
Duration
At least 1 week of observation and recording

Bibliography

Birkeland J (2010) Positive development: from vicious circles to virtuous cycles through built environment design. Earthscan, London
de Certeau M (1984) The practice of everyday life. University of California Press, Berkeley
Grbić M (2023) ElementarNo.1. Građevinski fakultet Subotica, Subotica
Gross M (2010) Ignorance and surprise: science, society, and ecological design. MIT Press, Cambridge
Hawley AH (1950) Human ecology: a theory of community structure. Ronald, New York
LeVine R (1982) Culture, behavior, and personality. Aldine Publishing, New York
Moran J (2005) Reading the everyday. Routledge, London and New York
Norberg-Šulc K (2002) Egzistencija, prostor i arhitektura. Građevinska knjiga, Beograd

Chapter 53
Spatial Storytelling to Mitigate Urban Sprawl

Markus Schaffert and Cemre Sahinkaya

53.1 Introduction

Storytelling is a scientific method that uses narrative techniques to convert complex scientific information and research findings into an engaging and understandable format. It serves a vital role in presenting scientific discoveries, capturing audience interest, and making content easily comprehensible. Narrating future scenarios and articulating our desired self-perceptions of the future is the crucial first step in turning these aspirations into reality. Digital storytelling combines text, images, videos, and interactive elements, while spatial storytelling incorporates geographical context using maps and spatial data.

Spatial storytelling can have a profound impact on urban planning, challenging entrenched assumptions. Geovisualizations simplify intricate spatial and temporal relationships, fostering better communication among diverse professionals. In global climate change discussions, the focus shifts from "what," "where," and "when" to the crucial question of "how." Visualizing climate change processes requires open dialogue and interactive engagement, with storytelling playing a pivotal role in fostering understanding (Veland et al. 2018).

M. Schaffert (✉)
i3mainz c/o Mainz University of Applied Sciences, Mainz, Germany
e-mail: markus.schaffert@hs-mainz.de

C. Sahinkaya
i3mainz c/o Mainz University of Applied Sciences, Mainz, Germany
e-mail: cemre.sahinkaya@hs-mainz.de

J. Bentz, J. Ristić Trajković (eds.), *Imagining, Designing and Teaching Regenerative Futures: Art-Science Approaches and Inspirations From Around the World*, Science for Sustainable Societies,
https://doi.org/10.1007/978-981-96-9029-9_53

53.2 Transformative Potential

Urban sprawl in rural Germany is marked by unsustainable sealing of soil, increased energy use, greenhouse gas emissions, and reduced green spaces (Privitera et al. 2018). Due to the negative impacts of urban sprawl, local sustainability targets have been set in Germany to reduce land consumption. The 30-hectare target was specified as part of the federal government's national sustainability strategy in 2002 and was later incorporated into the spatial planning laws of the federal states. It states that the daily nationwide land taken for settlement purposes in Germany should be limited to a maximum of 30 hectares per day. The current federal government has now even agreed to achieve a circular economy without new urban land take by 2050, aligning with European objectives. However, land consumption for residential and commercial purposes remains high, with municipalities often failing to implement federal objectives (Fina 2021). Despite hopes for growth in peripheral rural areas, these expectations are often misleading, and a change in approach shifting away from a focus on quantitative urban growth in planning and development is needed.

Therefore, alternative methods are necessary to make urban sprawl more visible and understandable. Here, storytelling serves as an important tool to address this issue with great potential, especially in spatial terms. First, spatial storytelling facilitates the documentation and communication of spatial and intangible data that were previously unrecorded, thereby revealing the true extent of urban sprawl in rural areas through a holistic approach. Second, storytelling simplifies the concept of urban sprawl prevention, making it more comprehensible for local communities compared to complex theoretical policies. This approach enhances awareness and cooperation, enabling a more participatory resolution of the issue. Finally, storytelling fosters community engagement and shared understanding, which is crucial for effective urban planning and sustainable development. For all these reasons, spatial storytelling can make a significant contribution to the transition to a new and more sustainable development by preventing the ongoing urban sprawl in rural areas and efficiently contributing to regenerative futures (Schaffert and Sahinkaya 2024).

53.3 Application

The importance of local people in municipal politics is often overlooked in the fight against urban sprawl. Although they readily organize against issues like wind turbine projects, such movement is uncommon for new residential or commercial developments. This disparity is likely due to the gradual nature of urban sprawl, which unfolds over generations rather than occuring abruptly. Spatial storytelling with the help of geovisualization offers a promising approach to make the impacts of urban sprawl in rural areas more apparent, emphasizing long-term consequences and advocating for sustainable development.

The first step is based on data preparation which involves gathering aerial photographs and quantitative data on population number and density, encompassing both historical and current datasets. Aerial images, both historical and current, vividly illustrate substantial land consumption since the 1950s with minimal effort. The use of a current aerial photograph and the marking of the area that has grown since the Second World War provide us with the extent of urban sprawl and the resulting difference in settlement structure between the town core and the later zones of expansion. Stakeholders such as municipalities, local administrations, residents, non-governmental organizations, and other organizations are contacted to ensure their participation in this phase.

The second step involves geospatial analyses and visualizations that integrate spatial and quantitative data to show urban expansion over time. More detailed insights can be generated using additional data (e.g., population and building density) and geographic information systems. Historical maps and animations can show settlement expansion over different periods, presenting temporal changes in an accelerated format.

In the third step, workshops are organized where stakeholders participate in story-telling interviews. These begin with an introduction to urban sprawl and use spatio-temporal visualizations to prompt personal reflections among participants from diverse backgrounds. During the workshop, local residents and other stakeholders are also encouraged to share their narratives and memories about significant landmarks and reference points in the selected area focusing on the changes over time in terms of urban development and urban sprawl. This fosters collective memory and objective discussion. Critical questions are posed, and new narratives emphasizing qualitative growth are developed with the community, addressing conflicts and featuring fictional or real protagonists. Examples of these questions could include: "*Why do you think urban sprawl has continued, even without significant population growth? What role can each of us play, individually or collectively, in shifting our growth patterns toward more sustainable practices? How do you envision growth that meets community needs without further outward expansion?*" The information is digitally processed, refined, and documented collaboratively by the project team to create an interactive, feedback-driven output involving local stakeholders.

As the fourth step, digitalization is performed. At this stage, the spatial data obtained from the story narratives of the interviews conducted in the previous step are transferred to maps with appropriate symbologies. This process allows for the creation of numerous up-to-date and comprehensive visualizations on various themes related to urban sprawl, enriched by the narratives and intangible data collected.

In the final step, an evaluation is made through visualizations created with spatial data reinforced with storytelling, so that the changing/increasing trends of urban sprawl over the years are revealed in a more clear and understandable way with both quantitative and qualitative/abstract data. By recording previously unrecorded abstract information, stakeholders from diverse backgrounds contribute to a deeper understanding of the urban sprawl process which makes the issue clearer, more

Table 53.1 Method overview

Main purpose
The main purpose is to visualize and present the urban sprawl in rural Germany over time and to make it more understandable, especially to the local population with geospatial information through the storytelling method
Gained competences
System thinking, collaborative thinking, and geospatial thinking
Educational setting
None
Space requirements/restrictions
Indoor/outdoor activities through a workshop
Resources and necessary materials
Digital storytelling tools, geovisualization software, indoor place for a workshop, hard copies of the visualizations, and drafting materials
Number of participants
10–100
Facilitator competences and skills
Persuasion, oratory and communication skills, basic geoinformatics, and mapping knowledge
Participants skills/age/competences
All skill levels and ages, living in the same area for a long time is a plus
Duration
Geovisualizations and digitalization would need from 1 to several weeks and the workshop can take a day or longer (if it is more complex)

visible, and more comprehensible to all stakeholders, facilitating collective action against urban sprawl and promoting the adoption of more sustainable policies (Table 53.1).

References

Fina S (2021) Nachhaltigkeitslücke Landschaftszersiedelung. Berichte Geographie und Landeskunde 94(4):278–299

Privitera R, Palermo V, Martinico F, Fichera A, La Rosa D (2018) Towards lower carbon cities: urban morphology contribution in climate change adaptation strategies. Eur Plan Stud 26(4):812–837

Schaffert M, Sahinkaya C (2024) Mit Geovisualisierungen zum Flächensparen? Eine Ideenskizze. In: Meinel G, Behnisch M (eds) Flächennutzungsmonitoring XVI—Flächenpolitik—Flächenanalysen—Methoden und Werkzeuge. Rhombos-Verlag (IÖR Schriften Band 82), Berlin

Veland S, Scoville-Simonds M, Gram-Hanssen I, Schorre AK, El Khoury A, Nordbø MJ et al (2018) Narrative matters for sustainability: the transformative role of storytelling in realizing 1.5 C futures. Curr Opin Environ Sustain 31:41–47

Chapter 54
Sustainable Business Model Canvas

Sabrina Schork

54.1 Introduction

The *Sustainable Business Model Canvas* is a strategic management tool that helps organizations develop, visualize, and innovate their business models with a focus on sustainability. It extends the traditional Business Model Canvas to include elements that specifically address environmental, social, and economic impacts. The *Sustainable Business Model Canvas* encourages companies to consider not only their profitability but also their responsibility to the planet and society. It consists of 13 elements and can be downloaded from the Threebility website (see references).

Inspiration for the *Sustainable Business Model Canvas* can be drawn from the Circular Pattern Cards and Business Model Navigator Cards created by the Business Model Innovation Lab (2024), a spin-off of the University of St. Gallen. The Circular Pattern Card Set (2024) is physical and can be ordered from a website listed in the references. The Business Model Navigator Cards (2024) are available online (free of charge) and can be accessed from the website listed in the references. The cards are designed to inspire participants to come up with innovative ideas for sustainable value creation and revenue streams. The *Sustainable Business Model Canvas* can be completed without the inspiration cards, but participants tend to generate less diverse or innovative solutions without the card sets.

Sustainable business models such as the one from Threebility (2024) are powerful because they enable companies to thrive economically while making a positive impact on society and the environment. This holistic approach to business ensures

S. Schork (✉)
Aschaffenburg University of Applied Sciences, Faculty of Economy and Law, Aschaffenburg, Germany
e-mail: sabrina.schork@th-ab.deTH

J. Bentz, J. Ristić Trajković (eds.), *Imagining, Designing and Teaching Regenerative Futures: Art-Science Approaches and Inspirations From Around the World*, Science for Sustainable Societies,
https://doi.org/10.1007/978-981-96-9029-9_54

long-term success, fosters innovation, enhances reputation, and contributes to a more sustainable and equitable world.

Sustainable business models can be applied in a variety of contexts and industries, including corporations, small- and medium-sized enterprises, start-ups, non-profit organizations, educational institutions, and the public sector. By adopting sustainable practices, organizations in these contexts can drive long-term success while making a positive contribution to society and the environment.

54.2 Transformative Potential

The broader impact of sustainable business models extends beyond individual companies to include environmental protection, economic resilience, and social equity. These models are powerful tools for creating a sustainable and equitable future.

Sustainable business models are powerful enablers of a regenerative future because they integrate environmental, social, and economic considerations into core business strategies, driving innovation and long-term resilience. By prioritizing resource efficiency, reducing waste, and promoting social equity, these models not only mitigate negative impacts but actively restore and enhance ecosystems and communities. This holistic approach not only ensures business viability but also contributes to a healthier planet and more equitable societies, making it a fundamental element in building a regenerative and sustainable future.

The transformative potential of sustainable business models lies in their ability to fundamentally reshape industries by embedding sustainability at the heart of economic activity. These models drive systemic change by fostering innovation, improving resource efficiency, and promoting social equity, resulting in more resilient and adaptable businesses. When companies adopt sustainable practices, they set new industry standards, influence consumer behavior, and catalyze broader societal shifts toward sustainability, ultimately transforming how value is created and shared in the economy. The environmental potential of sustainable business models is profound, as they prioritize reducing environmental impacts through efficient resource use, waste minimization, and the use of renewable energy sources. By designing products and processes that support circular economy principles, these models help conserve natural resources, reduce pollution, and protect biodiversity. This approach not only mitigates the negative impacts of industrial activities but also actively contributes to the restoration and regeneration of ecosystems, paving the way for a healthier and more sustainable planet.

54.3 Application

Target audiences:

- Students creating sustainable business models as part of an entrepreneurship and/or innovation course.

- Executives or business strategists who want to transform their business model into a sustainable one.
- Interdisciplinary teams from start-ups and medium, small, or large companies who want to create sustainable business models.

Preparation tips for facilitators:

- Set up a flexible room with round tables, seating four people per table (up to 100 participants).
- Prepare poster printouts of the *Sustainable Business Model Canvas.*
- Provide one set of Business Model Navigator and one set of Circular Pattern Cards for each table.
- If possible, provide Wi-Fi and tablets are available so that participants can access the *Sustainable Business Model Canvas* and Business Model Navigator cards online.

Duration: 4 hours.
Process: Complete each of the 13 *Sustainable Business Model Canvas* elements together.

- Empathize with the target audience: Divide the foursomes into twos and have them conduct a 1:1 design thinking interview empathizing with the target group. Each person steps into the target group once and is interviewed by their partner once for 4 minutes. The interviewer then spent 3 minutes delving deeper into the points where emotions were raised.
- After 15 minutes, stop the interviews and have the foursome create an empathy map together (Interaction Design 2024). It is very important to stay on schedule (30 minutes).
- Now the four identify the core problem of the sustainable value proposition. If the participants are blocked, they can use the Problem Statement Canvas (Disciplined Entrepreneurship, 2021) (30 minutes).
- Define the positive and negative impacts of your value proposition, using the 1–2-4-All method (Liberating Structures, 2024), and apply the pattern cards where necessary (30 minutes).
- Create a simple visual prototype of your solution. Test it with members of the target group or team members from other groups (what is possible at that time). Pitch the final solution to a business coach in 60 seconds and integrate feedback (1 hour).
- Specify the resources, activities, and partners to implement your solution, using the pattern cards as needed (30 minutes).
- Estimate the fixed and variable costs of your business model. Set realistic prices and forecast customer numbers for the first 5 years. Determine the break-even point. Validate business assumptions with a market expert (1 hour).
- Research subsidization programs that fit your sustainable solution (optional).

Case Studies:

- Master Course in Strategic Innovation Management: Creation of a *Sustainable Business Model Canvas* for self-selected societal problems in groups of four. Use of circular pattern and business model navigator cards.
- Bachelor Elective: Creating Sustainable Business Models for Digital Innovations: Creation of a *Sustainable Business Model Canvas* for a given creative question targeting the sustainable development goal "Quality of Education." Based on this, the development of a 5-year business case.
- Lisbon Summer School on Innovation, Creativity, and Entrepreneurship: Creation of a *Sustainable Business Model Canvas* in 10 groups of 6–7 people from 6 countries. Use the circular pattern and business model navigator cards as inspiration. Address United Nations Sustainable Development Goal 4, focusing on activating GenZ in sports with a sustainable solution.

Considerations:

- Participants should be at least 18 years of age.
- Provide adequate time allocation and a comfortable environment with all necessary materials (Table 54.1).

Table 54.1 Method overview

Main purpose
Development of socially and environmentally compatible business models
Gained competences
Entrepreneurial action/change management/problem solving/creativity/decision-making/cooperation/future design
Educational setting
Formal or informal setting (depending on the participants)/open and adaptable space (natural light, fresh air, nature-based materials, colorful design environment)
Space requirements/restrictions
Indoor activity/several tables for three to four people
Resources and necessary materials
1 circular pattern card set per table/1 business model navigator card set per table/1 *sustainable business model canvas* poster per table/post-its; pens/additional canvas where necessary (i.e., empathy map or problem statement canvas)/online timer
Number of participants
3–100
Facilitator competences and skills
Personal coaching for individual advice and liberating structures for bigger groups
Participants skills/age/competences
Communication/reflection / > 18
Duration
4 hours

References

Business Model Navigator, Business Model Navigator Cards, https://businessmodelnavigator.com/explore. Accessed 06 June 2024

Business Model Navigator, Circular Pattern Cards, https://bmi-lab-shop.myshopify.com/products/test-circular-economy-pattern-cards. Accessed 06 June 2024

Disciplined Entrepreneurship, Problem Statement Canvas, https://www.d-eship.com/articles/the-problem-statement-canvas-a-deep-dive-in-problem-definition-for-startups-and-innovation-teams/. Accessed 06 June 2024

Interaction Design, Empathy Map, https://www.interaction-design.org/literature/topics/empathy-mapping. Accessed 06 June 2024

Liberating Structures, 1-2-4-All, https://www.liberatingstructures.com/1-1-2-4-all/. Accessed 06 June 2024

Teaching Entrepreneurship, Design Thinking Interviews, https://www.teachingentrepreneurship.org/design-thinking/. Accessed 06 June 2024

Threebility, Sustainabile Business Model Canvas, https://www.threebility.com/_files/ugd/b67836_7bd275b345cf4f20bc560590113a8f47.pdf. Accessed 06 June 2024

Chapter 55
Integral Ecology: A Method for Education in Addressing the Climate Crisis

Zoran Turza

55.1 Introduction

Learning and leadership during times of global change require the active engagement of all participants in social processes. Social groups and associations play a significant role in communicating the importance of global changes to their members and encouraging them to adopt specific changes in habits to address global challenges better. This process involves adapting to climate change, reducing human impact on the environment, and directing human influence toward restoring the entire environment. To drive regenerative changes, active participation at different levels is essential. This involves acting at social life's micro, meso, and macro levels, enabling individuals to participate and actively initiate various regenerative changes.

Integral ecology is a method that enhances transformative and regenerative changes. While it targets Catholics explicitly, it may also apply to other Christian communities and all people, promoting a sense of community and shared responsibility. This method is based on Pope Francis' encyclical letter "Laudato Si′" (2015), which emphasizes the concept of "integral ecology." According to Pope Francis, we face not separate social and environmental crises but a single socio-environmental crisis. Integral ecology seeks to address not only individual social or ecological issues but also the broader issues stemming from this crisis, such as economic inequality, urban planning, sustainable development, and changes in behavior to reduce human ecological impact. This interconnectedness of social and environmental problems highlights the seriousness of the situation and the need for a comprehensive approach.

Z. Turza (✉)
Catholic University of Croatia, Zagreb, Croatia
e-mail: zoran.turza@unicath.hr

J. Bentz, J. Ristić Trajković (eds.), *Imagining, Designing and Teaching Regenerative Futures: Art-Science Approaches and Inspirations From Around the World*, Science for Sustainable Societies,
https://doi.org/10.1007/978-981-96-9029-9_55

To tackle the climate crisis effectively, we can follow three key stages: grasping the theoretical framework, raising awareness about the issue, and engaging in practical transformation.

The first stage provides insights into knowledge about the climate crisis, explores theoretical frameworks within the Catholic community, and examines the link between environmental and social issues. The second stage presents various examples and guidelines for changing lifestyles. The third stage motivates participants to design, create, and implement small projects in their local environments. This final stage fully realizes the method's transformative potential by connecting theoretical knowledge with practical application.

This method can be a valuable tool for addressing the climate crisis not just for Catholic institutions, parishes, educational institutions, and teachers of school and university subjects but also for all practitioners and educators.

55.2 Transformative Potential

The method's transformative potential is based on the idea that ecological engagement (such as reducing the use of plastic and saving water) is an integral part of Catholic identity (Turza and Jurić 2023). This is important because it raises awareness about Catholics' responsibility to care for the environment and emphasizes the connection between ecological engagement and Catholic identity. It can also influence and promote ecologically sustainable behaviors within Catholic communities. Given that there are 2.3 billion Catholics worldwide, this method has significant potential to impact society at large.

This method assumes that humans are in a relationship with God, their neighbors, and nature. Any disruption of these relationships is deemed undesirable.

Integral ecology opposes a tyrannical attitude toward nature involving unchecked natural resource exploitation. It criticizes interpretations of the Bible that seek to justify the idea that humans have absolute dominion over nature. The encyclical Laudato Si′ firmly rejects this viewpoint and emphasizes that caring for nature is integral to a believer's identity. This approach addresses these issues comprehensively: it seeks solutions for dealing with the impacts of accelerated climate change and aiding the most vulnerable in these processes simultaneously. Its goal is to restore damaged connections essential to human identity, including relationships with God, neighbors, and nature. It involves religious prayer and meditation practices, caring for the most vulnerable people (i.e., the people suffering most from the climate crisis's consequences), and making lifestyle changes, such as embracing sustainable and ecological practices.

Thus, this method of educating believers promotes various transformative practices as the core of this type of education.

First, all participants are encouraged to include in their prayer intentions, whether at liturgical celebrations or in their prayers, a prayer for the preservation of all creation, with a specific focus on all those affected by the impacts of climate change.

Second, participants are urged to support and take care of those affected by the consequences of climate change to the best of their abilities. If there are no such individuals in their immediate surroundings, participants can contribute to their care by making financial donations to organizations or individuals involved in their assistance.

Third, participants are encouraged to embrace ecologically sustainable behavior, such as reducing or eliminating the use of plastic, conserving water and electricity at home, minimizing car usage, and utilizing public transportation whenever possible.

55.3 Application

This method can be applied to education across different educational institutions, such as kindergartens, universities, parish communities, and families. It involves collaboration between various experts, the transfer of theoretical knowledge, and practical application.

Implementing the integral ecology method effectively requires following a structured process that ensures participants gain theoretical understanding and practical skills. This section provides detailed step-by-step instructions on applying the method in Catholic educational settings or other relevant contexts. The method is divided into three stages: Knowledge, Awareness, and Practice.

55.3.1 Knowledge Stage: Theoretical Foundations

Objective: Equip participants with the foundational knowledge of the climate crisis and its theological implications.

1. Preparation: Arrange a 2-hour session facilitated by:
   - A climate science expert will provide an overview of the current state of the climate crisis, including data on global warming, ecological degradation, and their societal impacts.
   - A theology expert to highlight the moral and spiritual responsibilities of individuals and communities, focusing on the teachings of the encyclical *Laudato Si'*.
2. Materials Needed:
   - Presentation slides summarizing key scientific data.
   - Excerpts from *Laudato Si'* and related theological resources.
3. Delivery:
   - Divide the session into two 60-minute segments for each expert.
   - Conclude with a 15-minute interactive Q&A session.

55.3.2 Awareness Stage: Learning from Best Practices

Objective: Inspire participants by showcasing successful projects and strategies for addressing the climate crisis.

1. Preparation: Arrange a 2-hour session with a practitioner or environmental project expert.
2. Content: Present at least three case studies, such as:

   - Community-based renewable energy initiatives.
   - Urban greening projects or zero-waste movements.
   - Faith-driven ecological campaigns, like the projects presented in the Laudato Si′ Action Platform (Laudato Si' Action (laudatosiactionplatform.org) Laudato Si' Movement (laudatosimovement.org).

3. Materials needed:

   - Videos, photographs, or reports of successful projects.
   - Practical guides outlining steps for implementing similar initiatives.

4. Activity:

   - Engage participants in small group discussions to brainstorm how such practices could be adapted to their local contexts.

55.3.3 Practice Stage: Project Development and Execution

Objective: Empower participants to design, implement, and present small-scale ecological projects.

1. Preparation:

   - Assign participants to pairs or small groups, ensuring diversity in skills and backgrounds.
   - Provide a project template, including sections for objectives, activities, expected outcomes, and evaluation.

2. Implementation:

   - Allocate a 2-week (or more if needed) period for participants to complete their projects in their families, parishes, or communities.
   - Encourage projects such as:

   - Reducing single-use plastics in local events.
   - Organizing tree-planting campaigns.
   - Hosting workshops on energy conservation in their neighborhoods.

Table 55.1 Method overview

Main purpose
Integral ecology
Gained competences
Acquiring basic knowledge of theology about the responsibility of Christians for all creation. Gaining basic knowledge in the field of natural sciences about the causes and consequences of the climate crisis. Understanding all the processes involved in preparing and implementing small projects
Educational setting
Formal (kindergarten, school, university, and parishes)
Space requirements/restrictions
Indoor (lectures) and outside activities (implementation of projects)
Resources and necessary materials
Technical equipment for lectures
Number of participants
Group size: 20–25
Facilitator competences and skills
Professional knowledge in the field of environmental sciences and theology. Experience in implementing projects
Participants skills/age/competences
Children of kindergarten, school, students, and believers in parish communities
Duration
8 hours over a month

3. Presentation:
   - Host a 3-hour group session for participants to present their projects.
   - Encourage feedback and discussion to refine ideas and share lessons learned.
4. Evaluation:
   - Instructors provide individual feedback based on specific criteria, such as feasibility, impact, and originality (Table 55.1).

References

Francis (2015) Laudato Si' (May 24, 2015), Laudato si' (24 May 2015) | Francis (vatican.va). Accessed 1 July 2024

Turza Z, Jurić AK (2023) Environmental engagement in catechesis: critical analysis of the apprehension of the relation between catechesis and the care for the environment by the "new directory for catechesis". Bogoslovni Vestnik 83(1):213–225

Chapter 56
Carbon Detox

Mithat Ekici

56.1 Introduction

Detoxification can be defined as the natural neutralization, transformation, or destruction of substances or toxins that the human body does not want (Lipman 2019). In other words, detox or purification is the process of cleansing the human body, affected by pollution due to environmental factors, malnutrition, unhealthy lifestyle, or polluted air and water.

It has been hypothesized that all of the living organisms in the Earth's biosphere may act as a single entity to regulate its chemical composition, surface pH, and climate. The idea that the biosphere is an active adaptive control system that can keep the Earth in homeostasis, (meaning that it maintains its stability while adjusting to conditions that are optimal for survival (Britannica 2024), is called the Gaia hypothesis (Lovelock and Lynn 1974). The Gaia theory emphasizes that the Earth system can act as an amplifier and small changes in both warming and cooling can be intensified, resulting in erratic temperature changes (Lovelock 2008). According to this hypothesis, the Earth is a super-organism. All components act together. Unfortunately, the Earth is rapidly being polluted and the need to purify it from this accumulated pollution and toxins—carbon and equivalent greenhouse gasses should be evaluated in this context—must be cleansed. The earth needs detox just like the human body.

The concept of Carbon Detox is a new concept that individuals can use to address climate change. It is possible to calculate the carbon footprint of individuals and consequently reduce CO_2 emissions in order to apply carbon detox. With this

M. Ekici (✉)
Ministry of Environment, Urbanization and Climate Change, Turkish State Meteorological Service, Ankara, Türkiye
e-mail: mekici@mgm.gov.tr

J. Bentz, J. Ristić Trajković (eds.), *Imagining, Designing and Teaching Regenerative Futures: Art-Science Approaches and Inspirations From Around the World*, Science for Sustainable Societies,
https://doi.org/10.1007/978-981-96-9029-9_56

method, people may become curious about their individual ecological footprint, realize it, calculate it, share it, and make it a part of their lives. They can then learn how to emit less emissions and adopt changes. If every person reduces carbon emissions by 10%, it will mean a huge emission reduction for the world.

56.2 Transformative, Regenerative Potential

The Carbon Detox method can help to focus people's attention at individual and group levels. Since it is a medical term, it can make it easier to directly relate climate change to their own lives. It may be transformative because it deals with the theme of taking better care of oneself, draws attention to individuals' ecological footprint calculations, and encourages less consumption.

56.3 Application

Almost all countries' environmental ministries or environmental agencies have websites on carbon footprint calculations or have suggestions for a page that calculates carbon footprints. For example, Free Carbon Calculators for Individuals and Small Businesses for the United States Environmental Protection Agency (https://www.carbonfootprint.com/calculator1.html) and (https://knowsdgs.jrc.ec.europa.eu/cfc) for EU can be given as examples. If we take an example from the United States Environmental Protection Agency's suggestion from these two sample calculation pages, individuals first question the type and amount of energy consumption in their homes in these calculation applications and include it in the calculation. These are questioned as to whether it is electricity, natural gas, heating oil, coal, liquefied petroleum gas, propane, or wood. Then, the number of times and where they have traveled by plane is entered into the calculation. The type of vehicle they use for transportation (public transportation, car, motorcycle), how many kilometers they have traveled, and the type of fuel used for the vehicle are calculated. Thanks to these, our ecological footprint related to daily, weekly, and monthly transportation can be calculated. Answers are given to questions about food and beverage habits as secondary consumption types. Our medicine consumption, clothing and shoe consumption, paper product consumption, computer and information technology product consumption, TV, radio, telephone product consumption, motor vehicle product consumption, furniture and manufactured goods consumption, hotel, restaurant and pub habits, mobile phone and phone calls, bank and finance expenses, insurance and education expenses, and recreation, cultural, and sports expenses will be calculated (Ekici 2019). As a result of the calculations, our CO_2 equivalent consumptions under these main headings will be revealed. It is possible to do this in any time measurement. With the carbon detox application, it can be revealed how much of these emission-emitting habits we can abandon and turn to less emission-emitting practices. In this way, awareness and habit change can be achieved. "Carbon Detox" can be turned into a lifestyle. Individual and group applications can be developed. Awareness can be increased and motivation can be increased with group applications. Groups and communities can calculate the group carbon footprint by calculating the individual carbon footprint first and adding it to the number of people in the group in cooperation with each other. Carbon detox can be fun and more productive with activities done in groups. All groups can do this application, such as family, school, neighbors, co-workers, sports groups, art groups, and so on. The elements required are a leader, an online carbon footprint calculator application (Table 56.1). Created by Mithat Ekici and Designed by Orhan Can Özcan using Adobe Illustrator.

Table 56.1 Method overview

Main purpose
Awareness building /behavior change; as a slowly developing disaster, it is very difficult for society to understand the negative effects of climate change. The aim is to raise awareness and benefit from the detox habit, which is a popular approach that attracts attention, and to change people's habits
Gained competences
Competences acquired through the method; with the method, people learn about human activities that change the climate and why to calculate the carbon footprint. In addition, they can learn that they can apply these competencies individually as well as with all communities they are involved in
Educational setting
Formal (level and format)/informal (format—Workshops, seminars, etc.); all environments where they can formally learn carbon footprint calculation methods in their country. They can also learn informally by participating in various NGOs and training courses
Space requirements/restrictions
Outside activities/indoor activities/etc.; preserving biodiversity by traveling in nature as an outdoor activity, protecting water resources, developing waste awareness, less consumption, walking culture, and cycling culture are important parts of this method. Indoors also reduce carbon print in the home, improving energy use, optimizing water use, using renewable energy in homes and workplaces, choosing electric cars, etc
Resources and necessary materials
Specify (materials, technical equipment, etc.); people or society and calculation of carbon footprints
Number of participants
Group or individual work, group size, number of groups; from one person to community
Facilitator competences and skills
Background, experience, skills; all persons who are unprejudiced, nature-loving, love the earth, and care about future generations
Participants skills/age/competences
Community/students/policymakers/researchers/designers/practitioners, etc.; all people of all ages and walks of life
Duration
Length of time required to apply this method; always, it is a lifestyle change

References

Britannica (2024) Homeostasis, physiology. https://www.britannica.com/science/homeostasis

Ekici M (2019) Resilience to the climate change: carbon detox for the earth. Resilience J 3(2):113–125

Lipman F (2019) What do you mean by detox? https://www.bewell.com/blog/what-do-you-mean-by-detox/. Accessed 17 Jan 2019

Lovelock JE, Margulis L (1974) Atmospheric homeostasis by and for the biosphere: the gaia hypothesis. Tellus 26(1–2):2–10. https://doi.org/10.3402/tellusa.v26i1-2.9731

Lovelock JE (2008) A geophysiologist's thoughts on geoengineering. Phil Trans R Soc 366:3883–3890. https://doi.org/10.1098/rsta.2008.0135

Chapter 57
More-Than-Human Future Ministries

Samuel Yu, Bem Le Hunte, Susanne Pratt, and Scott Matter

57.1 Introduction

Futures Ministries is a speculative world-building game, originally developed by Bem Le Hunte in 2015, in which participants collaborate to imagine a new government in a future scenario. Thinking through the lens of governance and policy, participants role-play as alternative government ministers to envision a new and better future for an interconnected society. The aim of the game is to suspend disbelief for change, re-writing narratives to envision new ways of thinking, governing, and living. This entry focuses on a regenerative extension of this game—More-Than-Human Future Ministries.

Introducing a more-than-human lens to Future Ministries encourages players to represent other ecological actors, not just focusing on humans. It encourages life-centered futures thinking and creativity, going beyond human-centered approaches to imagining new regenerative possibilities. It is an activity that encourages participants to be hopeful, radical, critical, empathetic, and collaborative while considering both human and wider ecological well-being.

S. Yu (✉) · B. Le Hunte · S. Pratt · S. Matter
Transdisciplinary School, University of Technology Sydney, Ultimo, NSW, Australia
e-mail: samuel.yu@uts.edu.au; bem.lehunte@uts.edu.au; susanne.pratt@uts.edu.au; scott.matter@uts.edu.au

J. Bentz, J. Ristić Trajković (eds.), *Imagining, Designing and Teaching Regenerative Futures: Art-Science Approaches and Inspirations From Around the World*, Science for Sustainable Societies,
https://doi.org/10.1007/978-981-96-9029-9_57

57.2 Transformative Potential

More-Than-Human Future Ministries is an open-ended role-play activity for imagining new forms of government and society through a range of alternative lenses. In an Australian context—where this activity originated—ministries are government departments headed by a minister. They are responsible for portfolios that can lead to siloed thinking and action—for example, departments for Defense, Finance, or Social Services. This activity asks—what kinds of ministries might better address the interconnected nature of the global challenges that we currently face?

Framed as an ecologically focused task, participants are directed to think through different aspects of government and more-than-human society to step beyond the restrictive remits of our current government ministries. It generates a forum to create new narratives for human relationships with nature, as nature—rather than positioning humans as separate. This contributes to the growing body of other futures-oriented role-play, simulation, and story-telling activities that invite participants to adopt the perspective of other more-than-human actors and think through the consequences of human impact and governance. For example, (The Treaty of Finsbury Park 2025) and Council of All Beings (Seed et al. 1988) which both present forms of creative play in which participants act as another entity (and not as humans) to discuss concerns and possibilities in fantastical ways.

57.3 Application

The activity can be applied in a range of different contexts, from classrooms to industry workshops. As an activity, it only requires some light facilitation and everyday workshop materials. The beauty of the activity is its flexibility and open-endedness and its ability to incorporate transdisciplinary perspectives. The instructions below outline our general procedure, but facilitators and participants are invited to frame the activity as they choose, as well as its process and outcomes. More-Than-Human Future Ministries could be used as an open-ended exploration to develop futures thinking or be used in an applied context with a more refined scope toward generating actionable insights. Additional resources such as costumes and props can add to this suspension of disbelief and create a more theatrical, creative, and engaging environment.

57.3.1 How to Play

The approach below provides a generic structure to More-than-human Future Ministries. This is the approach we have used within the Transdisciplinary School at the University of Technology Sydney, with undergraduate students in the Bachelor of Creative Intelligence and Innovation. We find that it typically works best with a group of 20–30 people and a timeframe of 90–120 minutes.

57.3.2 Framing the Activity

More-Than-Human Future Ministries begins with a short introduction that recognizes that new futures need to be imagined. Participants are prompted to dream big, think radically, and push boundaries. Their aim is to come up with a new society, through the lens of government and policy. The activity may be broadly framed as an open call to re-imagine any aspect of the world or be framed within specific contexts or topics. The activity works well in prompting participants to imagine their futures through the lens of regeneration. Engaging with ecological well-being from both human and more-than-human perspectives sets them up to imagine government, policy, and society that addresses well-being more holistically.

57.3.3 Forming Ministries

Forming groups of 4–6, participants select, or are assigned, a future government department from a descriptive list of different ministries prepared in advance. For example, the Ministry of Abundance, the Ministry of Tinkering, and the Ministry of Care. These future departments are not traditional government ministries as we currently know them but have been created to encourage new ways of thinking across silos. Each alternative ministry adopts a unique framing, perspective, and focus to prompt participants to imagine governance, actions, and policies in an interconnected way that extends to the more-than-human world. Participants are not required to stick to the established list and are free to propose their own ministry with a new agenda.

Participants should consider the membership of their ministries to adopt more-than-human roles. This promotes policy planning that considers constituents beyond human citizens and residents. More-than-human members may be relevant animals, plants, and other entities such as rivers, mountains, and forests, or even buildings and technologies. To support a life-centered, rather than only human-centered approach—more-than-human personas can be used to support participants' thinking (Lutz 2023).

57.3.4 Envisioning a New World and Future Policies

Participants begin by introducing themselves and discussing the ministry they have chosen. It is important to establish a collective understanding of the ministry's agenda within the framing of the activity. In establishing the future ministry's direction, scope, and responsibility, groups should collectively reflect on two key questions:

- What are your biggest wishes for the future of government?
- What are your biggest concerns for the way that government is currently conducted?

Discussing these questions and the group's responses is a mode of collective sense-making for everyone to critically reflect on what "government" means to them and how it could operate differently. This involves debating the current functions of government and the kinds of new inclusive and expansive policies that future departments would prefer to consider.

The objective is to imagine the workings of a new society and the role that the ministry plays in supporting it. To assist with creating future scenarios, facilitators can incorporate other futuring tools and methods.

Prompts for future ministries to consider:

- What is your department's vision for a more-than-human society? This could involve what kind of future the ministry wants to promote, as well as avoid.
- What are the key relationships between the human and more-than-human constituents? What role does each of them play in the ecosystem?
- What would your ministry look like in a networked, complex, open, and dynamic world?
- How might you engage with other ministries toward a shared vision?
- What are the public's greatest concerns/worries about the future to address?
- What is the narrative of this future? How did this new government form?

57.3.5 Presenting Future Visions and Policies

By the end of the activity, each ministry presents a summary of their vision and policy proposals to the other ministries. This may take many transdisciplinary and creative forms, such as a conventional presentation or a manifesto. To create parliamentary consensus, the ministries then work together to form a new constitution for their future society.

In our experience, students who have engaged with More-than-human Future Ministries have found this to be an engaging and insightful experience with a lasting impact on their understanding of complexity, more-than-human relations, and the importance of acting wisely to benefit all players in an interconnected world (Table 57.1).

Table 57.1 Method overview

Main purpose
More-than-human awareness-building and decision-making
Gained competences
More-than-human empathy and mindset, futures thinking, and collaborative decision-making
Educational setting
Formal (undergraduate and postgraduate classes) Informal (workshops)
Space requirements/restrictions
Indoors, but can be done outdoors
Resources and necessary materials
List of alternative ministries (please contact authors if you would like more information)
Number of participants
Groups of 4–6; approximately 20–30 people but can be more
Facilitator competences and skills
Basic workshop facilitation, understanding of more-than-human concepts, and understanding of basic government structures
Participants skills/age/competences
Students, policymakers, and community, but open to all with flexible framing
Duration
Approximately 90–120 minutes

References

Lutz D (2023) The non-human persona guide: how to create and use personas for nature and invisible humans to respect their needs during design. Damien Lutz, Sydney

Seed J, Macy J, Fleming P, Naess A (1988) Thinking like a mountain: towards a council of all Beings. New Society Publishers, British Columbia

The Treaty of Finsbury Park. The interspecies assemblies or the story so far. https://treaty.finsburypark.live/. Accesed 13 June 2025

Part VII
Innovative Pedagogies for Regeneration and Societal Transformation

To regenerate our societies, we must transform the ways we teach, learn, and communicate within educational spaces. This part brings together methods focused on pedagogical innovations that cultivate transformative competencies essential for navigating and reshaping a rapidly changing world. These include self-awareness, reflective and critical thinking, empathy, active listening, sustainability leadership, and a sense of collective responsibility. Through diverse approaches, these methods invite learners to question dominant assumptions, reframe inherited knowledge systems, and engage education as a site of possibility. Rather than treating education as a neutral process of knowledge transmission, these pedagogies embrace it as a deeply relational, ethical, and political practice that shapes not only what we know but also how we live and relate to others and the world. They create space for learners to participate actively in the co-creation of just, caring, and regenerative societies. These approaches foster the capabilities needed to think systemically, act collaboratively, and imagine alternative futures. They remind us that to teach for sustainability is also to teach for harmony, creativity, and ecological responsibility—and that education itself can be a powerful act of regeneration.

Chapter 58
Reflective Lifelines: Tracing the Path to Regenerative Horizons

Leonie Paul

58.1 Introduction

> Before we can be different, we need to meet ourselves where we are. (unknown)

This quote beautifully encapsulates the essence of the Lifeline exercise and underscores the profound importance of understanding our past as a catalyst for "growth" and change (McAdams 2008).

Our lives are a tapestry woven with unique stories, each representing memories that have shaped our minds, actions, and the world we inhabit (Prebble et al. 2013). These threads weave a tale of the personal meaning and significance we assign to events, people, and ourselves (Eysenck and Keane 2015).

In the quest for regenerative futures, we can find that the most profound steps begin within ourselves. I believe that this introspective journey exemplified by the Lifeline exercise is the cornerstone of meaningful change in the external world (see Frank et al. 2019). It is where we unearth the seeds of transformation that can sprout into purposeful actions, from inside us to the outside world.

Whether you are just beginning to explore your regenerative understanding or have been involved in the field for a long time, lifelines can be created repeatedly at different points in your journey. You can conduct this exercise individually for personal reflection or in small group settings as conversation starters to explore regenerative futures and visions. Sharing your lifeline with others can reinforce the concepts of inspiration, (un-)learning, and collectively evolving.

I invite you, dear readers, to share the "turning point" moments of your journeys with one another and to nurture the hope for a regenerative future where all beings

L. Paul (✉)
LUT University Lappeenranta Finland & RMIT University, Melbourne, Australia
e-mail: leonie.paul@lut.fi

J. Bentz, J. Ristić Trajković (eds.), *Imagining, Designing and Teaching Regenerative Futures: Art-Science Approaches and Inspirations From Around the World*, Science for Sustainable Societies,
https://doi.org/10.1007/978-981-96-9029-9_58

thrive. Your stories, your insights, and your care are the fuel that drives this collective endeavor.

58.2 Transformative Potential

This exercise is not a mere stroll down memory lane; it is a powerful tool for nurturing the regenerative futures we aspire to create. First, by dedicating time to unraveling our personal sustainability narratives, we gain profound insights into our ever-evolving relationship with the planet and our role within it. It serves as a reflective mirror, illuminating our individual journeys as well as the collective odyssey of humanity toward a more sustainable and regenerative world.

Second, understanding how our perceptions of sustainability have transformed over time empowers us to be deliberate, conscious, and effective change agents. It invites us to acknowledge and celebrate our progress while also discerning the areas in which potential shifts in our perception are needed. Through this exercise, we deepen our self-awareness and support a sense of agency and responsibility, propelling us toward a future where each of us actively participates as an integral part of nature.

Third, in sharing our personal sustainability stories, we contribute to a joint source of inspiration and wisdom that can illuminate the paths of others on their transformative journeys toward regenerative thinking and action (see Bentz et al. 2022). Thus, we collectively weave a narrative of hope and learning, guiding us toward the regenerative future we envision.

The exercise is refreshingly simple, demanding nothing more than your imagination. However, you are welcome to use pen and paper for visualization. So, let us begin…

58.3 Application

Through my own journey and interactions with others deeply engaged in sustainable development and regeneration, the Lifeline exercise has unveiled a mosaic of diverse experiences. Critical incidents that have forged a stronger connection or deeper understanding of oneself and regeneration often transcend traditional boundaries. The Lifeline exercise can be used as a prompt in interviews, classroom settings, intra- and inter-organizational workshops, interdisciplinary seminars, and many other contexts.

When done alone, a debriefing session with a small group or partner can be helpful to process the findings. The transcript describing the exercise can be found below and can be read out by a facilitator. Alternatively, you can follow along with the provided audio file. Feel free to draw your lifeline to bring reflections to life.

Creating a first lifeline draft will take about 15–20 minutes. Trust your instincts and initial thoughts; there are no right or wrong answers, and everyone's journey is unique. If you have the chance to reflect on your lifelines in small groups, I recommend scheduling ample time to openly share thoughts and ideas from the exercise.

Audio https://drive.google.com/file/d/1BXB40vkd0osMk7Tw3ttuwtobK-c11lad/view?usp=sharing

Transcript

Dear Listener/Dear Reader,

I would like to invite you to find a comfortable position. You might be seated or standing. Take a deep breath in and try to release any tension, physically and mentally. Before delving into the dreaming and designing of regenerative futures, I encourage you to have a look at your past and your own story.

Visualize your life's journey, from the day you were born to this very moment as you read/hear these words. Your life's path may take on various forms: a straight line, a spiral, a series of steps, or even a winding, curvy road. It is entirely your choice.

Now, traverse this lifeline, adding significant markers along the way. These markers could be your first day at school, a memorable family vacation, a life-altering accident, a true love story, your graduation, or the wedding of dear friends. Begin adding these marks to your imagined life course. These significant points in your life not only define your unique journey but also serve as crucial signposts guiding you through the rest of this exercise.

Next, let us come back to the present moment and begin walking backward along your lifeline. As you journey back in time, try to pinpoint the moment when "Sustainability" first made its entrance into your life's narrative. I purposefully keep the concept vague because this is your story, and you are the storyteller. Only you truly grasp what the term means to you. When did sustainability appear in the picture for the first time?

If you identify an occasion that you believe marked your initial encounter with sustainability, continue walking further back into your past. Search for additional instances, understanding that during those early moments, you may not have explicitly labeled them as "sustainable." Yet, armed with the knowledge you have gained over time, you can now see the connection. Can you identify specific incidents or multiple instances that signal your evolving perception of sustainability? What sparked these changes?

Take your time reflecting on these questions, and when you are ready, return to the *Reflective Lifelines* chapter and debrief the exercise.

Debrief

Having embarked on this introspective journey through your lifeline, you have unearthed the precious gems of your own sustainability narrative. Now, as we collectively aspire to craft regenerative futures, it is essential to harness the insights you have gained and channel them toward meaningful action.

For some, it is the transformative power of education, like a teacher's words igniting a lifelong passion for social justice or animal welfare. For others, it is the fascination of travel, where immersion in new environments sparks eco-consciousness.

Family can play a significant role; the tales of eco-savvy parents or the influence of a grandparent's regenerative farming practices can resonate across lifelines. Furthermore, global events can trigger eye-opening moments, whether experienced firsthand or solely witnessing the devastation of natural disasters on a TV screen. For others, the impact of being confronted with harsh realities through documentaries vividly sharing images of, for example, the meat industry can trigger shifts in awareness and action. Each lifeline is unique; however, I invite you to look out for commonalities.

After completing the exercise, consider taking the following steps:

Document and Act Record your reflections to capture the essence of your lifeline. Keep a journal for your thoughts, milestones, and realizations to deepen your insights. Identify factors that shaped your views on sustainability and consider how to transform these into actions. For example, if learning about regenerative farming from your grandparents inspired you to explore urban gardening, you might offer neighborhood workshops to share this knowledge.

Engage in Dialogue Share your experiences and insights with others, whether their journeys are similar or different. Meaningful conversations about personal transformations can foster new perspectives and collaborative initiatives, leading to broader change. Think together about the common determinators to strengthen sustainability awareness and ultimately develop a regenerative mindset. How can we invite others on this journey? (Table 58.1).

Table 58.1 Method Overview

Main purpose
Reflection on inner transformations
Gained competences
Self-awareness, reflective thinking, integrative thinking, agency, empathy
Educational setting
Formal (undergraduate and graduate courses; leadership and management programs, and adult education) Informal (workshops, seminars, retreats, and team building)
Space requirements/restrictions
Quiet reflection space
Resources and necessary materials
Pen and paper (optional)
Number of participants
Individual, pair, or group work (recommended) Smaller groups of up to 6 people
Facilitator competences and skills
Active listening Moderating diverse opinions
Participants skills/age/competences
Community, students, policymakers, researchers, and practitioners from age 16 General interest in learning about sustainability
Duration
30 minutes including some reflection time

References

Bentz J, do Carmo L, Schafenacker N, Schirok J, Corso SD (2022) Creative, embodied practices, and the potentialities for sustainability transformations. Sustain Sci 17(2):687–699

Eysenck MW, Keane MT (2015) Cognitive psychology: a student's handbook. Taylor & Francis Group, London

Frank P, Sundermann A, Fischer D (2019) How mindfulness training cultivates introspection and competence development for sustainable consumption. Int J Sustain High Educ 20(6):1002–1021

McAdams DP (2008) Personal narratives and the life story. In: John OP, Robins RW, Pervin LA (eds) Handbook of personality: theory and research. The Guilford Press, New York, pp 242–262

Prebble SC, Addis DR, Tippett LJ (2013) Autobiographical memory and sense of self. Psychol Bull 139(4):815–840

Chapter 59
Regenerative Practices, Teaching as Learning

Lígia Oliveira

59.1 Introduction

This approach aims at reconfiguring the relationship we have with Nature. This is seen as the key to regenerative practices to unfold, independent of scale, target groups, discipline, and place. As a designer trained in interdisciplinary approaches, I have been developing this as a means to approach challenges within the multiple fields of my expertise: design, arts, architecture; the urban realm; and any disciplines belonging to these, to the cultural and innovation frameworks.

The regenerative practice proposed here consists of regular visits to a specific place, other species, or element. However, this is not a method per se, as no exclusive approach can be considered regenerative. Instead, these practices emerge from a relational stance within a specific context, requiring our presence, self-inquiry, values, discipline, observation, and cooperation. Therefore, it is more fitting to refer to these as practices and approaches rather than a single method.

To reconfigure our relationship with Nature, we need a body–mind relational approach. Key aspects of this include mental knowledge, by having clear values and goals, and analytical thinking; listening and interpreting our bodies' sensory cues—including emotions, as these, aligned with our experiences, values, and current moods, provide constant information about our context and its unfolding. Finally, a relational approach acknowledges the quality of our engagement toward places, more than humans, elements: not as consumers, users, and visitors, but as friendly parts of a complex, interconnected system. This includes the awareness of these relational dynamics: how we impact other humans, other species, and ecosystems; the impact these have on us; and the interconnectedness among all. It also integrates

L. Oliveira (✉)
Independent Artist and Designer, Vila Real de Santo António, Portugal
e-mail: ligiaoliveirastudio@gmail.com

J. Bentz, J. Ristić Trajković (eds.), *Imagining, Designing and Teaching Regenerative Futures: Art-Science Approaches and Inspirations From Around the World*, Science for Sustainable Societies,
https://doi.org/10.1007/978-981-96-9029-9_59

an understanding of how our own biographies, character, and aspirations shape us, and by extension, define the quality of what we do in the world.

Observing the complexity of these, we can loop back to the mental knowledge and awareness of values and behaviors, apply those within this relational framework, and ask: What are the values and behaviors that make good relationships, and which ones can be applied to regenerative practices? The answer to this might be individual, circumstantial, and local, yet the questions themselves are a fundamental step in this practice.

59.2 Transformative Potential

We become full members of our society by being specialists on what is around us—people, places, and Nature—by being their students. This curiosity is a key feature in establishing positive relationships, done through engagement with them: by listening, observing, first. Our disposition to learn defines the quality and complexity of our analysis and of our relationships and largely indicates how good the results are. Getting to know the object of our analysis directly, the other, as often as possible; in different conditions; its patterns of behavior, of change; how it interacts with other elements; and its history. Where did it come from? How was it in the past? How does it relate to its system?

The power of the approach for regenerative futures presented here lies in its foundational framework toward deeply understanding the interconnectedness of all elements within ecosystems: how they relate to each other and with us. By recognizing that everything relates to its context and that we relate to everything, we develop a profound awareness of the interdependence of all elements. This perspective fosters a systemic understanding essential for addressing complex environmental challenges. It emphasizes the importance of both external observation and internal reflection. It encourages us to be "me-searchers," understanding our motivations and histories and how these relate to our contexts. This dual focus helps identify individual and collective shortcomings, frustrations, misguided actions, and opportunities, leading to more effective and meaningful actions.

By reframing our practices in response to social and environmental crises, we can foster a new imagination that drives positive change. This promotes a shift from a human-centric to a relational-centric approach, recognizing the interconnectedness of all life and the importance of listening to and respecting our environment. This framework is transformative because it acknowledges that we are also transformed by our methodologies and frameworks, not only the reality that we aim to regenerate. We transform, and all of these transform us too. Through integrating intellectual and emotional intelligence, transformation occurs within the author, relationship, and outcome. And by having a framework grounded in the relational through awareness, communication, and values such as respect and compassion, the intentions are toward the right course of action. By using this framework, we foster deeper relationships—with ourselves and with Nature.

59.3 Application

The work I do starts outside—getting to know my context and defining my commitment: I commit to respect and care. Not in an abstract way, but specifically: a tree, a river, a landscape, or a sea. The process of getting to know it—the ecosystem we are looking into—increases the commitment. Value it and make that the basis of your relational, loving practice: an action verb.

In this simple walking, sitting, and observing routine, no complicated materials or extensive preparation are needed. These are framed by repetition: regular, in-person visits. Being in nature, I engage emotionally and physically with the environment, and by asking questions like, "What would the trees or the river do?" I open myself to new perspectives and solutions from the natural world. In the process toward any outcome, object, or process, I keep contrasting it to the relational dimension: how would it affect this place: this tree or similar ones, this landscape or others I have never seen, or how does it affect the functioning of the ecosystem? Does it provide disruption, and to whom? Does it create entanglements, opportunities, benefits, and to whom? What would be necessary for regeneration to occur?

For this practice, you will need the following:

Time Commitment: Set aside regular time to visit your chosen natural place, element, or species. Even short, frequent visits can be powerful.

Materials: Minimal materials are needed—comfortable clothing, a notebook for observations, and perhaps a camera to document thoughts, metaphors, sensations, impressions, and moods.

Application: This method can be applied individually or in groups. Individual practice helps with personal growth and understanding, while group activities can foster community engagement and collective learning.

Preparation: No extensive preparatory work is required. Simply choose a place that resonates with you and begin your visits. Over time, your observations and interactions will deepen your understanding and connection.

Further Considerations: Emotional regulation is an important element when witnessing environmental changes, as these can be challenging. Practices of compassion, hope, and commitment, rather than avoidance, help manage discomfort and anxiety, as well as sharing concerns with a trusted friend and/or within your community. This practice encourages viewing problems from a broader perspective and understanding the interconnections within ecosystems. This comprehensive view is crucial for developing regenerative practices and aims to encourage a sense of belonging. And finally, combine physical presence in nature with self-reflection. Understanding your motivations and histories, and how these relate to the larger environmental context, contributes to weaving your individual narrative into the collective, planetary one. We are all in this together (Table 59.1).

Table 59.1 Method overview

Main purpose
Awareness building/decision-making/behavior change
Gained competences
Awareness: Of self, of the other/nature, of the relationship; systemic perspective
Educational setting
Informal and/or formal: Individual and collective decision-making processes and workshops
Space requirements/restrictions
Outside activities
Resources and necessary materials
Comfortable clothing, a notebook for observations (optional: Camera)
Number of participants
Individual and small groups (max. 25)
Facilitator competences and skills
Self-knowledge, compassion, and interpersonal skills
Participants skills/age/competences
Community, students, policymakers, researchers, and practitioners; university level and above
Duration
From 1 hour on, with benefits resulting from its repetitive use

References

Celidwen Y, Keltner D (2023) Kin relationality and ecological belonging: a cultural psychology of indigenous transcendence. Front Psychol 14:994508. https://doi.org/10.3389/fpsyg.2023.994508

Ekman P, Ekman E (2016) The Atlas of emotions. https://atlasofemotions.org/

Pallasmaa J (2005) The eyes of the skin: architecture and the senses. Wiley-Academy, Chichester

Waldinger RJ, Schulz MS (2023) The good life: lessons from the world's longest study of happiness. Simon and Schuster, New York

Watson J (2019) Lo-TEK: design by radical indigenism. Taschen

Whitburn J, Linklater W, Abrahamse W (2020) Meta-analysis of human connection to nature and proenvironmental behavior. Conserv Biol 34(1):180–193. https://doi.org/10.1111/cobi.13381

Yunkaporta T (2020) Sand talk: how indigenous thinking can save the world. Harper Collins

Chapter 60
Active Listening Workshop: The Art of Listening to Words and Emotions

Maria Bakatsaki

60.1 Introduction

Active listening is one of the most powerful communication skills, and it can be cultivated through daily practice. The proposed 2-hour experiential workshop uses role-playing to effectively develop active listening skills. This technique can be used for experiential teaching, either in-person in a classroom or via an online platform. Participants often reported quickly connecting with others, despite the brevity of the exercise. They also recognized shared challenges, highlighting active listening's potential to foster a sense of belonging and shift thoughts and behaviors.

Active listening is the art of listening to words and emotions. The ability to connect empathetically by understanding others' emotions, avoiding interruptions and distractions, while using techniques such as interpreting both verbal and non-verbal cues, reflective paraphrasing, summarizing, asking open-ended or clarifying questions and staying curious about the speaker's intentions and motivations. This entails questioning perceptions, empathizing, using warm and supportive vocal tone and showing genuine concern and full attention facilitating others to express themselves more openly and authentically (Tennant et al. 2023).

Stephen Covey (1990), author of the bestseller 'The 7 Habits of Highly Effective People', suggested as a tip that '*We should seek first to understand, then to be understood*'. Emotional regulation is essential to prevent anger outbursts and set aside judgment, manipulation, egocentric attitudes, stereotypes, and personal assumptions that hinder communication. This approach reduces misunderstandings and conflicts, facilitates information exchange, builds trust and mutual respect, and

M. Bakatsaki (✉)
School of Production Engineering and Management, Technical, University of Crete, Chania, Greece
e-mail: mbakatsaki@tuc.gr

J. Bentz, J. Ristić Trajković (eds.), *Imagining, Designing and Teaching Regenerative Futures: Art-Science Approaches and Inspirations From Around the World*, Science for Sustainable Societies,
https://doi.org/10.1007/978-981-96-9029-9_60

fosters fruitful dialogue and meaningful communication, ultimately creating a more positive and welcoming atmosphere in both personal, educational and professional settings.

60.2 Transformative Potential

The American psychotherapist Carl Rogers introduced active listening as a key communication tool in 1967, recognizing its significance in therapeutic contexts. He observed that allowing individuals to share what has happened to them and how they feel about it was a crucial step in facilitating changes in their thoughts and behaviors (Rogers and Farson 1987). While advice can help them, real change happens when they work through problems themselves. Having the space to speak openly with a compassionate listener is one of the most powerful ways to support the process (Tuhovsky 2018). Since then, active listening has proven valuable in fields such as counseling psychology, administration, education, and healthcare.

Human relationships are the sum of interactions. When people's concerns, efforts, and opinions are acknowledged, they feel seen and valued. This enhances communication, reduces misunderstandings, and facilitates conflict management by promoting a collaborative atmosphere where both parties feel respected. Open dialogue can challenge existing norms and paradigms, paving the way for innovative solutions and transformative actions. Under such conditions, fear and anxiety are reduced, while self-esteem increases, creating space for trust, mutual respect, and a sense of security. This fosters integrity and ethical behavior, empowering people to be authentic and innovative. Moreover, when open dialogue is paired with empathy, compassion, and openness, it strengthens social cohesion, resilience, and a sense of belonging, ultimately contributing to the overall well-being of society. This, in turn, encourages people to become engaged citizens who actively participate in civic life, fostering a more informed and democratic society .

In education, active listening empowers students to voice their opinions, ideas, and concerns, promoting critical thinking and creating a positive, inclusive learning environment. It strengthens teacher–student relationships, and when students feel heard and valued, they are more likely to engage in their academic community. This sense of belonging encourages students to work harder to achieve their learning goals (Oyeronke et al. 2024), significantly contributing to their educational success and well-being (Conner et al. 2022).

In the professional realm, active listening is an essential skill for effective leadership and team collaboration (Hersey et al. 2001). Leaders who actively listen to their team members demonstrate a commitment to inclusivity and employee well-being, which enhances workplace morale and fosters innovation as employees feel empowered to share their ideas. Moreover, by actively listening to stakeholders, businesses can develop practices that align with ecological principles, contributing to a more sustainable and regenerative economy.

Active listening also plays a key role in integrating cultural and Indigenous knowledge into ecological practices. In a broader context, the ability to understand, respect, and respond to diverse voices in environmental discussions is crucial for building a collective commitment to ecological sustainability. By fostering understanding, collaboration, and meaningful dialogue, active listening contributes to regenerative futures, promoting a holistic approach to problem-solving, decision-making, and innovation in sustainability and environmental stewardship (Taylor 2023).

It takes awareness, effort and can be cultivated through daily practice to become a more effective listener. The proposed Active Listening workshop, specifically using role-play, holds significant transformative potential for several reasons:

1. It is an effective conflict management practice, enabling two opposing sides (CONS and PROS) to reach a mutually acceptable solution quickly (within 5–7 minutes). During the group session, participants engage in self-assessment and peer-assessment techniques to enhance their self-awareness and active listening skills. Occasionally, I encourage members of the academic community, including administrative staff, professors, researcher and deans, to participate and interact with the students, as well as with each other. This amplifies the results of the workshop, strengthening the students' sense of belonging to the academic community, improving their relationships with their co-students and academic staff and boosts their motivation, peristence and academic performance (Silverman 2021; Wang et al. 2024). Such inclusive interactions in a non-hierachical setting foster mutual trust, respect and authentic dialogue among academic stakeholders, humanize institutional roles, demystify authority figures, which in turn can reduce social distance and power imbalance. Additionally, academic staff gain insight into students' perspective, concerns and communication styles, which can inform more empathetic and effective teaching (Flavia & Enachi-Vasluianu 2016) and administrative practices. Ultimately, this cross-participation improves community bonds, interpersonal relatiohships and contributes significantly to greater academic performance, personal well-being and a more cohesive and inclusive organisational culture.
2. Participants are tasked with supporting an opinion contrary to their own beliefs by "putting themselves in the shoes" of the opposing side. Our minds often justify our actions, causing us to overlook important factors. By adopting the opposite perspective, we can identify overlooked issues and better prepare for future actions. For example, consider interracial baby adoption: in the context of a hostile school environment, how prepared would I be as a parent to support my child against stereotyping and bullying from children unfamiliar with racial diversity? Would I be strong enough to confront close-minded teachers who contribute to a toxic environment?
3. The observer in this exercise is required to actively listen to the two participants without intervening, which puts them in a unique position. They gain a more objective perspective, observing the entire process from a distance. This practice promotes cognitive clarity, reduces the negative effects of stress on the brain, and

encourages a relaxed state. Observers also become more attuned to the "DOs" and "DON'Ts" of active listening, increasing their awareness of effective communication strategies.

60.3 Application

This experiential Active Listening workshop is based on a group exercise from Bateman and Snell (2017), and can be applied to students either in person in a classroom or online:

(a) Online version: An online platform (such as Zoom, Webex, or any platform that allows participants to be randomly and automatically divided into break-out rooms) is used. Each participant must have a camera and microphone. The presentation and instructions for the workshop can be sent via email before or at the beginning of the workshop using the chat function of the online platform. The advantages of the online version include: (1) the facilitator can quickly create break-out rooms and easily visit groups without interrupting them, (2) the groups can work without disturbances, (3) the workshop can accommodate many participants, and (4) it is easier to record the entire process (though consent from all participants is required). The disadvantages include: (1) recordings are only available in the break-out rooms visited by the facilitator, (2) despite cameras, many non-verbal cues are missed since only faces are visible, (3) there is always a risk of technical issues (e.g., internet or electricity failure), leading to participant drop-offs or disconnections, and (4) the cost of the online platform license must be considered.

(b) In-Person Version: For in-person workshops, a video projector is required to present the theory of active listening and the instructions for role-playing. Participants are randomly divided into groups of three and receive hard copies of the workshop instructions. The in-person version allows participants to observe full-body non-verbal cues and eliminates technical disruptions. However, the disadvantages are: (1) dividing participants into groups takes more time, (2) it can be very noisy when all groups are working in the same room, and (3) as the number of participants increases, more space is needed, making it harder for the facilitator to monitor group progress.

The workshop requires at least 100 minutes, though 2-2.5 hours is recommended to allow participants to fully experience all roles and still have time for discussion during the subsequent plenary session. The Active Listening Role-Play workshop is organized into five stages:

Stage 1 (25 Minutes) The workshop begins with a joint session where all participants are introduced to the theory of active listening (find helpful links below in the section of References), its advantages, and practical techniques. My own presentation can also serve as a helpful guide, which can be downloaded from the link: https://users.isc.tuc.gr/~mbakatsaki/Workshop_ActiveListening-EN.pptx.

Stage 2 (5 Minutes) The process of the exercise is explained to participants, as detailed in the following stages and more extensively in the presentation link (https://users.isc.tuc.gr/~mbakatsaki/Workshop_ActiveListening-EN.pptx).

Stage 3 (at Least 50 Minutes) Participants are randomly divided into groups of three to begin the role-playing exercise. Groups should be as diverse as possible (in terms of gender, nationality, discipline, age, etc.). Each participant is assigned a number (1, 2, or 3). In the first round: 1 takes the PROS position, 2 takes the CONS position, and 3 acts as the OBSERVER. A suggested list of "difficult" topics is provided (e.g., the death penalty for convicts, legalization of marijuana, etc.), and participants can propose alternative controversial subjects.

The PROS (1) and CONS (2) members have 1 minute to prepare their points of view, followed by 5–7 minutes to reach a mutually acceptable solution. The OBSERVER tracks time and notes the behaviors and comments of the conflicted parties, based on a list of active and less-active listening behaviors provided in Table 60.1.

After reaching a mutual (acceptable or non-acceptable) conclusion, team members share feedback on which behaviors worked well and which need improvement. The OBSERVER presents their own observations and experiences regarding how the PROS and CONS members practiced active listening.

If possible, two further rounds are conducted, each with a different topic, allowing everyone to play all roles. The facilitator during the role-play visits the breakout rooms (online format) or the groups (onsite format) and observes/facilitates the whole process.

Stage 4 (at least10 Minutes) In a plenary session, all groups exchange their experiences and reflect on the following questions:

1. How did you come to a mutually acceptable solution? What helped you get there?
2. Which factors hindered this process?
3. How comfortable did you feel supporting the PROS or CONS position? How did this affect your active listening ability?
4. If the given position (PROS vs. CONS) was opposed to your values, did you see the issue from a different perspective before the workshop?

Table 60.1 Information about listening

Indications of active listening
1. Asked questions for clarification
2. Paraphrased the opposite view
3. Reacted to non-verbal cues (body language, vocal tone)
4. Worked towards a mutually acceptable solution
Indication of less active listening
1. Interrupted his/her interlocutor instead of letting him/her finish
2. Became defensive about his/her position
3. Tried to dominate the conversation
4. Ignored non-verbal cues

What steps could you take to improve your ability to listen actively to your friends or collaborators, especially if you disagree with their opinions?

Stage 5 (10 Minutes) The workshop concludes with a discussion on how to further cultivate active listening. To deepen participants' learning and evaluate their experience, the link of an online short anonymous questionnaire (e.g. via Google Forms) can be shared (Table 60.2). The facilitator waits participants to complete the questionnaire and respond to any questions.

Table 60.2 Method overview

Main purpose
Experiential workshop using role-playing to practice of the fundamental communication skill of active listening. Active Listening is a powerful conflict management tool and can foster mutual trust, respect, empathy, and meaningful open dialogue, as well as a sense of belonging. This can subsequently lead to changes in behavior, encouraging more ethical and authentic behavior, engaged citizenship, and improvement in general social cohesion, resilience, and society's well-being. It can also contributes to building the foundations for a more informed and democratic society and promotes a more holistic approach to problem-solving, decision-making, and innovation in the context of sustainability and environmental stewardship.
Gained competences
Active listening and other verbal and non-verbal communication skills, educational & professional benefits (strengthen teacher-student relationships, sustainability leadership, decision-making, creativity, inclusivity and workplace effectiveness), Social and Relational skills (teamworking, building trust, conflict management) and Cognitive & Reflective skills (critical thinking, improved self-awareness, problem solving). It increases Behavioral & Transformational skills (compassion, empathy, self-esteem, ethical and responsible behavior, stress reduction and resilience).
Educational setting
This capacity-building workshop can be easily applied to all educational settings (primary, secondary and higher education) as well as informal vocational settings. I included this workshop as an experiential learning method in the "Human Resource Management" course). Occasionally, I encourage members of the academic community, including administrative staff, professors, researcher and deans, to participate and interact with the students, as well as with each other. This amplifies the results of the workshop, strengthening students' sense of belonging to the academic community, improving their relationships with their co-students and academic staff, and encouraging them to work harder to achieve their learning goals, which contributes significantly to their educational success and well-being.
Space requirements/restrictions
Indoor activities either: – Online format conducted via a video conferencing platforms, allowing participants to join from any location without physical space limitations. Web platforms have usually a limitation of 300 participants. The problem is that only the participant's upper body is visual and the non-verbal cues of the full body cannot be taken into account. – In person format in a physical space room size becomes a constraint. There is a limitation of 8 groups of 3 people = 24 people per 50 m^2; otherwise, it will be too noisy. The advantage is that since all participants are present, full-body non-verbal cues are able to be observed, which can significantly enhance the active listening experience.

(continued)

Table 60.2 (continued)

Resources and necessary materials
Material: – 1 presentation of the theory about active listening (my own presentation can be also used as a helpful guide, which can be downloaded from this link: https://users.isc.tuc.gr/~mbakatsaki/Workshop_ActiveListening-EN.pptx), – 1 page with printed directions of the role-play, which is shared to all participants for in person version. In the case of online version a Word file is shared via the chat of webplatform. Technical equipment: – Online version: (1) mobile phone or computer with microphone and video camera and (2) web platform for video conferencing, which has the possibility of break out rooms – With a physical presence in a room/class: 1 video projector
Number of participants
Groups of three participants: – The online version of the free account of webplatform has a limit of 100 participants. The paid version can host from 300 to 1000 participants (depended on the webplatform payment plan). The more participants there are the lower the quality of the connection. – With physical presence in a class/room (8 groups of 3 people = 24 people per 50 m^2)
Facilitator competences and skills
The workshop can easily applied even by primary/secondary/higher education teachers to pupils or students if they follow the directions of this chapter/tipsheet and are aware of verbal and non-verbal communication skills. The introductory presentation on the theory of active listening could be also a helpful guide, which is given at the following link: https://users.isc.tuc.gr/~mbakatsaki/Workshop_ActiveListening-EN.pptx.
Participants skills/age/competences
Anyone over the age of 10 can participate regardless of their scientific background. A diverse range of participants, including those from various countries, genders, ages, businesses, policymakers, citizens, and more, can make the workshop more effective and encourage meaningful dialogue.
Duration
At least 100 Minutes, preferably 2-2.5 hours (150 Minutes)

References

Bateman T, Snell S (2017) Management leading & collaborating in a competitive world. Mc-Graw-Hill, Boston

Conner J, Posner M, Nsowaa B (2022) The relationship between student voice and student engagement in urban high schools. Urban Rev 54(5):755–774. https://doi.org/10.1007/s11256-022-00637-2. Epub 2022 Jun 23. PMID: 35761953; PMCID: PMC9219399

Covey SR (1990). The 7 habits of highly effective people. New York: Simon & Schuster

Flavia M, Enachi-Vasluianu L (2016) The importance of elements of active listening in didactic communication: a student's perspective. CBU Int Conf Proc 4:332. https://doi.org/10.12955/cbup.v4.776

Hersey P, Blanchard K, Johnson D (2001) Management of organizational behavior. Leading Human Resources. Prentice-Hall Inc., New Jersey

Oyeronke P, Adeoye M, Arowosaye S, Yejide I (2024) The impact of active listening on student engagement and learning outcomes in educational settings. Int J Univ Educ 2:77–89. https://doi.org/10.33084/ijue.v2i2.8898

Rogers CR, Farson RE (1987) In: Newman RG, Danzinger MA, Cohen M (eds) Active listening. Communicating in Business. D.C. Heath & Company, Boston

Silverman A (2021) What universities can do to advance student belonging. EAB https://eab.com/resources/blog/student-success-blog/universities-advance-student-belonging/

Taylor L (2023) Five benefits of placing indigenous knowledge at the heart of environmental monitoring. Shared Values Solutions Ltd. https://info.sharedvaluesolutions.com/blog/environmental_monitoring_incorporating_ik-0?utm

Tennant K, Long A, Toney-Butler TJ (2023) Active listening. StatPearls Publishing LLC, Treasure Island

Tuhovsky I (2018) 21 days of effective communication: everyday habits and exercises to improve your communication skills and social intelligence. Positive Psychology Coaching Series

Wang G, Chen G, Li B (2024) The effects of teacher–student reciprocity on students' sense of belonging to the university; a moderated mediation model. Front Educ 8:1284177. https://doi.org/10.3389/feduc.2023.1284177

Chapter 61
Glocal Pedagogy

Munir Moosa Sadruddin

61.1 Introduction

Glocal Pedagogy combines local and global perspectives and is an integrative and inclusive approach that promotes a deeper understanding and appreciation of culturally relevant practices. It incorporates diverse contexts in teaching, ensuring that the learning process is culturally significant and ideologically relevant.

In educational contexts, "local" involves incorporating knowledge that reflects the country's/region's/community's unique cultural and social landscape. Meanwhile, "global" involves integrating international viewpoints and trends to provide a broader understanding of the subject matter.

Glocal Pedagogy is more relevant for the Global South. Many available teaching manuals are from the Global North, and how they fit the contexts in the Global South is not questioned, creating rejection and boycott. This pedagogy takes a reflected approach to bridge that gap of resistance.

Pedagogies greatly rely on educators' competencies, including content knowledge and teaching skills. However, inclusion and cultural responsiveness should be included, making their integration significant to avoid ideological clashes. Pedagogy is a multi-layered and dynamic process rather than a direct strategy for teaching a subject matter. There are three components of glocal pedagogy: (1) integration of local perspectives, (2) gathering global perspectives that fit into local cultural context, and (3) filtration and incorporation of culturally relevant elements.

The Glocal Pedagogy approach begins with understanding the country's educational system. Building on this foundation, educators can assess their understanding of human values through observations, understand local curriculum practices through content analysis, and recognize how it has influenced their behaviors and

M. M. Sadruddin (✉)
Sindh Madressatul Islam University, Karachi, Pakistan

J. Bentz, J. Ristić Trajković (eds.), *Imagining, Designing and Teaching Regenerative Futures: Art-Science Approaches and Inspirations From Around the World*, Science for Sustainable Societies,
https://doi.org/10.1007/978-981-96-9029-9_61

attitudes through surveys. Teachers reflect on whether their knowledge and the student's current knowledge are free from biases by asking questions such as: Do students bring preconceived ideas or misconceptions into the classroom? Do students' experiences and cultural backgrounds shape their interpretation of the material? Are there power imbalances that may silence some students' views or over-amplify others? Am I presenting diverse perspectives predominantly from a single cultural or ideological viewpoint? They then identify best practices in the existing curriculum and learning resources, selecting culturally appropriate content and activities. Additionally, teachers can gather local resources with the help of community members.

The approach then involves identifying global elements in the local curriculum and scrutinizing whether they are well-adjusted and culturally suitable. Following this, it is crucial to gather global content, perspectives, and practices on the subject matter within the framework of social contexts. Educators must connect with others to access open educational resources (curriculum, books, etc.) and gather others' resources and narratives.

Subsequently, through collective efforts, both local and global content is filtered, scrutinized, and adapted to suit local cultural context. This means selecting content that does not hurt the sentiments of local people and is culturally sensitive and respectful. Only content meeting these criteria is considered relevant.

This pedagogy harmonizes local and global knowledge, prioritizing local contexts without leaving culturally relevant global practices.

61.2 Transformative Potential

This transformative pedagogy balances local and global knowledge and harmonizes diverse educational perspectives, fostering an inclusive and culturally relevant learning environment. It focuses on creating a well-rounded educational experience that is both globally informed and locally relevant. It emphasizes local contexts while incorporating global perspectives to enhance relevance, promote inclusivity, contextualize global knowledge, and address local needs. It supports sustainable learning by continuously integrating and updating content, ensuring that education remains relevant and dynamic. Furthermore, employing this pedagogical approach mitigates the risk of prejudice and nurtures resilience.

Pedagogy significantly influences learners' participation and understanding of concepts. However, it is often imposed as non-negotiable. Graeber (2015) uses the term pedagogical violence and views teachers as bureaucrats rather than pedagogues.

The rationale behind introducing Glocal Pedagogy is to balance the influence of global practices and content that often evoke resistance among young people. Unfortunately, this occurs primarily because of misinterpretation and biased ideological integration into teaching praxis. In many cultural settings, some elements of global content are often included without considering the socio-cultural and political context. Conversely, some aspects of global content relate well with the local content, but because the global perspective usually dominates, learners are unwilling to pay attention to them.

Global content often overshadows local knowledge content, such as information about indigenous cultures, local leaders, and examples of local issues. This, in turn, needs to be clarified regarding concepts. Every educational system possesses unique qualities that are closely tied to the culture to which it belongs. These aspects should be noticed and addressed in pedagogical practices. With growing misperceptions and misinterpretations among learners, educators must respect the sensitivity of many topics and pay more attention to their interactions with learners. They need to revisit pedagogies that may raise unanswered questions in students' minds and potentially provoke aggression, emotional outbursts, rejection, or interpersonal conflict.

Glocal pedagogy does not impose personal ideologies but rather invites educators to participate in reflective practices, thereby reducing chances of conflicts when teaching complex or debatable concepts, for example, the rights of persons with disabilities and issues of migrants and refugees.

61.3 Application

Glocal Pedagogy is suitable when teaching complex, global topics such as human rights education, global citizenship education, multiculturalism, and climate change at the university level in various settings and contexts.

Here, I use an example to illustrate the use of Glocal Pedagogy. It involved education on human rights for future teachers pursuing a Bachelor of Education. Earlier, lecture, project-based, and peer teaching were used to teach human rights. At times, I failed to address complex topics. For example, when teaching about the Universal Declaration of Human Rights, participants rejected learning about it. They showed resistance and anger toward the global convention, stating that some rights were not in line with their culture while others that should be part of the document were missing. However, I insisted on learning about it, claiming that it was significant and part of the course outline; thus, it could not be omitted. Similarly, when I informed them about female genital mutilation and child abuse issues, participants refused to learn. They avoided attending class because the pedagogy used by the educators was traditional and caused distress. I then applied glocal pedagogy to the new batch of learners to see if it helped improve their understanding of topics.

I first studied the social and cultural practices and their impact on shaping the country's educational system. It was followed by evaluating the learners' knowledge of human rights education and examining how it influenced their behaviors and attitudes. For this purpose, activities, including discussions, focus groups, and case studies, were arranged. I revisited the existing human rights education curriculum and identified sensitive and culturally biased elements. I realized many things that the learners needed to clarify regarding human rights. These misconceptions should have been addressed during their previous education and influenced their attitudes and behaviors toward others. For example, a few students believed that global people were the worst and that their cultural practices were a curse. They were considered insensitive to the problems faced by poor people. I also reflected on my competence and biases followed by students' competence and biases, I or they might bring into the classroom. I then shared the course outline with the students and asked them to review the topics they wished to learn and topics they wanted to omit from the course outline with justification. After gathering insights, I identified good practices and gaps in the curriculum and books while collecting student feedback. I found some topics in the course outline needed to be culturally appropriate to teach.

Furthermore, content written by local writers was limited to the subject matter. To address this, I consulted community members to gather insights. Only culturally appropriate content was selected with the help of experts, and activities were designed accordingly. Next, I scrutinized global elements in the local curriculum through content analysis. Later, I referred to the online content on the subject matter, selecting many Open Educational Resources (OER). Students were also engaged in adding content to the list during the activity. Furthermore, I connected with global scholars working in human rights education to gather additional resources and understand best practices. I then built a network of likely-minded educators, filtering, scrutinizing, and adapting content to suit local contexts. Any element that could hurt the sentiments of local culture or biased content was omitted. For example, women's rights that were compatible with cultural values were added instead of rights that did not fit in the local context.

My prime focus was to develop OER as collaborative content on topics unavailable in the local context, involving the participants and other teachers. For example, we developed case studies together, wrote an open educational storybook, participated in community work, and filtered global resources that aligned with local culture.

Glocal pedagogy is a transformative tool that imparts knowledge and fosters a sense of responsibility and activism among prospective teachers. It is a continuous process that does not confine itself to any specific timeframe but is developmental and cyclic and can be applied within a group (Table 61.1).

Table 61.1 Method overview

Main purpose
To promote a deeper understanding and appreciation of local and culturally relevant global practices; balances local and global tensions and harmonizes diverse educational perspectives, fostering an inclusive and culturally relevant learning environment.
Gained competences
Critical thinking, decision making, ethical thinking, collaboration, ownership, resilience, diversity, and pluralism
Educational setting
Higher educational institutions
Space requirements/restrictions
Indoor
Resources and necessary materials
Readings and activities
Number of participants
5–100
Facilitator competences and skills
Reflective practice, personal identity exploration, pedagogical skills, digital literacy, ownership, multiculturalism, resilience, and competence to teach and research
Participants skills/age/competences
It suits anyone above 16 years
Duration
1–6 months (cyclic)

Reference

Graeber D (2015) The utopia of rules: on technology, stupidity, and the secret joys of bureaucracy. London, Melville House

Chapter 62
Asking Questions to Co-create Future

Dagmar Nolden and Pia Rox

62.1 Introduction

As change agents, community workers, and educators, we, like other people, often face complex challenges in diverse fields such as education, peace work, or climate action. We tend to fall back on known solutions, yet what if we embraced complexity and uncertainty?

The "Wicked Question Game" introduces a playful approach to these seemingly unsolvable problems, exploring the underlying questions and uncovering a wide range of perspectives and new courses of action in a short space of time. The game is about the questions behind the question and draws attention to the multitude of related issues that often go unnoticed.

In a nutshell: The game questions a problem until it bursts.

Useful for in-person and online settings, it supports peer counseling, project work, and meetings. The only requirements are (1) everyone's willingness to engage in the process with an open mind and heart and (2) a specific initial question that gets to the heart of the challenge at hand.

D. Nolden (✉)
Berghof Foundation Operations gGmbH, Tübingen, Germany
e-mail: D.Nolden@berghof-foundation.org

P. Rox
University of Tübingen, Tübingen, Baden-Württemberg, Germany
e-mail: pia.rox@uni-tuebingen.de

J. Bentz, J. Ristić Trajković (eds.), *Imagining, Designing and Teaching Regenerative Futures: Art-Science Approaches and Inspirations From Around the World*, Science for Sustainable Societies,
https://doi.org/10.1007/978-981-96-9029-9_62

62.2 Transformative Potential

62.2.1 Creating Transformative Spaces Through Art of Hosting

Addressing the challenges of our time demands more than individual effort, it requires spaces that foster encounter, participation, and collaboration (Brocchi 2022). Innovation seldom arises within rigid, rule-bound environments; rather, it thrives in the liminal space between order and chaos (Pogatschnigg 2021). Embracing this requires a fundamental willingness to explore uncharted territory, to accept not knowing, and to be comfortable with the uncertainty and ambiguity that accompany it. The methods and mindset of Art of Hosting support these processes within the space between chaos and order. They allow us to create and sustain spaces of trust and connection, where slowing down, honest listening, and focusing on the essentials become possible. Art of Hosting taps into a group's collective intelligence and capacity for self-organization (ibid).

62.2.2 Inviting Curiosity and Playfulness

Aligned with the principles of the Art of Hosting (Pogatschnigg 2021), the Wicked Question Game encourages participants to embrace a playful and curious mindset, shifting from solution-oriented thinking to emergence and co-creation. Various adaptations of the method exist, such as Liberating Structures. This brief chapter introduces the Wicked Question Game as it is presented in one of the modules of the online training program *Hosting for Cultural Change*, offered by the non-profit organization Pioneers of Change, where it is an integral part. Further contextual and audience-specific adaptations are possible.

62.2.3 Harnessing the Power of Questions

The strength of this method lies in the power of impactful questions. Each question carries an implicit statement that can disrupt habitual thinking and offer an alternative perspective (e.g., Erbring 2022; Lindemann 2023; von Schlippe 2019). While questions in everyday life aim to gather information, the questions in the Wicked Questions Game are designed to generate insight by redirecting attention. During the game, the person who initiates with the first question is encouraged to focus their attention on the feelings, sensations, and thoughts that arise in response to the questions. The other group members open up the space by responding with further questions, rather than immediate solutions or concrete ideas. In particular, open-ended W-questions such as What (for/if)...? To whom? When? Why? Where...? appeal to our autonomy and invite introspection, reflection, and exploratory curiosity—even toward our inner resistances (e.g., Kolodej 2019; Lindemann 2023). This inquisitive attitude of all participants allows for an appreciative and respectful approach, which fosters a pressure-free environment where the questioner selects which insights to take forward. Thus, the Wicked Question Game unfolds its full potential through the attitude with which it is used. The core values of this mindset are openness, appreciation, and trust.

62.3 Application

62.3.1 Basic Principles of the Wicked Question Game

1. *Ask questions.* Reply to the initial question by asking more questions, rather than giving answers or good advice.
2. *Prefer open questions.* Ask (W-questions) and avoid closed questions (decision/yes-no-questions).
3. *Be present.* Ground yourself, practice self-empathy, and empathize with other group members.
4. *Be open.* Adopt an open, curious, experimental attitude. Allow your mind not to have an answer to every question and not to have to understand everything.
5. *Trust the flow.* Allow the pace of questions to change as the process unfolds—sometimes fast, sometimes slow, sometimes even with short moments of silence between questions.

62.3.2 What Roles Are There?

Person A Initiates a Round of the Wicked Question Game by Posing the First Question The person then actively listens to the questions posted by the other group members and directs their attention to the feelings, sensations, and thoughts that arise in response. Person A decides when the round concludes—whether the initial question has been sufficiently explored, a breakthrough has been reached, or a new question emerges that may offer further direction.

Sample question:

- "What could help me feel more connected to myself and my colleagues at our next team meeting, so that we can manage the substantial workload constructively before the upcoming deadline?"

Person B, Who Holds the Space and Witnesses the Process, Stands Up This person is one of the "other group members." Standing allows a different perspective on the process so that often entirely different questions emerge. When person B feels the impulse to ask a question, they sit down to do so, while another person from the group takes the place of person B standing.

Other Group Members, Persons C and D (at Least Two People), Listen Actively to Person A's Question They then explore person A's question further by asking additional questions, one after another. Following their intuition and associative thinking, group members say questions as they think of them. There are no right or wrong questions—experience shows that questions that seem paradoxical at first can have a profound impact.

Sample questions:

- "Who do you feel connected to?"
- "What one thing would make you aware of your available resources and constraints during the meeting?"
- "What would you need to do to avoid feeling connected to yourself and your team members?"
- "What should the workload be for the project to be completed on time?"

What Else Is Needed?

- *Location:* a virtual or physical space where a group can work undisturbed
- *Time:* 30–60 min per session
- *Materials:* for offline sessions: paper, pen, markers, blue card(s) for initial question(s), orange card(s) for in-depth/concluding question(s)
- *Participants:* 3–100 people

62.3.3 Experiences Implementing the Wicked Question Game

In the *Hosting for Cultural Change* online course, the Wicked Questions game supports participants in reflecting on their role as hosts and facilitators. Their questions relate to current challenges in their hosting or facilitation, such as "How will I, as a host/facilitator, manage uncertainty and chaos when facilitating a session on a controversial and emotionally charged topic for 200 people?" Participants frequently state that the game generates entirely new lines of thought and ideas, or that it

Table 62.1 Sample session plan for the application of the Wicked Question Game

Duration	Content
05′	*Introduction*: Set the scene by giving a short introduction, for example, on the art of asking effective questions and the function and effect of different types of questions. Perhaps also set an intention such as curiosity or openness
05′	*Preparing the group*: Each participant formulates their own initial question and writes it down on a small card. Sometimes, it is helpful to formulate several variations of the same question until the most appropriate one is found. Alternatively, a few minutes of "stream of consciousness" journaling beforehand can be helpful in identifying a question
05′–10′	Explain and demonstrate the method if it is new to the participants
20′–40′	*Play the Wicked Question Game*: In situ: Participants meet in groups of 5 and form a (chair) circle Online: Participants meet in groups of 5 in a breakout session *Participants assign roles for the first round*: Participant A poses their question. Participant B stands up to observe the process. C, D, and E, or all remaining group members, ask open-ended questions, for example, about the relationship between the questioner and their question When the round is over, that is, when the person who asked the first initial question says that it is enough, the roles are redistributed. Depending on the time available, it may not be possible for everyone to explore their initial question. In this case, it is helpful to announce this beforehand and appeal to the participants' sense of responsibility
05′	*Personal harvesting*: *In situ* and online: Invite participants to individually write down on a (colored) piece of paper their initial question and the insight or essence that emerged from the process
05′	*Collective harvesting*: Gather (a few) voices from the group: "What did I experience/observe/perceive/realize? What was surprising?" Encourage participants to share the essence rather than long stories

reveals personal ambivalences, preferences, or insecurities. In the example above, the participant discovered that she felt comfortable facilitating such a large group of people and that neither the controversial topic nor the potential for chaos was of concern. However, she felt overwhelmed by the idea that these 200 people would be physically in the same room and not just connected online. As a result, she was able to focus her preparation specifically on this aspect of facilitation (Table 62.1 and 62.2).

Table 62.2 Method overview

Main purpose
To explore various perspectives on an issue or question To generate new ideas/solutions/ways forward for a regenerative and transformative future To move and liquefy blockages and uncertainties
Gained competences
Take different perspectives Create new solutions and ideas to an unresolved problem Become aware of things in the unconscious Learn to make hypotheses Become aware of the potential of asking questions
Educational setting
Informal (e.g., workshops, seminars, team meetings, and communities of practice) Formal (e.g., classrooms, university courses, and colloquia)
Space requirements/restrictions
A space where you can work undisturbed in a small group In- and outdoors In situ and online
Resources and necessary materials
In situ: paper, pens, markers, blue card(s) for initial question(s), and orange card(s) for follow-up question(s) at the end Online: a software application to set up breakout rooms (e.g., breakout sessions in Zoom)
Number of participants
100 or more people: Minimum 3 people per group, ideally 5–6 people per group Works equally well for larger groups up to 10 people Groups of more than 10 should be divided into smaller units to play the Wicked Question Game.
Facilitator competences and skills
Hosting skills: creating and maintaining open and confidential spaces for dialogue in which all participants feel comfortable Questioning skills: how to frame challenges or problems into a good question
Participants skills/age/competences
Suitable for different audiences, for example, communities/students/policymakers/researchers/designers/practitioners, etc. Age: suitable for different age groups from 14 years onward Skills: curiosity, creativity, ability to abstract, challenge and change perspectives, and formulate (open ended) questions
Duration
5–15 min for one round of the Wicked Question Game, slightly longer if an introduction to the Wicked Question Game is required 30–60 min for a session with several rounds of the Wicked Question Game, allowing all group members to contribute their topics in the form of a starting question

References

Brocchi D (2022) By disaster or by design? Transformative Kulturpolitik: Von der multiplen Krise zur systemischen Nachhaltigkeit. Springer VS, Wiesbaden
Erbring S (2022) Systemische Beratung für eine inklusive Gesellschaft. Vandehoeck & Ruprecht, Göttingen
Kolodej C (2019) Strukturaufstellungen für Konflikte, Mobbing und Mediation. Vom Sichtbaren zum Unsichtbaren. Springer Gabler, Wiesbaden
Liberating Structures https://liberatingstructures.de/liberating-structures-menue/wicked-questions/. Accessed 15 Jan 2024
Lindemann H (2023) Einführung in die systemisch-lösungsorientierte Gesprächsführung und Beratung. Vandehoeck & Ruprecht, Göttingen
Pogatschnigg IM (2021) The art of hosting. Wie gute Gespräche Führung und Zusammenarbeit verbessern. Franz Vahlen GmbH, München
von Schlippe J (2019) Systemische Interventionen. Vandehoeck & Ruprecht, Göttingen

Chapter 63
Prebunking of Climate Change Conspiracy Theories

Colăcel Onoriu

63.1 Introduction

Prebunking seeks to prevent climate change conspiracy theories from gaining traction. This can be achieved by educating people about the discourse of conspiratorial thinking. Helping audiences identify the specifics of language use that contribute to anti-climate change narratives rooted in conspiracy theories means understanding the use value of conspiracy theories. Improving the persuasiveness of arguments for human-caused climate change is possible. Teaching how conspiratorial rhetoric is used to craft anti-climate change narratives and addressing cultural differences in opinions are key methods.

The approach *Prebunking Climate change Conspiracy* theories consists of analyzing the rhetoric of conspiracy theories, in order to engage diverse audiences and foster transformative action.

1. *Using key terminology* that underpins the conceptualization of conspiracy theory ideation to calibrate attention;
2. *Deconstructing the rhetoric* of critical thinking behind conspiracy theorizing, by understanding and replicating the common narrative and discursive tactics used by conspiracy theorists;
3. *Testing* different *communication strategies* for prebunking, by experimenting with the same conspiratorial tactics of presenting information to see what works best;
4. *Humanizing transformative climate practices* with narratives that connect human experiences with environmental interactions to climate issues more relatable;

C. Onoriu (✉)
University of Suceava, Suceava, Romania
e-mail: onoriucolacel@litere.usv.ro

J. Bentz, J. Ristić Trajković (eds.), *Imagining, Designing and Teaching Regenerative Futures: Art-Science Approaches and Inspirations From Around the World*, Science for Sustainable Societies,
https://doi.org/10.1007/978-981-96-9029-9_63

5. *Advocating for climate action as social justice*, by framing climate justice as aligned with regenerative futures approaches.

The overall goal is to provide audiences with the skills to identify and label climate change conspiracy theories before they become entrenched. Regenerative sustainability approaches face considerable challenges due to opposition fueled by the idea that human-caused carbon emissions are not the sole driver of climate change (Tenbrink and Willcock 2023). While there is consensus that greenhouse gas emissions contribute to global warming, the communication of climate science is often undermined by both well-intentioned and self-serving arguments that distort risk perception and indicate a cultural divide in opinions on climate change (Tenbrink and Willcock 2023). Conspiracy theories about the environment hinder efforts to combat climate change, as they appear plausible and deter understanding, influencing behaviors toward rejecting carbon footprint reduction (Jolley and Douglas 2014). These climate-related conspiracy theories also cast doubt on the potential for human adaptation to global warming. Consequently, educating the public through critical thinking could be a user-friendly approach, but it risks leading to a culture of relentless questioning that can contribute to the spread of misinformation (Boyd 2017). A more realistic perspective on renewable energy's potential would acknowledge the substantial societal shifts (Floyd et al. 2020) that transformative practices may entail (Davidson 2010).

63.2 Transformative Potential

Prebunking climate change conspiracy theories can promote acceptance of climate science by engaging diverse audiences and fostering transformative action. Although it has had limited success, debunking conspiracy theories and providing evidence-based information can equip individuals with the practical understanding to recognize and label conspiracy theories. Prebunking focuses on addressing the underlying motivations and misleading claims, rather than individual conspiracy theories, to challenge the rhetoric that perpetuates conspiracy ideation (Traberg et al. 2022). This requires engaging in lengthier conversations, with patience and a willingness to listen to those who feel left behind by climate change adaptation. By addressing the concerns of those affected and building their rhetorical competence, this approach can help mitigate the proliferation of climate change conspiracy theories and facilitate empathy-centered practices through inclusive dialogue. The goal is to equip the public commonly exposed to climate change conspiracy theories with the practical understanding to recognize and label them as such, as well as to foster their engagement with transformative action on climate change.

63.3 Application

In what follows, a strategy for addressing climate change conspiracy theories through the use of empathy-instilling narratives in public communication is delineated. The approach can highlight a transformative potential to address the genre of climate change conspiracy theories, as well as conspiracy theorizing. By blending human experiences with environmental interactions, this approach makes climate-related issues more relatable to diverse audiences, fostering empathy and collective consciousness toward climate change challenges. It addresses the underlying motivations and rhetoric of conspiracy theorists engaging audiences skeptical of climate change (rather than climate deniers), through patient and empathy-centered dialogue. Essentially, it aims to develop rhetorical competence to understand and counter the persuasive language of conspiracy theories and equip the public with practical skills to recognize and label conspiracy theories based on cultural contexts.

The Radar Chart of Prebunking presents the comparison of five different steps to achieve prebunking in order to counter climate disinformation before it spreads. Each approach is evaluated across five key dimensions: Complexity, Importance, Time Required, Impact, and Difficulty.

Humanizing transformative practices is most comprehensive but demanding. Calibrating attention and deconstructing rhetoric show high impact with varying levels of complexity and time requirements. Testing communication strategies offers a balanced approach, while advocating for climate action is complex and time-consuming but low in impact for prebunking purposes.

This recommendation suggests starting with low-complexity, high-impact methods like "Calibrate attention" and "Deconstruct rhetoric," and then progressively incorporating more complex strategies as resources and expertise allow. The tiered approach ensures that even organizations with limited resources can begin effective prebunking efforts, while also providing a roadmap for scaling up as capacity increases.

The approach requires several steps:

- Step 1: Identify common climate change conspiracy theories
- Research and list prevalent conspiracy theories related to climate change.
- Example: "Climate change is a hoax invented by scientists to secure research funding."
- Step 2: Analyze the rhetoric and structure of the conspiracy theory
- Break down the argument structure and identify persuasive techniques used.
- Example: This theory uses appeal to motive (questioning scientists' intentions) and false dichotomy (presenting only two options: truth or hoax).
- Step 3: Develop weakened versions of the conspiracy theory
- Create milder forms of the argument that maintain its essence but are easier to debunk.
- Example: "Some people think scientists might exaggerate climate data to get more attention."

- Step 4: Craft evidence-based counterarguments
- Compile factual information that directly addresses the conspiracy theory's claims.
- Example: Provide data on the peer-review process, diverse funding sources for climate research, and the global scientific consensus on climate change.
- Step 5: Design empathy-centered communication strategies
- Create narratives that address the emotional appeals of the conspiracy theory.
- Example: Share stories of scientists' personal motivations for studying climate change, emphasizing their concern for the planet and future generations.
- Step 6: Tailor the prebunking approach to specific audiences
- Adapt language, examples, and presentation style to resonate with different demographics.
- Example: For a youth audience, use social media formats and relatable climate impacts on their future. For older audiences, focus on long-term changes they have observed in their lifetime.
- Step 7: Implement inoculation theory
- Expose the audience to the weakened version of the conspiracy theory, followed immediately by its debunking.
- Example: Present the mild version: "Some think scientists might exaggerate climate data." Then immediately explain how peer-review and diverse funding sources prevent such exaggeration.
- Step 8: Engage the audience in critical thinking
- Encourage active participation in analyzing and debunking the conspiracy theory.
- Example: Host a workshop where participants identify logical fallacies in climate change conspiracy theories and construct counterarguments.
- Step 9: Provide tools for resilience
- Equip the audience with strategies to identify and resist future conspiracy theories.
- Example: Teach the CRAAP test (Currency, Relevance, Authority, Accuracy, Purpose) for evaluating information sources.
- Step 10: Monitor and adapt
- Assess the effectiveness of your prebunking efforts and adjust as needed.
- Example: Conduct surveys to measure changes in beliefs and attitudes toward climate science after prebunking interventions.

Resources:

- Online platforms for storytelling and narrative development (e.g., StoryCenter and The Moth)
- Tools for audience analysis and feedback collection (e.g., SurveyMonkey and Google Forms)
- Access to scientific research and credible sources (e.g., IPCC reports and NASA climate data) (Table 63.1 and Fig. 63.1)

Table 63.1 Method overview

Main purpose
Awareness building/behavior change
Gained competences
Rhetorical media literacy, communication skills, academic literacy in climate science, collaborative learning, rhetoric and evidence-based decision making, communication skills for policymaking, and advocacy
Educational setting
Both formal (Science and Communication Studies Conferences and Classes) and informal (workshops, seminars, local meetups, etc.)
Space requirements/restrictions
Indoor activities
Resources and necessary materials
Workshop/Conference Room Setup, *Workshop Materials*, *Collaborative Software* (Google Suite, Microsoft Office 365, or Slack), *Software for Content Creation and Editing* (Adobe Creative Suite, etc.), *Online Tools for Note-Taking and Mind Mapping* (Evernote, Obsidian, etc.)., and *Feedback and Assessment Tools* (surveys, online quizzes, or feedback forms)
Number of participants
Individual and Group work. Ideal group size under 12 people
Facilitator competences and skills
Content curation, *empathy and cultural competence*, *educational pedagogy*, *conflict resolution*, and *digital literacy*
Participants Skills/Age/Competences
Community, students, policymakers, and researchers Of all ages and abilities
Duration
Community members Anywhere from 2 to 4 h initially, with possibly one follow-up session after a month if tailored activities are designed for reinforcement Students An initial session lasting about 3–4 h might be sufficient, followed by in-class discussions or assignments throughout their academic year to reinforce learning Policymakers/researchers A more comprehensive approach, potentially spanning several days to accommodate in-depth discussions and presentations

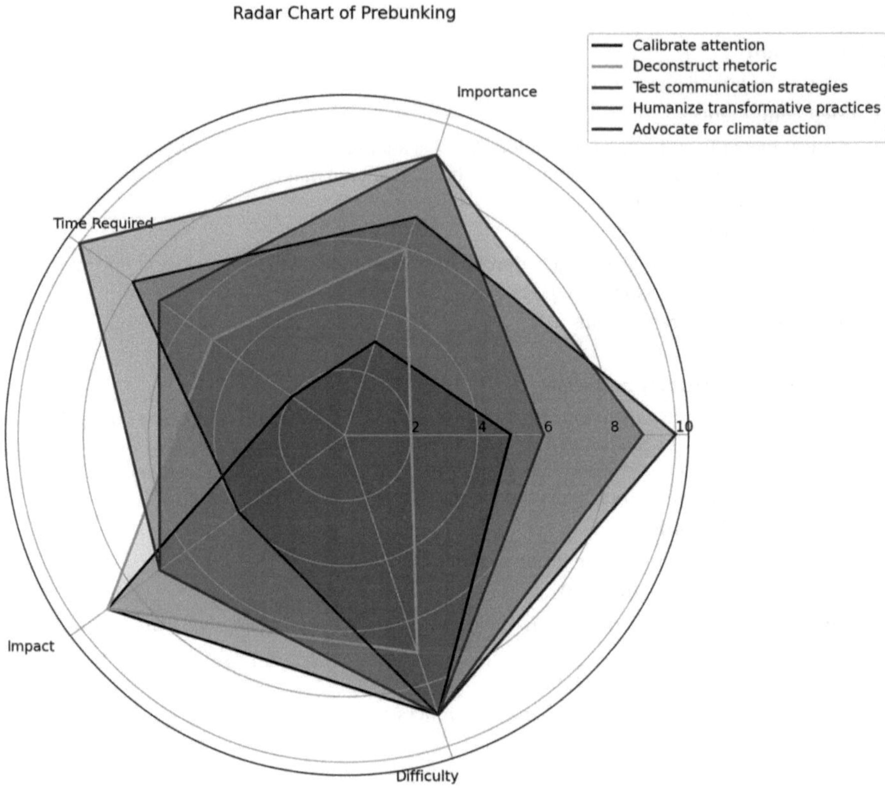

Fig. 63.1 Radar chart of prebunking

References

Boyd D (2017) Did media literacy backfire? J Appl Youth Stud 1(4):83–89. https://doi.org/10.3316/informit.607936397466888

Davidson M (2010) Hacking away at sustainability: science, ideology and cynical blockage. Hum Geogr 3(2):83–90

Floyd J et al (2020) Energy descent as a post-carbon transition scenario: how 'knowledge humility' reshapes energy futures for post-normal times. Futures 122:102565. https://doi.org/10.1080/14797585.2021.1886424

Jolley D, Douglas KM (2014) The social consequences of conspiracism: exposure to conspiracy theories decreases intentions to engage in politics and to reduce one's carbon footprint. Br J Psychol 105(1):35–56. https://doi.org/10.1111/bjop.12018

Tenbrink T, Willcock T (2023) Place attachment and perception of climate change as a threat in rural and urban areas. PLoS One. https://doi.org/10.1371/journal.pone.0290354

Traberg CS, Roozenbeek J, van der Linden S (2022) Psychological inoculation against misinformation: current evidence and future directions. Ann Am Acad Polit Soc Sci 700(1):136–151. https://doi.org/10.1177/00027162221087936

Chapter 64
Effective Innovation Leadership Framework: Self-Reflection Method

Sabrina Schork

64.1 Introduction

The *Effective Innovation Leadership framework* is a self-reflection method designed to help especially individuals understand and harness their values, strengths, and practices to become more effective innovation leaders. This framework emphasizes self-awareness and self-leadership as the foundation for leading others and driving innovation.

The *Effective Innovation Leadership framework* can be used individually or in small groups. It reflects on three core aspects:

1. Values: Participants identify which values guide their actions, using the seven values provided by the *Effective Innovation Leadership framework*. They can add personal values using resources like the ten broad personal values recognized across cultures such as self-direction, stimulation, hedonism, achievement, power, security, conformity, tradition, benevolence, and universalism (Sagiv and Schwartz 2022). Values act as a compass navigating decision.
2. Strengths: Participants determine their unique strengths by asking, "What am I better at than others without much effort?" They can start with the six strengths in the *Effective Innovation Leadership framework* and refer to the VIA Institute on Character's (2024) list of character strengths for additional insights.
3. Practices: Participants reflect on the practices they use in their daily innovation work. The *Effective Innovation Leadership framework* provides six practices, and further practices can be explored in the methods manual edited by van

S. Schork (✉)
Aschaffenburg University of Applied Sciences, Faculty of Economy and Law, Aschaffenburg, Germany
e-mail: sabrina.schork@th-ab.deTH

J. Bentz, J. Ristić Trajković (eds.), *Imagining, Designing and Teaching Regenerative Futures: Art-Science Approaches and Inspirations From Around the World*, Science for Sustainable Societies,
https://doi.org/10.1007/978-981-96-9029-9_64

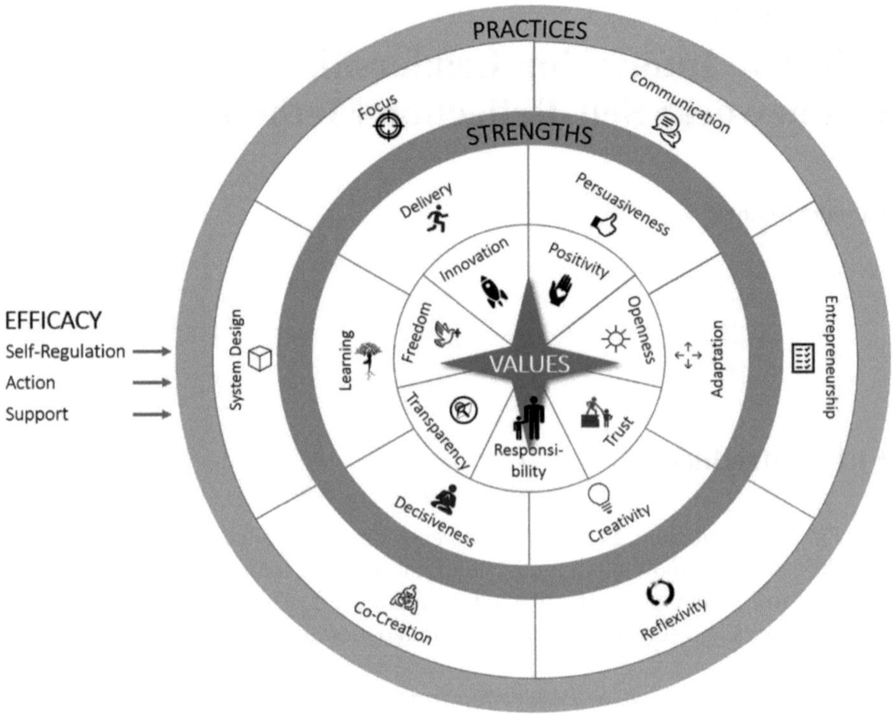

Fig. 64.1 Effective Innovation Leadership framework

Aerssen and Buchholz (Burkhardt et al. 2018) or the Innovator's Toolkit (Silverstein et al. 2012).

The *Effective Innovation Leadership framework* empowers individuals to lead themselves, which is essential for leading others. This self-leadership is crucial in innovation, where bringing new ideas to life requires clarity of purpose and confidence in one's abilities. The values of a person are also very important in the context of sustainability. The inventor of the framework is driving the development of sustainable or circular innovation (Schork et al. 2016, 2023; Schork 2021a, b). The framework encourages individuals to share their reflections with two others who have engaged in the same process. This three-way dialogue enhances insights and fosters a supportive environment for personal and collective growth. The dialogue can be self-organized or moderated in a bigger group (up to 100 participants), amplifying the framework's impact (see Fig. 64.1).

64.2 Transformative Potential

The Effective Innovation Leadership framework is a powerful tool for contributing to regenerative futures due to its emphasis on self-awareness, values alignment, and holistic leadership development. Here is why this method stands out:

- Fostering Self-Awareness and Personal Growth: The Effective Innovation Leadership framework helps individuals understand their values, strengths, and practices through reflection. This self-awareness supports personal growth and ensures leaders act with authenticity and purpose, crucial for sustainable practices.
- Building Resilient and Adaptive Leaders: The framework enhances resilience and adaptability by helping people identify and use their unique strengths. These qualities are essential for tackling complex challenges like climate change and inspiring others to adopt regenerative practices.
- Encouraging Reflective and Values-Driven Leadership: By emphasizing values-driven leadership, the framework ensures that decisions prioritize long-term well-being over short-term gains. This shift is key to creating regenerative futures, moving from exploitative to restorative practices.

Catalyzing Systemic Change: The Effective Innovation Leadership framework develops leaders who are self-aware, resilient, and values-driven. These leaders can challenge the status quo and create innovative solutions for ecological balance and social equity. Their ability to inspire others multiplies their impact, leading to widespread positive change.

Creating a Culture of Continuous Improvement: The framework promotes ongoing reflection and dialogue, fostering a culture of continuous learning and improvement. This is crucial for transformative change, helping organizations and communities adapt to new challenges and opportunities.

Enhancing Collaboration and Collective Impact: The Effective Innovation Leadership framework's group reflection process encourages collaboration and collective impact. By sharing insights and learning from diverse viewpoints, it helps create inclusive solutions for complex, interdependent issues that need collective action.

Promoting Sustainable Practices: Leaders who understand their values and strengths are more likely to champion sustainable practices within their organizations and communities.

Aligning Personal and Organizational Goals: The framework helps align personal and organizational goals with broader ecological and social objectives. This alignment ensures that actions taken at individual and organizational levels contribute to larger regenerative efforts, creating a synergistic impact.

The Effective Innovation Leadership framework empowers individuals to lead themselves and others more effectively, fostering a culture of reflection, resilience, and values-driven action. Its potential to create regenerative and transformative change makes it a powerful tool for leaders committed to building sustainable and thriving futures. By embracing this method, leaders cultivate the self-awareness and adaptive capacity needed to navigate and shape a rapidly changing world, ensuring their contributions are impactful and regenerative.

64.3 Application

Target audiences:

- Pupils that are deciding what to do after school, for example, study, career, stay abroad, or social year
- Students who want to shape their career actively
- Young professionals who want to shape their career proactively

Preparation advice for facilitators:

- Arrange an adaptable room with private niches and stable Wi-Fi.
- Prepare printouts of the Effective Innovation Leadership framework, personal values, and core strengths lists if Wi-Fi is unreliable.
- Have the book "Das große Handbuch Innovation" (German) and the "Innovator's Toolkit" (English) online and/or physically available.
- Prepare printouts of the values and strengths.
- Notify participants to bring tablets or laptops.

Individual Sessions:

- Duration: 45–60 min
- Process: Participants reflect on three questions about their values, strengths, and practices.

Group Sessions:

- Group Size: Up to 100 people
- Duration: approximately 1.5 h

Process:

- Individual Reflection (45 min): Participants consider the three core questions.
- Small Group Discussion (15 min): Sharing reflections in threes
- Group Synthesis (30 min): Facilitator-led discussion on common themes and intersections

Case Studies:

- THNK Creative Leadership Program: The Effective Innovation Leadership framework was used to challenge ten participants individually, fostering self-awareness and leadership effectiveness.
- Innovation Management and Design Study Program: The Effective Innovation Leadership framework helped a group of six students build self-knowledge for their innovation studies and career path.

Considerations:

- Allocate appropriate time and ensure a comfortable environment with the necessary materials (Table 64.1).

Table 64.1 Method overview

Main purpose
Reflection of personal values/strengths/innovation practices
Gained competences
Self-efficacy/self-responsibility/self-awareness/reflection skills/future-design/leadership skills/innovation practices
Educational setting
Informal workshop setting with movable chairs and tables as well as niches for privacy
Space requirements/restrictions
Sufficient space for quiet individual thinking or joint reflection in groups of three; indoor or outdoor activity
Resources and necessary materials
Wi-Fi, tablet, paper, online timer, and printed *Effective Innovation Leadership framework*
Number of participants
1–100 (ideally between 15 and 30)
Facilitator competences and skills
Personal coaching for individual advice and liberating structures for bigger groups
Participants skills/age/competences
Self-reflection/minimum 18 years old/communication in groups
Duration
45 min to 1.5 h depending on the group size

References

Burkhardt N, Ernst A, Rings J, Rings S, Schobloch A, Spicker M et al (2018) Das große Handbuch innovation: 555 Methoden und Instrumente für mehr Kreativität und Innovation im Unternehmen. Vahlen, Munich

Sagiv L, Schwartz SH (2022) Personal values across cultures. Ann Rev Psychol 73:517–546

Schork S (2021a) Circular innovator education. In: Ability to act. BoekenGilde, Enschede, pp 105–124

Schork S (2021b) Innovation leadership in the digital enterprise: lessons from pioneers. In: Handbook of research on management and strategies for digital enterprise transformation. IGI Global, Hershey, pp 86–109

Schork S, Heblich B, Terzidis O (2016) Effective innovation leadership. In: University-industry-innovation-network conference, Amsterdam, June 2016. University Industry Innovation Network B.V, Amsterdam, pp 1–16

Schork S, Özdemir-Kaluk D, Zerey C (2023) An integrative definition of digital, innovation, and sustainable: a bridge between disciplines, pp 1–37. Preprint

Silverstein D, Samuel P, DeCarlo N (2012) The innovator's toolkit: 50+ techniques for predictable and sustainable organic growth. Wiley, Hoboken

VIA Institute on Character, The 24 Character Strengths. https://www.viacharacter.org/character-strengths. Accessed 06 June 2024

Chapter 65
Holistic Approach to Environmental Law Education

Zerrin Savaşan

65.1 Introduction

While there are rich resources available separately on environmental education or law education, there are limited resources for Environmental Law Education (ELE). Current approaches address the subject in different scopes and dimensions, but not holistically. Therefore, I aim to provide a holistic approach that includes a wheel of different teaching tools based on different methods. By encouraging the effective use, sharing, and dissemination of original, innovative teaching materials and methods in a holistic way, by relating education/trainings to real-life problems, this method, the ELE wheel, aims to prepare future academics (educators, teachers, trainers, and professors) working on environmental law and policies to have a holistic understanding of teaching methods when they become part of the academy. It consists of two main activity phases for 10 days with additional collaboration and networking phases.

65.2 Transformative Potential

Due to the unique characteristics of environmental law, it is essential that environmental lawyers are able to view events from a holistic and multidisciplinary perspective and have the necessary practical skills built on the knowledge and simulations they acquire in class, such as building relationships with clients, fellow

Z. Savaşan (✉)
Department of International Relations, Sub-Department of International Law, Faculty of Economics and Administrative Sciences, Selçuk University, Konya, Türkiye
e-mail: szerrin@selcuk.edu.tr

J. Bentz, J. Ristić Trajković (eds.), *Imagining, Designing and Teaching Regenerative Futures: Art-Science Approaches and Inspirations From Around the World*, Science for Sustainable Societies,
https://doi.org/10.1007/978-981-96-9029-9_65

lawyers, and decision-makers, managing a case, observing how strategic decisions and ethical judgments are made, interviewing experts, obtaining relevant information, and facilitating negotiations. Environmental lawyers trained from this perspective can learn and adopt skills that are not readily apparent in classroom work and thus be more successful in reaching solutions to deal with real-world problems.

The aim of this approach is to support the transformation of environmental lawyers' perspectives on environmental problems into a holistic and interdisciplinary framework in the short term.

- By promising to reflect a holistic and interdisciplinary perspective on ELE
- By contributing to the training of future educators/lecturers/trainers and the development of the ELE curriculum in line with this approach
- By accelerating the quality, capacity, and number of academics teaching environmental law in line with this approach

In the long term, it prepares the ground for people to evaluate their behaviors regarding environmental problems with a holistic approach and aims to achieve sustainable behavioral change regarding environmental problems and environmental protection.

65.3 Application: ELE Wheel

Main Purpose To enable participants (educators, teachers, trainers, and professors) working on environmental law and policies to learn, think, and discuss about the holistic approach and apply it in their courses in higher education

Target Audience Graduate students (Master, Doctorate, Post-Doctoral levels) and research assistants/academicians teaching or aiming to teach environmental law (12 participants, 4 groups)

Phases of the Training The training program involves two basic activity phases for 10 days with additional collaboration and networking phases:

1. *Activity Phase-I:* The first activity program lasts 5 days involving a group discussion and simulation phases.

    1.a. *Holistic Approach—2 days:* This session includes discussions on different teaching methods and tools. Since participants are already researchers in the field, they can discuss, with the help of a facilitator, the strengths/weaknesses/limitations and benefits of implementing different approaches in law courses in higher education, for a minimum of 1.5 h for each tool group (narrative-interactive-collaborative-experiential).

1.b. *How to Apply a Holistic Approach in Class ?—3 days*: This session involves a simulative class that will work on a sample case study and a defense section.

Sample Case Study: An Environmental non-governmental organization asks for help and legal assistance from the Law Clinic of the Faculty, because a Building Construction Materials Company discharges waste, such as sand, gravel, rubble, garden soil, and fertilizer, which are the result of its activities, to the land next to their workplace.

1.b.1. *Simulative Class—1 day:* Through an interactive participation of the participants, a simulation class will be formed. A sample case with a real client (like mentioned above) will be provided to the class (experiential tools). On the basis of that case, the main legal problems to deal with, and applicable legislation, procedure will be explained just to draw a framework and guide the class (narrative tools). Then, the class in groups will be asked to conduct interviews with the client (experiential tools), seek for and find out solutions to the case, and discuss on different possibilities to lead to the optimal solution (based on both non-litigation and litigation tools) (interactive and collaborative tools).

1.b.2. *Defense and Justification—2 days:* The legal responses provided by the groups will be evaluated with constructive feedback and the participants will be asked to defend their solutions through alternative teaching methods such as showing related documentaries, referring to creative drama techniques, storytelling techniques, moot court, experts' views from other relevant disciplines, making field visits, or preparing posters calling for students to choose their environmental law course, thinking about how they can implement a holistic approach in their course through alternative ways, and so on.

2. *Activity Phase-II:* It is for 5 days involving two main sub-sessions: simulation and evaluation.

2.a. *Simulation—4 days:* The participants present their work specifying the tools they use, and the reasons behind them. If they do not use the whole wheel, they are expected to justify the reasons.

Those questions are asked to make the participants re-think holistically: What did you like? What didn't you like? What have you learned? Can such activities/tools be used in class as an effective way of education? What can be the weaknesses? What can be the strengths? What are their suggestions, if they have? What place do the tools/activities of each method used play into the weight ratio of the wheel? Which one is more important and why? How much impact does the topic being discussed have on this? Why?

2.b. *Evaluation—1 day:* While the participants evaluated each other, and the trainers evaluated the groups, the whole training program was evaluated by the participants in general by the survey study. In addition, an overall assessment was made by the facilitator with the contribution of the trainers through an observation form after the completion of the whole training program. This last phase can be considered to determine whether the objectives have been achieved and to measure and evaluate change/development in the participants thanks to the training program.

3. *Collaboration Phase:* In this phase, the aim is to evaluate, assimilate, and simulate the knowledge and skills acquired by the participants in the activity phases. Participants will be able to experience and evaluate what they have learned through group work and decide what to do for the next stages. Encouraging autonomy, independency, and flexibility, four groups are asked for a collaborative development of a course outline for dealing with a specific Environmental Law topic. They are expected to be based on the ELE wheel, and for each week to include at least one tool/activity of each method holistically. They will be monitored online and asked to report on their status and progress levels every 15 days (six times in total).
4. *Networking Phase:* In the network creation and maintenance phase, which is planned to continue after the training and observation phases, participants are expected to stay in touch with each other, share their knowledge and experiences, and organize similar trainings to ensure sustainability (Table 65.1).

Table 65.1 Method overview

Main purpose
Training environmental lawyers based on new approaches in ELE
Gained competences
Holistic perspective on ELE Effective use, sharing, and dissemination of original, innovative teaching materials and methods on ELE Associating education with real-life issues A network of professionals on ELE
Educational setting
Informal (training)
Space requirements/restrictions
Depends on group works (can include both outside/indoor activities, etc.)
Resources and necessary materials
Presentation materials and equipment Flip chart/board, poster papers, markers, pictures, texts, stories on the topic, crayons, etc. Depending on group work, more can be added.
Number of participants
Totally 12 (four groups with three members)
Facilitator competences and skills
Experienced in ELE, or law education expertised on new teaching methods
Participants skills/age/competences
Graduate students (Master, Doctorate, Post-Doctoral levels)/research assistants/academicians
Duration
10 days for activities/3 months for monitoring

Further Readings

Boud D, Cohen R, Walker D (1993) Using experience for learning. The Society for Research into Higher Education and Open University Press, Oxford

Clover DE, Bruno De OJ, Hall BL, Follen S (2013) The nature of transformation, environmental adult education. Sense Publishers, Rotterdam

Hecht SB, Weıs JL, Amos AL (2007) Environmental law teachers' perspectives on preparing new lawyers for practice. American Bar Association 36th Conference on Environmental Law Proceedings, March 2007. Laying the Foundation, American Bar Association Section of Environment, Energy, and Resources

Kugel P (1993) How professors develop as teachers? Stud High Educ 18(3):315–328

Mintz JA (1983) Teaching environmental law: some observations on curriculum and materials. J Legal Educ 33(1):94–110

Robertson HG (1998) Methods for teaching environmental law: some thoughts on providing access to the environmental law system. Colum J of Envtl L 23:237–298

Robinson-Dorn M (2007) Teaching environmental law in the era of climate change: a few whats, whys, and hows. Wash Law Rev 82:619–649

Yang Y (2013) Analysis of solution to environmental law education for college students. Beijing Law Rev 4(3):100–102

Chapter 66
3S: Seed, Save, Share

Maria João Fonseca and Rebeca Mateus

66.1 Introduction

Sunlight, dappled by a Tipuana tree, illuminates the school's transformed vegetable garden. Once underutilized, it now thrives with diverse greens. Laughter drifts from a nearby amphitheater, hinting at the program's engaging nature. The 3S (Seed, Save, Share) approach, rooted in the humble seed, cultivates environmental stewardship in students, particularly in urban areas where a disconnect from nature is prevalent.

3S echoes an ancient tradition. The practice of saving seeds has been handed down through generations, ensuring food security and adaptation to specific environments. This act embodies an understanding of ecological interconnectedness, fosters community, and recognizes the web of life.

The 3S approach is an educational initiative that reconnects students with nature by (1) transforming school spaces into agroecology laboratories with hands-on learning activities and (2) creating and offering a set of sessions to learn about gardens and how to care for them, using non-formal education methodologies.

By bridging this past to a sustainable future, 3S transforms classrooms and outdoor school areas into laboratories for hands-on learning in agroecology and ecosystems. Students explore the seed's lifecycle, historical significance, and role in shaping human evolution. The program emphasizes seed diversity conservation for ecological balance.

The core principles of 3S are not mere abstractions. Students actively participate in seed-saving techniques throughout the year, connecting historical practices to

M. J. Fonseca (✉)
HTC – CFE – NOVA FCSH, Lisbon, Portugal

R. Mateus
HortaFCUL, University of Lisbon, Lisbon, Portugal

J. Bentz, J. Ristić Trajković (eds.), *Imagining, Designing and Teaching Regenerative Futures: Art-Science Approaches and Inspirations From Around the World*, Science for Sustainable Societies,
https://doi.org/10.1007/978-981-96-9029-9_66

present solutions. These hands-on activities instill a sense of responsibility and an understanding of the interconnectedness between economic, social, and natural systems. The joy of cultivating and sharing harvests fosters a sense of community within the classroom and beyond. Events like open days, ethnobotanical walks, and seed-sharing meetings create spaces for students, families, neighbors, and local farmers to connect, building a network of environmental stewards.

3S transcends single schoolyards. The approach can be replicated in other contexts, acting on a larger scale. Through partnering with local organizations, the program can develop seed banks and national seed exchange initiatives, fostering a robust approach to biodiversity conservation. This collaborative spirit empowers students to see themselves as active participants in a global movement toward a sustainable future.

The 3S approach is more than just an educational program; it is a seed of change. It gives students the knowledge, skills, and confidence to become engaged citizens with an increased environmental consciousness. Through nurturing, they cultivate a sense of connection with the natural world and deepen their understanding of the delicate ecological balance. From school gardens, 3S empowers the next generation to embrace a sustainable future, one seed at a time.

66.2 Transformative Potential

The 3S approach addresses critical issues such as biodiversity loss and climate change. Promoting seed saving and sharing contributes to:

1. *Embodying Sustainability Values:* The 3S approach empowers students to become environmentally conscious leaders by raising awareness about topics such as food sovereignty, the environmental costs of agriculture, and how to act toward solutions, from daily practices to collective approaches. Seed saving in public schools fosters critical reflection on sustainability (cognitive, emotional, practical), building accessible student agency and collaboration.
2. *Embracing Complexity:* Hands-on activities and discussions addressing ecological and community perspectives equip students with critical thinking for problem-solving. Facilitators guide them toward solutions and group work activities, encourage healthy debates, stimulate imagination, and raise awareness of individual choices.
3. *Envisioning Sustainable Futures:* Envisioning a sustainable alternative future of biodiversity and food production encourages youth to practice behaviors that can make a positive change. Activities such as plant intercropping encourage students to explore future scenarios and sustainable practices.
4. *Acting for Sustainability:* Students translate learning into action by engaging in on-site theoretical-practical activities with observable short- and long-term results. Moreover, they are empowered to take action beyond school walls, fostering collaboration and a sense of community.

66.3 Application

The 3S approach can empower learners of all ages to become active environmental stewards. This highly adaptable program can be implemented in schools and other educational settings through projects like "Germinar um Banco de Sementes" (Sprouting a Seed Bank).

The entire project requires regular sessions throughout the academic year. However, individual seed-saving activities can be adapted for shorter time frames. Materials include seeds, educational resources, and potentially seed storage containers. The "Manual em Dispersão" [Handbook in Dispersion] (Fonseca et al. 2022) and "Manual em Germinação" [Handbook on Germination] (Fonseca et al. 2024) handbooks provide detailed instructions and activities for various age groups.

Steps for Implementation

1. *Preparation (2–4 weeks):*

   - *Curriculum Design:* Integrate seed-saving concepts and agroecological topics related to the seed life cycle into existing subjects (e.g., science and social studies).
   - *Seed Selection:* Choose regionally appropriate seeds suitable for student skill levels (for example, using large seeds such as beans or other legumes with very young children and including smaller seeds, which are more challenging to handle with older students).
   - *Community Engagement:* Partner with local organizations (e.g., gardens and seed banks) for support and outreach.

2. *Implementation (Throughout the program):*

   - *Seed-Saving Activities:* Engage students in nurturing (watering, fertilizing with organic compost, managing "weeds"), collecting, cleaning, and storing seeds, as well as introducing ecosystems and agroecological concepts and practices along the entire cycle "from seed to seed."
   - *Learning Activities:* Use the open-source handbooks (Fonseca et al. 2022, 2024), which serve as a guide for replicating the project in school environments for transformative action.
   - *Interdisciplinary Learning:* Link seed saving to art, history, and natural science for a holistic learning experience, for example, by planting through ancient practices such as seed bombs or the "3 sisters"—growing beans, corn, and pumpkin together. Moreover, there is an effort to engage in landscaping and artistic activities in the garden that are not always directly related to the growth of plants but also to creating a hosting environment.

3. *Community Engagement:*

   - *Public Events (optional):* Organize seed exchanges within the school or with the broader community. Host open days, talks, or ethnobotanical walks to raise awareness about seed diversity and to bridge the gap between schools and the local community.

Table 66.1 Method overview

Main purpose
Environmental awareness building/behavior change/non-formal education
Gained competences
Learning by doing/agroecology/biology
Educational setting
Non-formal (school sessions)/informal (walks, meetings, and talks)
Space requirements/restrictions
Outside activities
Resources and necessary materials
Seeds/soil/gardening equipment and structural conditions
Number of participants
In groups of 15–30 participants
Facilitator competences and skills
Agroecology/biology/scientific illustration/education
Participants skills/age/competences
Community/students
Duration
From one session to an academic year

- *Digital Platform (optional):* Create a website or online forum for free seed sharing and information exchange (as www.germinar.pt), raising awareness about the importance of seed diversity and conservation at a national level (Table 66.1).

References

Fonseca MJ, Alves R, Manta A, Peres J, Passos L, Alexandre A (2022) Manual em dispersão. Germinar um banco de sementes, Lisabon

Fonseca MJ, Peres J, Alexandre A, Passos L, Andrade J, Guia I (2024) Manual em disseminação. Germinar um banco de sementes, Lisabon

Correction to: Imagining, Designing and Teaching Regenerative Futures: Art-Science Approaches and Inspirations From Around the World

Julia Bentz and Jelena Ristić Trajković

Correction to: Chapters 26 and 50 in: J. Bentz, J. Ristić Trajković (eds.), *Imagining, Designing and Teaching Regenerative Futures: Art-Science Approaches and Inspirations From Around the World*, Science for Sustainable Societies, https://doi.org/10.1007/978-981-96-9029-9

The original version of the book was inadvertently published with errors. The following corrections have been made after the original publication.

Chapter 26:
The original version of the chapter "The BEEHIVE: Method to Explore Alternative Futures" was inadvertently published without the below graph. This has now been updated in the chapter.

The updated versions of these chapters can be found at
https://doi.org/10.1007/978-981-96-9029-9_26
https://doi.org/10.1007/978-981-96-9029-9_50

J. Bentz, J. Ristić Trajković (eds.), *Imagining, Designing and Teaching Regenerative Futures: Art-Science Approaches and Inspirations From Around the World*, Science for Sustainable Societies,
https://doi.org/10.1007/978-981-96-9029-9_67

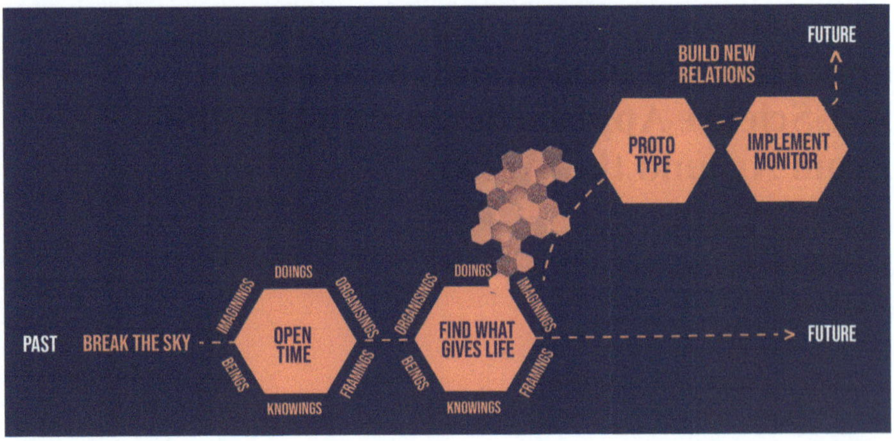

Chapter 50:
The last name of Karla Santos Zambrano was unfortunately published with an error. The initially published version has now been corrected.

Conclusion: Imagining, Designing, and Teaching Regenerative Futures as a Creative Journey—Reflections for Moving Forward

Julia Bentz and Jelena Ristić Trajković

Delving into complex social-environmental issues is rarely a linear or straightforward endeavor. It often involves the interplay of multiple elements and processes that converge to generate fresh insights and reshape existing knowledge. Imagining, designing, and teaching regenerative futures as a creative journey requires us to release preconceived notions of content and move beyond one-way knowledge transfer. It invites both learners and educators to become co-travelers, engaging together in a shared process of discovery, a creative journey.

Embarking on this journey reveals an extraordinary path involving imagination and transforming how we perceive the world and broadening our horizons. Along the way, we come to understand that the pursuit of meaningful transformation is not simply about accumulating scientific facts or data. Instead, it resembles a diverse set of experiences and perspectives and involves creativity, hope, and imagination. As we navigate this journey, we tune into the world around us, listening to shared aspirations for the future and different forms of knowledge and wisdom.

Art-science approaches are increasingly seen as effective means of developing both passion and an emotional connection with sustainability issues. Art can help to expand the learners' visions of the future, opening up their minds to alternative, regenerative scenarios and more-than-human worlds. Art-science approaches offer unique opportunities to foster critical thinking, empathy, and

J. Bentz
Center for Ecology, Evolution and Environmental Change, University of Lisbon, Lisboa, Portugal
e-mail: jhbentz@fc.ul.pf

J. Ristić Trajković
University of Belgrade – Faculty of Architecture, Belgrade, Serbia
e-mail: jelena.ristic@arh.bg.ac.rs

J. Bentz, J. Ristic Trajkovic (eds.), *Imagining, Designing and Teaching Regenerative Futures: Art-Science Approaches and Inspirations From Around the World*, Science for Sustainable Societies,
https://doi.org/10.1007/978-981-96-9029-9

imagination—qualities essential for addressing complex, interconnected global issues. Through storytelling, visual arts, performance, and participatory design, learners can experience diverse perspectives and envision alternative futures that transcend conventional boundaries. Integrating arts into educational contexts has been shown to increase engagement and motivation among learners. For example, exploring regenerative perspectives of the future and learning through art-science tools can be inspiring and empowering for (young) people. This can activate a desire to engage and contribute to alternative, regenerative futures (Bentz 2020, 2023; Bentz et al. 2022; Ristić Trajković et al. 2021).

Exploring regenerative futures as a creative journey invites educators, learners, and society as a whole to approach today's global challenges through a lifelong learning approach asking critical questions such as: What lessons might we uncover about ourselves, our relationships with others, our connection to nature, and our vision for the future? The ability to ask "What if?" lies at the heart of fostering empathy and imagining better ways of living. The lack of widespread action toward climate resilience and regenerative futures can be seen as a failure of imagination—a missed opportunity to craft shared visions of alternative futures that inspire transformation. By collectively envisioning desirable futures, we create both the motivation and the roadmap for meaningful change (Hopkins 2020).

To realize the transformative potential of art-science approaches, educators need support, training, and access to resources. Collaborative networks and open-source toolkits can equip teachers and change-makers with practical strategies to integrate arts-based learning into diverse educational contexts. There is a growing need to move beyond traditional, fact-based curricula toward more holistic, transdisciplinary, and participatory models. Policymakers and educational institutions generally still do not recognize enough the value of creativity and the arts as core components of education for sustainability, resilience, and regeneration. Cross-sector collaboration—between artists, educators, scientists, and communities—can spark innovation and ensure that creative tools are relevant, inclusive, and adaptable. Sharing best practices and lessons learned across regions and cultures will help scale successful approaches and inspire new ones.

Sustainability science emphasizes interdisciplinarity, transdisciplinarity, systems thinking, participatory engagement, and a focus on real-world problem-solving (Clark and Dickson 2003; Komiyama and Takeuchi 2006). The methods collected in this book align with these core tenets. They foster holistic perspectives, support collective knowledge production, and nurture transformative capacities such as imagination, empathy, critical reflection, and collaboration, which are crucial for enabling the societal changes sustainability science envisions (Wiek et al. 2011; Ristić Trajković et al. 2021, 2024).

By weaving together artistic experimentation, systems thinking, community collaboration, and transformative ways of teaching and learning, the approaches in this volume invite educators, learners, and change-makers to explore and grow essential capacities for shaping regenerative futures. They foster a deeper understanding of interconnected systems, the ability to anticipate and envision change, the strategic skills to navigate complexity, as well as the wisdom to collaborate across boundaries, and the critical reflection needed to reimagine our relationship with the world.

This book presents a broad array of art-science methods and educational practices that aim to go beyond a mere summation of artistic and scientific knowledge but weave together and work across different ways of knowing through participatory, imaginative, and transformative approaches. The diverse contributions in *Imagining, Designing, and Teaching Regenerative Futures* demonstrate how art-science methods can foster alternative imaginaries and ways of knowing which are required for navigating complexity and catalyzing societal change.

In doing so, this book responds to calls to move beyond traditional discipline-based models and toward holistic, solutions-oriented approaches (Clark and Dickson 2003; Kajikawa 2008). Importantly, it extends the ambitions of sustainability science by foregrounding regeneration as a vital educational and societal goal. Moving beyond sustainability's focus on minimizing harm, regenerative approaches seek to actively restore and enhance the health, vitality, and resilience of human and natural systems (Wahl 2016). By embedding regenerative thinking into educational practice, the methods collected here open new pathways for cultivating thriving, flourishing futures.

This compendium reminds us that regenerative learning is not just about what we teach but also about how, where, and with whom we create the conditions for change. By involving 120 contributors from 30 countries working across cultures, disciplines, practices, and sectors, the book constitutes a global, inclusive, and collaborative approach to knowledge production. The collection of 65 methods shared in this volume reflects a diverse tapestry of pedagogical possibilities united by a shared commitment to transformation. Whether mainly grounded in community engagement, systems thinking, and experiential learning, each method carries the potential to open new forms of learning and relating: to ourselves, to others, and to the living world. It invites readers to apply methods not as a fixed solution or finished formula, but as a flexible entry point into deeper research and educational experimentation. The methods are meant to be explored, transformed, and re-contextualized—shaped by the specific needs, aspirations, and wisdom of each learning setting. They offer tools not only for building agency and imagination but also for holding space: for reflection, emergence, and regenerative possibility. In this sense, regenerative learning is itself a living system—adaptive, evolving, and relational. It mirrors the ecological processes we seek to understand and restore: diverse, dynamic, and rooted in context.

Imagining, Designing, and Teaching Regenerative Futures affirms that art-science approaches are not peripheral but central to regeneration and sustainability transformation. To realize societies that live in harmony with nature, promote circular economies, and embrace inclusive governance, we must continue to imagine boldly, collaborate across boundaries, and teach for transformation.

As we move forward, this book is intended to serve not only as a resource but also as an invitation—an invitation to explore new forms of teaching and learning, to embrace uncertainty with curiosity, to imagine boldly and design compassionately, and to build futures that are not just sustainable but also regenerative, in which both people and planet can thrive. Further research is needed to understand the long-term impacts of arts-based education on learners' attitudes, behaviors, and capacities for transformative action. Rigorous evaluation and reflective practice will help refine methodologies and demonstrate the value of creative learning for addressing global challenges.

The concept of *Imagining, Designing, and Teaching Regenerative Futures* is about embracing the creative, hopeful journey itself—a journey toward shared aspirations for regenerative and thriving futures. Ultimately, creative and arts-based education can cultivate hope, agency, and a sense of belonging, empowering young people to become active participants in shaping regenerative futures. By nurturing imagination and creativity, we wish to strengthen all kinds of initiatives that collectively envision and work toward desirable futures as our contribution to a more just, resilient, and thriving world.

References

Bentz J (2020) Learning about climate change in, with and through art. Clim Change 162:1595–1612. https://doi.org/10.1007/s10584-020-02804-4

Bentz J (2023) Creative approaches to climate and peace education: an educator's guide to using storytelling and art. Leibniz Institute for Educational Media | Georg Eckert Institute, Leibniz

Bentz J, do Carmo L, Schafenacker N, Schirok J, Corso SD (2022) Creative, embodied practices, and the potentialities for sustainability transformations. Sustain Sci 17:687–699. https://doi.org/10.1007/s11625-021-01000-2

Clark WC, Dickson NM (2003) Sustainability science: the emerging research program. Proc Natl Acad Sci 100(14):8059–8061. https://doi.org/10.1073/pnas.1231333100

Hopkins R (2020) From what is to what if: unleashing the power of imagination to create the future we want. Chelsea Green Publishing

Kajikawa Y (2008) Research core and framework of sustainability science. Sustain Sci 3(2):215–239. https://doi.org/10.1007/s11625-008-0053-1

Komiyama H, Takeuchi K (2006) Sustainability science: building a new discipline. Sustain Sci 1:1–6. https://doi.org/10.1007/s11625-006-0007-4

Ristić Trajković J, Milovanović A, Nikezić A (2021) Reprogramming modernist heritage: enhancing social wellbeing by value-based programming approach in architectural design. Sustainability 13(20):11111. https://doi.org/10.3390/su132011111

Ristić Trajković J, Krstić V, Milovanović A, Calheiros CSC, Ćujić M, Karanac M, Kazak JK, Di Lonardo S, Pineda-Martos R, Garcia Mateo MC et al (2024) Moving towards a holistic approach to circular cities: obstacles and perspectives for implementation of nature-based solutions in Europe. Sustainability 16(16):7085. https://doi.org/10.3390/su16167085

UNESCO (2017) Guidelines on sustainability science in research and education. UNESCO, Paris

Wahl DC (2016) Designing regenerative cultures. Triarchy Press

Wiek A, Withycombe L, Redman CL (2011) Key competencies in sustainability: a reference framework for academic program development. Sustainability Science 6:203–218. https://doi.org/10.1007/s11625-011-0132-6

Appendix: Comprehensive Overview of All Methods by Tag Groups

Transformation involves the ability to act and to translate vision into practice and intention into experience. This final part of the book offers a reflection and practical guidance for selecting appropriate art-science methods introduced throughout the volume. It brings together all 65 contributions and organizes them into thematic clusters and core sustainability competencies, such as systems thinking, climate action, community empowerment, well-being, cultural perspectives, and regenerative design. This curated overview offers educators, facilitators, and change-makers a structure for navigating the book's methods. Through tables, tags, and thematic and competence-based groupings, it enables readers to select, compare, and adapt approaches based on their learning goals, audience, and educational context. By linking methods with key transformative competencies, this section supports the practical implementation of regenerative learning. It reflects the idea that meaningful education is not prescriptive or fixed, but adaptive, evolving, place-based, and rooted in context. This overview invites readers to treat each method as a flexible entry point into deeper research and educational experimentation.

J. Bentz, J. Ristic Trajkovic (eds.), *Imagining, Designing and Teaching Regenerative Futures: Art-Science Approaches and Inspirations From Around the World*, Science for Sustainable Societies,
https://doi.org/10.1007/978-981-96-9029-9

Collaborative Approaches

01_Participatory action research
02_Collaborative workshops
03_Community engagement

	01	02	03
Deliberative futures workshops for transformative sustainability	■	■	■
Collaborative maps of curiosity	■	■	■
Future and nature stakeholder integration in climate deliberation	■	■	■
Communities in action	■	■	
Community creativity booster method	■	■	■
Scale shift and mapping drift	■	■	
Enspirited envisioning approach			
Creative partnerships for the regeneration of urban rivers	■	■	■
Foot(notes): a walking diary of learning	■		
Embodied transformation theater	■	■	■
A-lab performed facilitating aesthetic exchange and performance	■	■	
Circle of life workshop		■	■
Herbarium with poetry: how to connect people and plants		■	■
Ecotheater as a tool of environmental imaginary	■	■	■
Ecostories for children: an art-science approach	■	■	■
The inner beauty of an interdisciplinary lab		■	■
Photography with intervention	■		
The living canvas: community mural creation		■	■
The echographic poiesis approach		■	■
Overcoming otherness: an artistic approach			
Future-oriented creative workshops		■	
Narrating new future visions		■	■
Regenerative futures with children and arts: the radio approach	■	■	■
Innovation know-how workshop		■	
The beehive—method to explore alternative futures		■	
Tales of transformation		■	■
The future you: visioning and empathizing for futures literacy		■	■
Imagination activation exercise: what if…?		■	
Future visioning as a tool for imagining regenerative futures			
Sensing the city			
Deep time walk			
Building an umwelt apparatus			
Eco-sensory learning pathways for neurodiverse children			■
The explorers in the garden	■	■	
Connecting to self and to nature			
Walking as relational aesthetics	■		
Developing daily sustainable habits		■	■

	01	02	03
Perception walks			■
The permaculture approach			■
Web-based application to foster pro-environmental behavior	■		
Vroom: a group activity for sharing energy		■	■
Dancing in the ruins of the future: an interplay intervention		■	■
Arosa: argumentative role play	■		
Circular families game			■
Climate community street play	■		■
REGEN-D: template-based games for sustainability and regeneration		■	■
Creating social change through urban gaming			■
A listening exercise in care and forgiveness			
Reframing perspectives: a systems mapping exploration		■	■
Systems sensing approach	■	■	■
Exploring the regenerative potential of the micro level	■	■	■
Spatial storytelling to mitigate urban sprawl	■	■	■
Sustainable business model canvas		■	■
Integral ecology: a method for catholic education	■	■	■
Carbon detox			■
More-than-human future ministries		■	
Reflective lifelines: tracing the path to regenerative horizons			
Regenerative practices, teaching as learning		■	■
Active listening workshop		■	
Glocal pedagogy	■		■
Asking questions to co-create future		■	■
Prebunking of climate change conspiracy theories			
Effective innovation leadership framework: self-reflection method		■	
Holistic approach to environmental law education	■		■
3S: seed, save, share			■

Creative Artistic Approaches and Experimentation

01_Art-based practices
02_Creative experimentation

	01	02
Deliberative futures workshops for transformative sustainability		■
Collaborative maps of curiosity	■	■
Future and nature stakeholder integration in climate deliberation		■
Communities in action		
Community creativity booster method	■	■
Scale shift and mapping drift	■	■

	01	02
Enspirited envisioning approach	■	■
Creative partnerships for the regeneration of urban rivers		■
Foot(notes): a walking diary of learning		■
Embodied transformation theater		■
A-lab performed facilitating aesthetic exchange and performance	■	■
Circle of life workshop	■	
Herbarium with poetry: how to connect people and plants	■	■
Ecotheater as a tool of environmental imaginary	■	■
Ecostories for children: an art-science approach	■	■
The inner beauty of an interdisciplinary lab	■	■
Photography with intervention	■	■
The living canvas: community mural creation	■	■
The echographic poiesis approach	■	
Overcoming otherness: an artistic approach	■	■
Future-oriented creative workshops		
Narrating new future visions	■	■
Regenerative futures with children and arts: the radio approach	■	■
Innovation know-how workshop		
The beehive—method to explore alternative futures		
Tales of transformation	■	■
The future you: visioning and empathizing for futures literacy	■	
Imagination activation exercise: what if…?		
Future visioning as a tool for imagining regenerative futures		
Sensing the city		
Deep time walk		
Building an umwelt apparatus		■
Eco-sensory learning pathways for neurodiverse children		
The explorers in the garden		
Connecting to self and to nature		■
Walking as relational aesthetics	■	
Developing daily sustainable habits		
Perception walks	■	
The permaculture approach		
Web-based application to foster pro-environmental behavior		
Vroom: a group activity for sharing energy	■	■
Dancing in the ruins of the future: an interplay intervention	■	■
Arosa: argumentative role play	■	
Circular families game		
Climate community street play	■	■
REGEN-D: template-based games for sustainability and regeneration		
Creating social change through urban gaming		
A listening exercise in care and forgiveness		
Reframing perspectives: a systems mapping exploration		
Systems sensing approach		■

	01	02
Exploring the regenerative potential of the micro level		■
Spatial storytelling to mitigate urban sprawl		
Sustainable business model canvas		■
Integral ecology: a method for catholic education		■
Carbon detox		■
More-than-human future ministries	■	■
Reflective lifelines: tracing the path to regenerative horizons		
Regenerative practices, teaching as learning		■
Active listening workshop		
Glocal pedagogy		
Asking questions to co-create future		
Prebunking of climate change conspiracy theories		
Effective innovation leadership framework: self-reflection method		
Holistic approach to environmental law education		
3S: seed, save, share		■

Future Visioning and Scenario Building

01_Future visioning exercises
02_Scenario planning
03_Visioning for sustainable futures
04_Foresight methods

	01	02	03	04
Deliberative futures workshops for transformative sustainability			■	
Collaborative maps of curiosity	■			
Future and nature stakeholder integration in climate deliberation	■	■	■	■
Communities in action	■			
Community creativity booster method	■		■	
Scale shift and mapping drift			■	
Enspirited envisioning approach	■	■	■	■
Creative partnerships for the regeneration of urban rivers	■		■	
Foot(notes): a walking diary of learning				
Embodied transformation theater	■	■	■	
A-lab performed facilitating aesthetic exchange and performance				
Circle of life workshop				
Herbarium with poetry: how to connect people and plants				
Ecotheater as a tool of environmental imaginary			■	
Ecostories for children: an art-science approach	■		■	
The inner beauty of an interdisciplinary lab		■	■	
Photography with intervention				

	01	02	03	04
The living canvas: community mural creation	■		■	
The echographic poiesis approach	■		■	
Overcoming otherness: an artistic approach				
Future-oriented creative workshops	■	■	■	■
Narrating new future visions	■		■	■
Regenerative futures with children and arts: the radio approach				■
Innovation know-how workshop		■	■	
The beehive—method to explore alternative futures	■			
Tales of transformation			■	
The future you: visioning and empathizing for futures literacy	■	■	■	■
Imagination activation exercise: what if…?	■		■	
Future visioning as a tool for imagining regenerative futures	■		■	■
Sensing the city			■	
Deep time walk				
Building an umwelt apparatus				■
Eco-sensory learning pathways for neurodiverse children				
The explorers in the garden				
Connecting to self and to nature				
Walking as relational aesthetics				
Developing daily sustainable habits			■	
Perception walks				
The permaculture approach	■		■	
Web-based application to foster pro-environmental behavior				
Vroom: a group activity for sharing energy	■		■	
Dancing in the ruins of the future: an interplay intervention	■		■	■
Arosa: argumentative role play		■	■	■
Circular families game				
Climate community street play			■	■
REGEN-D: template-based games for sustainability and regeneration				
Creating social change through urban gaming				
A listening exercise in care and forgiveness				
Reframing perspectives: a systems mapping exploration				
Systems sensing approach				
Exploring the regenerative potential of the micro level		■	■	
Spatial storytelling to mitigate urban sprawl	■		■	
Sustainable business model canvas			■	
Integral ecology: a method for catholic education				
Carbon detox			■	
More-than-human future ministries	■		■	
Reflective lifelines: tracing the path to regenerative horizons				
Regenerative practices, teaching as learning				
Active listening workshop				
Glocal pedagogy			■	■
Asking questions to co-create future				
Prebunking of climate change conspiracy theories				

	01	02	03	04
Effective innovation leadership framework: self-reflection method	■			
Holistic approach to environmental law education				
3S: seed, save, share			■	

Empathy and Perspective-Taking

01_Empathy-building activities
02_Role-playing and simulation
03_Storytelling approaches
04_Literacy workshops

	01	02	03	04
Deliberative futures workshops for transformative sustainability	■			
Collaborative maps of curiosity	■			
Future and nature stakeholder integration in climate deliberation	■		■	
Communities in action				
Community creativity booster method			■	
Scale shift and mapping drift	■	■		
Enspirited envisioning approach				
Creative partnerships for the regeneration of urban rivers	■	■	■	■
Foot(notes): a walking diary of learning	■			
Embodied transformation theater	■	■		
A-lab performed facilitating aesthetic exchange and performance	■	■		
Circle of life workshop	■			
Herbarium with poetry: how to connect people and plants			■	
Ecotheater as a tool of environmental imaginary		■	■	
Ecostories for children: an art-science approach	■		■	■
The inner beauty of an interdisciplinary lab	■	■		■
Photography with intervention	■			
The living canvas: community mural creation				
The echographic poiesis approach	■			
Overcoming otherness: an artistic approach	■	■		
Future-oriented creative workshops				
Narrating new future visions	■	■	■	
Regenerative futures with children and arts: the radio approach	■	■	■	
Innovation know-how workshop		■		
The beehive—method to explore alternative futures	■			
Tales of transformation			■	
The future you: visioning and empathizing for futures literacy	■	■		■
Imagination activation exercise: what if…?				
Future visioning as a tool for imagining regenerative futures				

	01	02	03	04
Sensing the city		■		
Deep time walk				
Building an umwelt apparatus	■	■		
Eco-sensory learning pathways for neurodiverse children	■			
The explorers in the garden				
Connecting to self and to nature	■			
Walking as relational aesthetics	■			
Developing daily sustainable habits	■			
Perception walks	■			
The permaculture approach				
Web-based application to foster pro-environmental behavior	■			
Vroom: a group activity for sharing energy	■	■	■	
Dancing in the ruins of the future: an interplay intervention	■	■	■	
Arosa: argumentative role play	■	■		
Circular families game				
Climate community street play	■	■	■	
REGEN-D: template-based games for sustainability and regeneration	■			
Creating social change through urban gaming		■	■	
A listening exercise in care and forgiveness	■	■		
Reframing perspectives: a systems mapping exploration				
Systems sensing approach	■			
Exploring the regenerative potential of the micro level				
Spatial storytelling to mitigate urban sprawl			■	
Sustainable business model canvas				
Integral ecology: a method for catholic education				
Carbon detox	■			
More-than-human future ministries		■		
Reflective lifelines: tracing the path to regenerative horizons			■	
Regenerative practices, teaching as learning	■	■	■	
Active listening workshop		■		
Glocal pedagogy				
Asking questions to co-create future	■			
Prebunking of climate change conspiracy theories	■			■
Effective innovation leadership framework: self-reflection method	■			
Holistic approach to environmental law education		■	■	
3S: seed, save, share				

Regenerative Design and Imagination

01_Design thinking
02_Visualization
03_Imagination
04_Urban regeneration

	01	02	03	04
Deliberative futures workshops for transformative sustainability				
Collaborative maps of curiosity		■	■	■
Future and nature stakeholder integration in climate deliberation	■	■	■	■
Communities in action	■		■	
Community creativity booster method	■	■	■	
Scale shift and mapping drift			■	■
Enspirited envisioning approach		■	■	■
Creative partnerships for the regeneration of urban rivers	■			■
Foot(notes): a walking diary of learning			■	
Embodied transformation theater			■	
A-lab performed facilitating aesthetic exchange and performance	■		■	
Circle of life workshop		■	■	
Herbarium with poetry: how to connect people and plants			■	
Ecotheater as a tool of environmental imaginary			■	
Ecostories for children: an art-science approach		■	■	
The inner beauty of an interdisciplinary lab				
Photography with intervention			■	
The living canvas: community mural creation			■	■
The echographic poiesis approach		■	■	
Overcoming otherness: an artistic approach				
Future-oriented creative workshops				
Narrating new future visions		■	■	
Regenerative futures with children and arts: the radio approach			■	
Innovation know-how workshop	■			
The beehive—method to explore alternative futures			■	
Tales of transformation			■	
The future you: visioning and empathizing for futures literacy		■	■	
Imagination activation exercise: what if…?	■	■	■	
Future visioning as a tool for imagining regenerative futures		■	■	
Sensing the city		■	■	■
Deep time walk				
Building an umwelt apparatus	■		■	
Eco-sensory learning pathways for neurodiverse children		■	■	
The explorers in the garden	■			
Connecting to self and to nature				
Walking as relational aesthetics			■	
Developing daily sustainable habits				■
Perception walks				
The permaculture approach	■			■
Web-based application to foster pro-environmental behavior				
Vroom: a group activity for sharing energy		■	■	
Dancing in the ruins of the future: an interplay intervention		■	■	■
Arosa: argumentative role play			■	■
Circular families game				■

	01	02	03	04
Climate community street play		■	■	■
REGEN-D: template-based games for sustainability and regeneration		■		
Creating social change through urban gaming			■	
A listening exercise in care and forgiveness				
Reframing perspectives: a systems mapping exploration		■		
Systems sensing approach				
Exploring the regenerative potential of the micro level	■	■	■	■
Spatial storytelling to mitigate urban sprawl		■	■	
Sustainable business model canvas				
Integral ecology: a method for catholic education				
Carbon detox			■	
More-than-human future ministries			■	
Reflective lifelines: tracing the path to regenerative horizons		■	■	
Regenerative practices, teaching as learning		■	■	
Active listening workshop				
Glocal pedagogy				
Asking questions to co-create future				
Prebunking of climate change conspiracy theories				
Effective innovation leadership framework: self-reflection method				
Holistic approach to environmental law education				
3S: seed, save, share				■

Game-Based Learning

01_Serious games
02_Learning games

	01	02
Deliberative futures workshops for transformative sustainability		
Collaborative maps of curiosity		
Future and nature stakeholder integration in climate deliberation		■
Communities in action		
Community creativity booster method		
Scale shift and mapping drift	■	
Enspirited envisioning approach		
Creative partnerships for the regeneration of urban rivers	■	■
Foot(notes): a walking diary of learning		
Embodied transformation theater		
A-lab performed facilitating aesthetic exchange and performance		
Circle of life workshop		■
Herbarium with poetry: how to connect people and plants		

	01	02
Ecotheater as a tool of environmental imaginary		
Ecostories for children: an art-science approach		
The inner beauty of an interdisciplinary lab		
Photography with intervention		
The living canvas: community mural creation		
The echographic poiesis approach		
Overcoming otherness: an artistic approach		■
Future-oriented creative workshops		
Narrating new future visions		
Regenerative futures with children and arts: the radio approach		
Innovation know-how workshop		
The beehive—method to explore alternative futures		
Tales of transformation		
The future you: visioning and empathizing for futures literacy		
Imagination activation exercise: what if…?		
Future visioning as a tool for imagining regenerative futures		
Sensing the city		
Deep time walk		
Building an umwelt apparatus	■	
Eco-sensory learning pathways for neurodiverse children		
The explorers in the garden		
Connecting to self and to nature		
Walking as relational aesthetics		
Developing daily sustainable habits		■
Perception walks		
The permaculture approach		
Web-based application to foster pro-environmental behavior	■	
Vroom: a group activity for sharing energy		■
Dancing in the ruins of the future: an interplay intervention		■
Arosa: argumentative role play		■
Circular families game	■	■
Climate community street play	■	
REGEN-D: template-based games for sustainability and regeneration	■	■
Creating social change through urban gaming	■	■
A listening exercise in care and forgiveness		
Reframing perspectives: a systems mapping exploration		
Systems sensing approach		
Exploring the regenerative potential of the micro level		
Spatial storytelling to mitigate urban sprawl		
Sustainable business model canvas		
Integral ecology: a method for catholic education		
Carbon detox		■
More-than-human future ministries	■	■
Reflective lifelines: tracing the path to regenerative horizons		

	01	02
Regenerative practices, teaching as learning		
Active listening workshop		
Glocal pedagogy		
Asking questions to co-create future		
Prebunking of climate change conspiracy theories		
Effective innovation leadership framework: self-reflection method		
Holistic approach to environmental law education		
3S: seed, save, share		

Climate Action and Resilience

01_Learning by doing
02_Climate action
03_Resilience-building activities

	01	02	03
Deliberative futures workshops for transformative sustainability			
Collaborative maps of curiosity			
Future and nature stakeholder integration in climate deliberation	■	■	■
Communities in action			
Community creativity booster method	■	■	■
Scale shift and mapping drift		■	■
Enspirited envisioning approach			
Creative partnerships for the regeneration of urban rivers	■		■
Foot(notes): a walking diary of learning	■		
Embodied transformation theater			
A-lab performed facilitating aesthetic exchange and performance	■		
Circle of life workshop	■		
Herbarium with poetry: how to connect people and plants			
Ecotheater as a tool of environmental imaginary		■	■
Ecostories for children: an art-science approach	■		
The inner beauty of an interdisciplinary lab	■	■	
Photography with intervention			
The living canvas: community mural creation			
The echographic poiesis approach			
Overcoming otherness: an artistic approach	■		
Future-oriented creative workshops			
Narrating new future visions		■	■
Regenerative futures with children and arts: the radio approach		■	

	01	02	03
Innovation know-how workshop			
The beehive—method to explore alternative futures			
Tales of transformation		■	
The future you: visioning and empathizing for futures literacy	■	■	■
Imagination activation exercise: what if…?			
Future visioning as a tool for imagining regenerative futures			
Sensing the city	■	■	■
Deep time walk			
Building an umwelt apparatus			
Eco-sensory learning pathways for neurodiverse children	■	■	■
The explorers in the garden	■		■
Connecting to self and to nature			
Walking as relational aesthetics	■		
Developing daily sustainable habits	■	■	■
Perception walks	■		■
The permaculture approach	■	■	■
Web-based application to foster pro-environmental behavior		■	
Vroom: a group activity for sharing energy	■	■	■
Dancing in the ruins of the future: an interplay intervention			■
Arosa: argumentative role play	■		
Circular families game			
Climate community street play		■	■
REGEN-D: template-based games for sustainability and regeneration			
Creating social change through urban gaming			
A listening exercise in care and forgiveness			
Reframing perspectives: a systems mapping exploration		■	
Systems sensing approach			
Exploring the regenerative potential of the micro level	■		
Spatial storytelling to mitigate urban sprawl		■	■
Sustainable business model canvas		■	
Integral ecology: a method for catholic education	■	■	■
Carbon detox	■	■	■
More-than-human future ministries			
Reflective lifelines: tracing the path to regenerative horizons			
Regenerative practices, teaching as learning			
Active listening workshop			
Glocal pedagogy			
Asking questions to co-create future			
Prebunking of climate change conspiracy theories			

	01	02	03
Effective innovation leadership framework: self-reflection method			
Holistic approach to environmental law education			
3S: seed, save, share	■		■

Community Empowerment and Social Change

01_Community-based initiatives
02_Social justice-oriented approaches
03_Participatory decision-making

	01	02	03
Deliberative futures workshops for transformative sustainability			■
Collaborative maps of curiosity	■		
Future and nature stakeholder integration in climate deliberation	■	■	■
Communities in action			■
Community creativity booster method	■		■
Scale shift and mapping drift	■		
Enspirited envisioning approach			■
Creative partnerships for the regeneration of urban rivers	■		■
Foot(notes): a walking diary of learning	■		
Embodied transformation theater	■		
A-lab performed facilitating aesthetic exchange and performance			
Circle of life workshop	■		
Herbarium with poetry: how to connect people and plants			
Ecotheater as a tool of environmental imaginary	■		
Ecostories for children: an art-science approach	■	■	
The inner beauty of an interdisciplinary lab			■
Photography with intervention	■	■	■
The living canvas: community mural creation	■		
The echographic poiesis approach	■		
Overcoming otherness: an artistic approach			
Future-oriented creative workshops			■
Narrating new future visions			
Regenerative futures with children and arts: the radio approach			
Innovation know-how workshop			
The beehive—method to explore alternative futures			■
Tales of transformation	■	■	
The future you: visioning and empathizing for futures literacy	■	■	■
Imagination activation exercise: what if…?			
Future visioning as a tool for imagining regenerative futures			
Sensing the city			

	01	02	03
Deep time walk			
Building an umwelt apparatus			
Eco-sensory learning pathways for neurodiverse children	■	■	
The explorers in the garden			■
Connecting to self and to nature			
Walking as relational aesthetics			
Developing daily sustainable habits	■	■	■
Perception walks			
The permaculture approach	■		
Web-based application to foster pro-environmental behavior			
Vroom: a group activity for sharing energy		■	
Dancing in the ruins of the future: an interplay intervention			■
Arosa: argumentative role play			■
Circular families game	■		
Climate community street play	■	■	
REGEN-D: template-based games for sustainability and regeneration	■		
Creating social change through urban gaming			
A listening exercise in care and forgiveness			
Reframing perspectives: a systems mapping exploration	■	■	■
Systems sensing approach	■		
Exploring the regenerative potential of the micro level			■
Spatial storytelling to mitigate urban sprawl			■
Sustainable business model canvas			■
Integral ecology: a method for catholic education	■	■	
Carbon detox	■	■	■
More-than-human future ministries			■
Reflective lifelines: tracing the path to regenerative horizons			
Regenerative practices, teaching as learning			
Active listening workshop			
Glocal pedagogy	■	■	■
Asking questions to co-create future			
Prebunking of climate change conspiracy theories			
Effective innovation leadership framework: self-reflection method			
Holistic approach to environmental law education	■		
3S: seed, save, share			■

Transformative Learning and Pedagogy

01_Experiential learning methods
02_Transformative learning
03_Critical thinking

	01	02	03
Deliberative futures workshops for transformative sustainability			
Collaborative maps of curiosity	■		
Future and nature stakeholder integration in climate deliberation		■	■
Communities in action		■	■
Community creativity booster method	■	■	
Scale shift and mapping drift	■		
Enspirited envisioning approach	■	■	
Creative partnerships for the regeneration of urban rivers	■	■	
Foot(notes): a walking diary of learning	■		
Embodied transformation theater			
A-lab performed facilitating aesthetic exchange and performance	■		
Circle of life workshop	■		
Herbarium with poetry: how to connect people and plants			
Ecotheater as a tool of environmental imaginary			■
Ecostories for children: an art-science approach	■	■	■
The inner beauty of an interdisciplinary lab			■
Photography with intervention			
The living canvas: community mural creation	■	■	
The echographic poiesis approach	■	■	
Overcoming otherness: an artistic approach		■	■
Future-oriented creative workshops			
Narrating new future visions	■	■	■
Regenerative futures with children and arts: the radio approach	■	■	■
Innovation know-how workshop		■	■
The beehive—method to explore alternative futures		■	■
Tales of transformation	■		■
The future you: visioning and empathizing for futures literacy	■	■	■
Imagination activation exercise: what if…?			
Future visioning as a tool for imagining regenerative futures	■	■	
Sensing the city	■	■	■
Deep time walk	■	■	■
Building an umwelt apparatus	■	■	
Eco-sensory learning pathways for neurodiverse children	■	■	■
The explorers in the garden	■	■	■
Connecting to self and to nature	■	■	
Walking as relational aesthetics	■		
Developing daily sustainable habits	■		■
Perception walks	■	■	
The permaculture approach	■	■	■
Web-based application to foster pro-environmental behavior		■	■
Vroom: a group activity for sharing energy			
Dancing in the ruins of the future: an interplay intervention	■	■	■
Arosa: argumentative role play	■	■	■
Circular families game			

	01	02	03
Climate community street play	■	■	■
REGEN-D: template-based games for sustainability and regeneration	■		
Creating social change through urban gaming	■	■	■
A listening exercise in care and forgiveness		■	
Reframing perspectives: a systems mapping exploration			
Systems sensing approach	■	■	
Exploring the regenerative potential of the micro level	■	■	■
Spatial storytelling to mitigate urban sprawl			■
Sustainable business model canvas			
Integral ecology: a method for catholic education		■	
Carbon detox		■	■
More-than-human future ministries	■		
Reflective lifelines: tracing the path to regenerative horizons		■	■
Regenerative practices, teaching as learning	■	■	■
Active listening workshop	■	■	
Glocal pedagogy		■	■
Asking questions to co-create future	■	■	
Prebunking of climate change conspiracy theories			■
Effective innovation leadership framework: self-reflection method		■	■
Holistic approach to environmental law education	■		
3S: seed, save, share	■	■	■

Interdisciplinary Collaboration

01_Transdisciplinary methods
02_Cross-disciplinary dialogue
03_Integration of multiple perspectives

	01	02	03
Deliberative futures workshops for transformative sustainability			■
Collaborative maps of curiosity			■
Future and nature stakeholder integration in climate deliberation	■	■	■
Communities in action	■		
Community creativity booster method		■	■
Scale shift and mapping drift	■		■
Enspirited envisioning approach	■	■	
Creative partnerships for the regeneration of urban rivers	■		■
Foot(notes): a walking diary of learning			
Embodied transformation theater			■
A-lab performed facilitating aesthetic exchange and performance	■	■	■
Circle of life workshop	■		

	01	02	03
Herbarium with poetry: how to connect people and plants	■	■	
Ecotheater as a tool of environmental imaginary			
Ecostories for children: an art-science approach	■	■	
The inner beauty of an interdisciplinary lab		■	
Photography with intervention			
The living canvas: community mural creation			■
The echographic poiesis approach			
Overcoming otherness: an artistic approach	■		■
Future-oriented creative workshops			■
Narrating new future visions	■		■
Regenerative futures with children and arts: the radio approach			
Innovation know-how workshop		■	
The beehive—method to explore alternative futures	■		
Tales of transformation	■		■
The future you: visioning and empathizing for futures literacy	■		■
Imagination activation exercise: what if…?			■
Future visioning as a tool for imagining regenerative futures			
Sensing the city			■
Deep time walk	■	■	■
Building an umwelt apparatus			■
Eco-sensory learning pathways for neurodiverse children	■	■	■
The explorers in the garden			■
Connecting to self and to nature			■
Walking as relational aesthetics		■	■
Developing daily sustainable habits		■	
Perception walks			
The permaculture approach			■
Web-based application to foster pro-environmental behavior	■		
Vroom: a group activity for sharing energy			
Dancing in the ruins of the future: an interplay intervention	■		■
Arosa: argumentative role play			■
Circular families game			
Climate community street play	■	■	■
REGEN-D: template-based games for sustainability and regeneration	■		
Creating social change through urban gaming			
A listening exercise in care and forgiveness			
Reframing perspectives: a systems mapping exploration			
Systems sensing approach			■
Exploring the regenerative potential of the micro level		■	
Spatial storytelling to mitigate urban sprawl		■	■
Sustainable business model canvas			■
Integral ecology: a method for catholic education	■	■	■
Carbon detox			■
More-than-human future ministries	■		■

	01	02	03
Reflective lifelines: tracing the path to regenerative horizons			
Regenerative practices, teaching as learning			■
Active listening workshop			
Glocal pedagogy		■	■
Asking questions to co-create future			■
Prebunking of climate change conspiracy theories		■	
Effective innovation leadership framework: self-reflection method			
Holistic approach to environmental law education		■	
3S: seed, save, share	■	■	■

Well-Being and Connection to Nature

01_Learning from nature
02_Mindfulness and nature connection
03_Well-being practices in nature
04_Embodiment practices

	01	02	03	04
Deliberative futures workshops for transformative sustainability				
Collaborative maps of curiosity				
Future and nature stakeholder integration in climate deliberation	■	■	■	■
Communities in action				
Community creativity booster method	■	■	■	
Scale shift and mapping drift	■	■	■	■
Enspirited envisioning approach				■
Creative partnerships for the regeneration of urban rivers	■	■	■	■
Foot(notes): a walking diary of learning				■
Embodied transformation theater	■			■
A-lab performed facilitating aesthetic exchange and performance				■
Circle of life workshop	■	■		
Herbarium with poetry: how to connect people and plants	■	■	■	
Ecotheater as a tool of environmental imaginary	■	■	■	■
Ecostories for children: an art-science approach		■		
The inner beauty of an interdisciplinary lab	■	■		
Photography with intervention				
The living canvas: community mural creation		■		
The echographic poiesis approach	■	■		
Overcoming otherness: an artistic approach		■		■
Future-oriented creative workshops				
Narrating new future visions				
Regenerative futures with children and arts: the radio approach	■	■	■	■

	01	02	03	04
Innovation know-how workshop				
The beehive—method to explore alternative futures				
Tales of transformation				■
The future you: visioning and empathizing for futures literacy	■	■	■	
Imagination activation exercise: what if…?				
Future visioning as a tool for imagining regenerative futures				
Sensing the city	■	■	■	■
Deep time walk	■	■	■	■
Building an umwelt apparatus	■			■
Eco-sensory learning pathways for neurodiverse children	■	■	■	■
The explorers in the garden	■	■	■	■
Connecting to self and to nature		■		■
Walking as relational aesthetics	■	■	■	■
Developing daily sustainable habits	■	■	■	
Perception walks	■	■	■	■
The permaculture approach	■		■	
Web-based application to foster pro-environmental behavior				
Vroom: a group activity for sharing energy				■
Dancing in the ruins of the future: an interplay intervention	■	■	■	■
Arosa: argumentative role play				
Circular families game				
Climate community street play	■	■		■
REGEN-D: template-based games for sustainability and regeneration				
Creating social change through urban gaming				
A listening exercise in care and forgiveness		■		■
Reframing perspectives: a systems mapping exploration				
Systems sensing approach				■
Exploring the regenerative potential of the micro level				
Spatial storytelling to mitigate urban sprawl				
Sustainable business model canvas				
Integral ecology: a method for catholic education		■		
Carbon detox	■	■	■	
More-than-human future ministries		■		■
Reflective lifelines: tracing the path to regenerative horizons				
Regenerative practices, teaching as learning	■			■
Active listening workshop				
Glocal pedagogy				
Asking questions to co-create future				
Prebunking of climate change conspiracy theories				
Effective innovation leadership framework: self-reflection method				
Holistic approach to environmental law education				
3S: seed, save, share	■			

Systems Thinking and Management

01_Systems mapping and analysis
02_Complexity awareness

	01	02
Deliberative futures workshops for transformative sustainability		
Collaborative maps of curiosity		
Future and nature stakeholder integration in climate deliberation		■
Communities in action	■	■
Community creativity booster method		
Scale shift and mapping drift	■	■
Enspirited envisioning approach		■
Creative partnerships for the regeneration of urban rivers	■	
Foot(notes): a walking diary of learning		■
Embodied transformation theater	■	
A-lab performed facilitating aesthetic exchange and performance		
Circle of life workshop		■
Herbarium with poetry: how to connect people and plants		
Ecotheater as a tool of environmental imaginary		
Ecostories for children: an art-science approach	■	■
The inner beauty of an interdisciplinary lab		■
Photography with intervention		
The living canvas: community mural creation		
The echographic poiesis approach		■
Overcoming otherness: an artistic approach		
Future-oriented creative workshops		
Narrating new future visions	■	■
Regenerative futures with children and arts: the radio approach		
Innovation know-how workshop		■
The beehive—method to explore alternative futures	■	■
Tales of transformation		
The future you: visioning and empathizing for futures literacy		■
Imagination activation exercise: what if…?		
Future visioning as a tool for imagining regenerative futures		
Sensing the city		■
Deep time walk		
Building an umwelt apparatus		
Eco-sensory learning pathways for neurodiverse children		■
The explorers in the garden		■
Connecting to self and to nature		
Walking as relational aesthetics		
Developing daily sustainable habits		
Perception walks		■

	01	02
The permaculture approach	■	
Web-based application to foster pro-environmental behavior		
Vroom: a group activity for sharing energy		
Dancing in the ruins of the future: an interplay intervention		■
Arosa: argumentative role play	■	■
Circular families game	■	
Climate community street play	■	■
REGEN-D: template-based games for sustainability and regeneration		
Creating social change through urban gaming		
A listening exercise in care and forgiveness		
Reframing perspectives: a systems mapping exploration	■	■
Systems sensing approach		■
Exploring the regenerative potential of the micro level	■	■
Spatial storytelling to mitigate urban sprawl	■	
Sustainable business model canvas		■
Integral ecology: a method for catholic education		
Carbon detox		■
More-than-human future ministries		■
Reflective lifelines: tracing the path to regenerative horizons		
Regenerative practices, teaching as learning		■
Active listening workshop		
Glocal pedagogy		
Asking questions to co-create future		
Prebunking of climate change conspiracy theories		
Effective innovation leadership framework: self-reflection method		
Holistic approach to environmental law education		
3S: seed, save, share		

Cultural Perspectives and Alternative Ways of Knowing

01_Sustainability and culture
02_Local knowledge
03_Indigenous knowledge systems

	01	02	03
Deliberative futures workshops for transformative sustainability			
Collaborative maps of curiosity			
Future and nature stakeholder integration in climate deliberation	■	■	■
Communities in action		■	
Community creativity booster method	■	■	
Scale shift and mapping drift		■	

	01	02	03
Enspirited envisioning approach	■		
Creative partnerships for the regeneration of urban rivers		■	■
Foot(notes): a walking diary of learning		■	
Embodied transformation theater	■	■	
A-lab performed facilitating aesthetic exchange and performance			
Circle of life workshop	■	■	
Herbarium with poetry: how to connect people and plants	■	■	■
Ecotheater as a tool of environmental imaginary	■		
Ecostories for children: an art-science approach	■	■	■
The inner beauty of an interdisciplinary lab	■	■	
Photography with intervention			
The living canvas: community mural creation			
The echographic poiesis approach	■	■	■
Overcoming otherness: an artistic approach	■		
Future-oriented creative workshops			
Narrating new future visions	■		
Regenerative futures with children and arts: the radio approach	■		
Innovation know-how workshop			
The beehive—method to explore alternative futures		■	
Tales of transformation			
The future you: visioning and empathizing for futures literacy	■	■	■
Imagination activation exercise: what if…?			
Future visioning as a tool for imagining regenerative futures			
Sensing the city	■	■	
Deep time walk			
Building an umwelt apparatus			
Eco-sensory learning pathways for neurodiverse children	■	■	■
The explorers in the garden		■	■
Connecting to self and to nature			
Walking as relational aesthetics		■	
Developing daily sustainable habits	■		
Perception walks			
The permaculture approach	■		■
Web-based application to foster pro-environmental behavior			
Vroom: a group activity for sharing energy			
Dancing in the ruins of the future: an interplay intervention	■		
Arosa: argumentative role play		■	
Circular families game			
Climate community street play	■	■	
REGEN-D: template-based games for sustainability and regeneration		■	
Creating social change through urban gaming	■	■	
A listening exercise in care and forgiveness			
Reframing perspectives: a systems mapping exploration			
Systems sensing approach			■

	01	02	03
Exploring the regenerative potential of the micro level	■	■	■
Spatial storytelling to mitigate urban sprawl		■	■
Sustainable business model canvas			
Integral ecology: a method for catholic education			
Carbon detox	■	■	
More-than-human future ministries	■		
Reflective lifelines: tracing the path to regenerative horizons			
Regenerative practices, teaching as learning			
Active listening workshop			
Glocal pedagogy	■	■	■
Asking questions to co-create future			
Prebunking of climate change conspiracy theories	■		
Effective innovation leadership framework: self-reflection method			
Holistic approach to environmental law education			
3S: seed, save, share	■	■	

Innovation and Technology for Sustainability

01_Digital tools for environmental education
02_Innovation

	01	02
Deliberative futures workshops for transformative sustainability		
Collaborative maps of curiosity		
Future and nature stakeholder integration in climate deliberation		■
Communities in action		
Community creativity booster method		■
Scale shift and mapping drift		
Enspirited envisioning approach		
Creative partnerships for the regeneration of urban rivers	■	
Foot(notes): a walking diary of learning		■
Embodied transformation theater		■
A-lab performed facilitating aesthetic exchange and performance		■
Circle of life workshop		
Herbarium with poetry: how to connect people and plants		■
Ecotheater as a tool of environmental imaginary		
Ecostories for children: an art-science approach		
The inner beauty of an interdisciplinary lab		
Photography with intervention		
The living canvas: community mural creation		
The echographic poiesis approach		
Overcoming otherness: an artistic approach		
Future-oriented creative workshops	■	■

	01	02
Narrating new future visions		
Regenerative futures with children and arts: the radio approach		
Innovation know-how workshop		■
The beehive—method to explore alternative futures		
Tales of transformation		
The future you: visioning and empathizing for futures literacy		
Imagination activation exercise: what if…?		
Future visioning as a tool for imagining regenerative futures		
Sensing the city		
Deep time walk		
Building an umwelt apparatus		
Eco-sensory learning pathways for neurodiverse children	■	■
The explorers in the garden		
Connecting to self and to nature		
Walking as relational aesthetics		
Developing daily sustainable habits	■	■
Perception walks		
The permaculture approach		■
Web-based application to foster pro-environmental behavior		
Vroom: a group activity for sharing energy		
Dancing in the ruins of the future: an interplay intervention		
Arosa: argumentative role play		
Circular families game		■
Climate community street play		■
REGEN-D: template-based games for sustainability and regeneration	■	
Creating social change through urban gaming	■	■
A listening exercise in care and forgiveness		
Reframing perspectives: a systems mapping exploration		
Systems sensing approach		
Exploring the regenerative potential of the micro level		■
Spatial storytelling to mitigate urban sprawl	■	
Sustainable business model canvas		■
Integral ecology: a method for catholic education		
Carbon detox		
More-than-human future ministries		■
Reflective lifelines: tracing the path to regenerative horizons		
Regenerative practices, teaching as learning		
Active listening workshop		
Glocal pedagogy		
Asking questions to co-create future		
Prebunking of climate change conspiracy theories		
Effective innovation leadership framework: self-reflection method		
Holistic approach to environmental law education		
3S: seed, save, share		

MIX
Papier aus verantwortungsvollen Quellen
Paper from responsible sources
FSC® C105338

If you have any concerns about our products,
you can contact us on
ProductSafety@springernature.com

In case Publisher is established outside the EU,
the EU authorized representative is:
Springer Nature Customer Service Center GmbH
Europaplatz 3, 69115 Heidelberg, Germany

Printed by Libri Plureos GmbH
in Hamburg, Germany